Climate Risks and Solutions

Series Editor

Ernesto D. R. Santibanez Gonzalez, University of Talca, Talca, Chile

This visionary series illuminates the symbiotic relationship between climate challenges and cutting-edge technologies such as artificial intelligence (AI) and mathematical modelling. The series examines how AI empowers decision-makers, enhances predictive capabilities, and guides proactive adaptation strategies. The series aim is to encourage and foster a transformative journey toward a low-carbon future, where academics explore, entrepreneurs innovate, and leaders advance resilience.

With its multidisciplinary approach, spanning fields like engineering, economics, environmental studies, computer science, management science, and operations management, this series offers applications across major industrial sectors such as mining, agriculture, energy, forestry, and finance.

Submissions of high-quality research in the form of monographs, edited volumes, conference proceedings, and advanced-level textbooks will be considered for publication. All submissions will be peer-reviewed.

The series will be of great interest to a broad audience, including scholars, researchers, undergraduate, and graduate students. Decision-makers aiming to lead society towards a low-carbon, resilient future will acquire the necessary tools for informed action. Those keen on discovering pathways to sustainable innovation will find invaluable insights.

Shashi Kant Tiwari · Atul Fegade · Mustafa Kamal
Editors

AI-Driven Environmental Pollution Management

Springer

Editors
Shashi Kant Tiwari
Balaji Institute of Modern Management
Sri Balaji University
Pune, Maharashtra, India

Atul Fegade
Balaji Institute of Modern Management
Sri Balaji University
Pune, Maharashtra, India

Mustafa Kamal
Department of Basic Sciences
Saudi Electronic University
Dammam, Saudi Arabia

ISSN 3005-1665 ISSN 3005-1673 (electronic)
Climate Risks and Solutions
ISBN 978-3-031-96242-4 ISBN 978-3-031-96243-1 (eBook)
https://doi.org/10.1007/978-3-031-96243-1

This Springer imprint is published by the registered company Springer Nature Switzerland AG
The registered company address is: Gewerbestrasse 11, 6330 Cham, Switzerland

Foreword 1: The Promise of AI in Tackling Environmental Crises

In an era where environmental degradation poses an existential threat to life on Earth, the need for innovative solutions has never been more urgent. Climate change, air and water pollution, and the relentless exploitation of natural resources are the biggest challenges that are affecting the health, economy, and survival of societies worldwide. Addressing these complex and interrelated issues requires a multidimensional approach that integrates technology, policy, and human ingenuity. The recent advances in artificial intelligence (AI) can be a game-changer and act as a transformative force that can revolutionize how we understand, monitor, and address environmental challenges.

This book, *AI-Driven Environmental Pollution Management*, is a timely and essential contribution to this critical field. By focusing on the intersection of AI and environmental management, it provides a roadmap for leveraging advanced technologies to combat pollution and foster sustainable development. The editors have curated a rich collection of chapters that related to the role of AI in monitoring air, water, and noise pollution, predicting environmental trends, and optimizing policy decisions. What sets this volume apart is its comprehensive approach—combining theoretical insights with practical applications and addressing both technical innovations and their socio-economic implications.

One of the book's key strengths lies in its ability to present complex ideas in an accessible manner, making it a valuable resource not only for researchers and academics but also for policymakers, environmentalists, and industry leaders. Each chapter offers a blend of rigorous analysis, practical case studies, and forward-looking perspectives. From exploring the latest advancements in IoT-based monitoring systems to evaluating the economic viability-driven solutions, the book provides a holistic view of the challenges and opportunities in this field.

The editors deserve commendation for their vision and dedication in bringing this project to completion. By highlighting the potential of AI to drive sustainable innovation, they have provided a ray of hope for those seeking to address the pressing environmental challenges of our time. It is my firm belief that this book will inspire collaborative efforts and innovative thinking, empowering a new generation of leaders to harness technology for the betterment of our planet.

The readers will not only gain a deeper understanding of AI's capabilities but also be inspired to act. This book is more than a guide—it is a call to use and leverage technology to create a cleaner, healthier, and more sustainable future. Let us embrace this opportunity with the urgency and determination it deserves.

Dr. Naveen Garg
Principal Senior Scientist
CSIR—National Physical Laboratory
New Delhi, India

Foreword 2: A Vision for Sustainable Innovation

With rising pollution levels, the depletion of natural resources, and the adverse impacts of climate change, humanity finds itself at a crossroads with the environment. The path we choose will determine the fate of future generations. In this context, the role of technology—particularly artificial intelligence can be of great help. AI has the potential to redefine how we address environmental issues, offering innovative solutions that are not only efficient and scalable but also economically viable and globally impactful.

This book, *AI-Driven Environmental Pollution Management*, is an extraordinary endeavor that seeks to bridge the gap between technology and sustainability. The editors have brought together a diverse group of authors to explore how AI can be harnessed to mitigate pollution, optimize resource utilization, and drive policy decisions that promote sustainable development. The book is a testament to the power of interdisciplinary collaboration, combining insights from environmental science, data analytics, and policy research to create a comprehensive guide for tackling one of the most pressing issues of our time.

What makes this book particularly remarkable is its balance between depth and breadth. It begins by laying the foundation—introducing readers to the sources and impacts of pollution and setting the stage for technological interventions. Subsequent chapters focus on advanced monitoring techniques, AI-driven solutions, and innovative case studies that demonstrate the real-world impact of these technologies. The book also addresses the economic and sustainable dimensions of AI, offering a nuanced perspective on its potential to drive both environmental and economic progress.

The editors' meticulous approach is evident in every aspect of this work. They have not only curated a wealth of knowledge but have also structured it in a way that facilitates understanding and application. It is a resource that will benefit a wide audience, from students and researchers to industry professionals and policymakers.

As we face the daunting task of addressing environmental degradation, this book serves as a reminder of what is possible when innovation is guided by purpose. It challenges us to rethink traditional approaches, embrace technological advancements, and work collaboratively toward a common goal. By doing so, it offers hope for a future where technology and sustainability are not at odds but are mutually reinforcing.

I have no doubt that this book will inspire its readers to act with urgency and creativity. It is a call to envision a world where AI not only mitigates the impacts of pollution but also fosters a culture of sustainability and innovation. Together, we can build a future that honors our responsibility to the planet and to each other.

Dr. Anand Kulkarni
Research Professor and Associate Director
Institute of Artificial Intelligence
Pune, India

Dr. Vishwanath Karad
MIT World Peace University
Pune, India

Preface

The alarming rise in environmental pollution poses one of the most critical challenges of our time, demanding innovative solutions that leverage cutting-edge technology. Artificial intelligence (AI) has emerged as a transformative tool capable of addressing these challenges with unprecedented precision, efficiency, and scalability. From monitoring air, water, and noise quality to formulating data-driven policies and optimizing sustainable practices, AI offers a new horizon in environmental management. This book, *AI-Driven Environmental Pollution Management,* is an ambitious endeavor to explore the multifaceted applications of AI in mitigating the impacts of pollution and fostering sustainable development.

This edited volume brings together contributions from leading researchers, academics, and industry experts, offering a comprehensive perspective on the interplay between AI and environmental management. The chapters span a wide range of topics, starting with an understanding of pollution sources and impacts, delving into advanced monitoring technologies, and emphasizing the critical role of AI and machine learning in pollution prediction, management, and policy-making. The book also examines the economic viability of AI-driven solutions, their potential for driving sustainable innovations, and their role in advancing global environmental cooperation.

The organization of this book reflects a logical progression of ideas. The first unit introduces readers to the fundamental issues of environmental pollution, setting the stage for technological interventions. The subsequent units explore the latest advancements in pollution monitoring and AI-driven solutions, bridging the gap between traditional practices and modern innovations. The final units highlight the economic and sustainable dimensions of these technologies, culminating in discussions on groundbreaking innovations in engineering and construction.

Each chapter presents in-depth analyses, practical case studies, and forward-looking insights, making this book a valuable resource for students, researchers, practitioners, and policymakers. It aims to inspire collaborative efforts and innovative thinking, ultimately contributing to a cleaner and healthier planet.

As the editor of this book, I express my sincere gratitude to the authors for their invaluable contributions and dedication to this project. I also thank the reviewers for their insightful feedback and the publishing team for their unwavering support. This book would not have been possible without their collective efforts.

It is our hope that this book serves as a guide and inspiration for leveraging AI-driven solutions to tackle the pressing challenges of environmental pollution. Let us embark on this journey together, using technology as a force for positive environmental change.

Pune, India Dr. Shashi Kant Tiwari
Pune, India Dr. Atul Fegade
Dammam, Saudi Arabia Dr. Mustafa Kamal

Contents

Editors and Contributors

About the Editors

Dr. Shashi Kant Tiwari is a highly accomplished Professor and a dedicated researcher, holding a bachelor's degree in mechanical engineering, a master's in Computer Aided Design from the Vellore Institute of Technology, Vellore (2017), and Ph.D. from the Indian Institute of Technology (Indian School of Mines) Dhanbad. His pioneering research areas lie in artificial intelligence, machine learning, deep learning, environmental acoustics, and data science. Dr. Tiwari has an extensive publication record in esteemed journals Environmental Science and Pollution Research, Applied Acoustics, Noise Mapping, NCEJ, MAPAN, IEEE Explore, serves as a reviewer for reputable publications such as Environmental Research, Scientific Report, ESPR, Measurement, and holds two registered design patents. Additionally, he is also associated in two projects of Saudi Electronics University, Saudi Arabia. He currently imparts his knowledge as Senior Assistant Professor at Balaji Institute of Modern Management, Sri Balaji University, Pune, in subjects like data science, and introduction to Python, machine learning, and deep learning inspiring the next generation of pioneers with his holistic perspective and impressive expertise.

Dr. Atul Fegade holds a Ph.D. in Marketing Management from Savitribai Phule Pune University (2018), an MBA in Marketing Management from the same university (2004), and a B.Sc. in Chemistry from Fergusson College, Pune (2002). He is UGC NET qualified and has over 15 years of experience in management education at both undergraduate and postgraduate levels. In addition, he has 4.5 years of industry experience, having held sales positions at reputable companies such as ICICI Lombard General Insurance Ltd., Novar India Pvt. Ltd. (part of the Honeywell Group), and GSK Pharmaceuticals Ltd.

Dr. Fegade's contributions to research and consultancy include authoring and presenting two papers at national and international conferences, as well as publishing seven research papers in reputable journals, including one in an ABDC B category publication and two indexed in Scopus. He has also published two book chapters indexed in Scopus.

His consultancy work includes a marketing research project aimed at improving packaged cow milk sales for Kolhapur Zilla Doodh Utpadak Sangh (Gokul Milk) and completing a minor research project funded by BCUD, Savitribai Phule Pune University.

Dr. Mustafa Kamal currently employed as an associate professor in the Department of Basic Sciences, College of Science and Theoretical Studies, at Saudi Electronic University in Saudi Arabia. He received his Ph.D. in statistics from Aligarh Muslim University in India in 2013. Prior to assuming his current position, Dr. Kamal worked as Research Project Fellow (From April 2012 to August 2014) in a Research Project (Code: 1206) of University Grants Commission SAP (DRS-I), Department of Statistics and Operations Research, Aligarh Muslim University, Aligarh.

Over the years, Dr. Kamal has established himself as a prolific researcher, with an impressive record of more than 75 research papers published in high-impact, peer-reviewed international journals and conference proceedings. His main research interest includes reliability theory, accelerated life testing, survey sampling, and distribution theory. He has also contributed to interdisciplinary fields, exploring the integration of statistical techniques with emerging areas like artificial intelligence, environmental sciences, statistical economics, and sustainable developments. His work emphasizes the application of statistical tools to address global challenges such as environmental sustainability and economic growth and reflects his deep expertise and innovative approach to addressing complex problems across various domains.

Contributors

Shoaib Ansari Department of Economics, AKI's Poona College of Arts, Science and Commerce, SPPU, Pune, Maharashtra, India

Gautam S. Bapat Dr. Vishwanath Karad MIT World Peace University, Pune, India

Alok Bhadauria Department of Mechanical and Industrial Engineering, Manipal Institute of Technology Bengaluru, Manipal Academy of Higher Education, Manipal, Karnataka, India

Shruti Bharadwaj Department of Computer Science & Engineering, Sharda University, Greater Noida, India;
Centre for SeNSE, Indian Institute of Technology, New Delhi, India

Susham Biswas Department of Computer Science Engineering, RGIPT Amethi, Uttar Pradesh, Amethi, India

Rajib Ghosh Chaudhuri Department of Chemical Engineering, National Institute of Technology, Durgapur, India

Parikesh Dhal GITA Autonomous College, BBSR, Gadajagasora, India

Rakesh Dubey Department of Mechanical Engineering, SITRC Nasik, Nasik, Maharashtra, India

Sachin Ghai Department of Management Studies, Graphic Era Deemed to be University, Dehradun, Uttarakhand, India

Manish Ghalwan Uttaranchal Institute of Technology, Uttaranchal University, Dehradun, Uttarakhand, India

Ashulekha Gupta Department of Management Studies, Graphic Era Deemed to be University, Dehradun, Uttarakhand, India

Ahsanuddin Haider College of Administrative and Financial Sciences, Saudi Electronic University, Riyadh, Saudi Arabia

Mohammad Haseeb School of Economics and Management, and Center for Industrial Economics, Wuhan University, Wuhan, China;

Department of Management Studies, Graphic Era Deemed to be University, Dehradun, Uttarakhand, India

Joseph T. Jose NLC India Ltd., Neyveli, Tamil Nadu, India;
Coal India Limited, Kolkatta, India

Mustafa Kamal Department of Basic Sciences, College of Science and Theoretical Studies, Saudi Electronic University, Dammam, Saudi Arabia

Aman Kumar Indian Institute of Management–Ranchi (IIM–Ranchi), Ranchi, India

Ankesh Kumar Department of Environmental Science and Engineering, Indian Institute of Technology, (Indian School of Mines), Dhanbad, India

Rakesh Kumar PP Savani University, Dhamdod, Gujarat, India;
Department of Mechanical Engineering, School of Engineering, P P Savani University, Surat, India

Vivek Kumar Department of Environmental Science and Engineering, Indian Institute of Technology, (Indian School of Mines), Dhanbad, India

Mousumi Maji Department of Chemical Engineering, National Institute of Technology, Durgapur, India

Mrinal Kanti Mandal Department of Chemical Engineering, National Institute of Technology, Durgapur, India

Raziqa Masood Department of Computer Science & Engineering, Integral University, Lucknow, India

Supriya Paigude Dr. Vishwanath Karad MIT World Peace University, Pune, India

Namrata Pandey Faculty of Civil Engineering, Shri Ramswaroop Memorial University, Barabanki, India

Rohit Patel Department of Environmental Science and Engineering, Indian Institute of Technology (Indian School of Mines), Dhanbad, India

Md. Mominur Rahman Bangladesh Institute of Governance and Management, Dhaka, Bangladesh

R. Rahul Department of Mechanical Engineering, Birla Institute of Technology and Science, Pilani, India

Deepak Kumar Rakesh University of Economics and Health Sciences, Warsaw, Poland

Prabhat Ranjan Padmabhooshan Vasantdada Patil Institute of Technology, Pune, India;
NLC India Ltd., Neyveli, Tamil Nadu, India

Masood U. Rehman College of Computing and Informatic, Saudi Electronic University, Dammam, Saudi Arabia

Shivam Saw Department of Chemical Engineering, National Institute of Technology, Durgapur, India

Deependra Sharma Dr. Vishwanath Karad MIT World Peace University, Pune, India

Mohd Shuaib School of Economics and Management, and Center for Industrial Economics, Wuhan University, Wuhan, China;
Department of Management Studies, Graphic Era Deemed to be University, Dehradun, Uttarakhand, India

Neeru Sidana Amity School of Economics, Amity University, Noida, India

Prasoon Kumar Singh Department of Environmental Science and Engineering, Indian Institute of Technology (Indian School of Mines), Dhanbad, India

Anubhava Srivastava Department of Computer Science & Engineering, Sharda University, Greater Noida, India

Sivarao Subramonian Faculty of Manufacturing Engineering, University Teknikal Malaysia Melaka, Durian Tunggal, Melaka, Malaysia

Varun Tripathi Uttaranchal Institute of Management, Uttaranchal University, Dehradun, Uttarakhand, India

Himanshu Vashishtha Department of Materials Science and Metallurgy, University of Cambridge, Cambridge, UK

Gyan Wrat Aalborg University, Aalborg, Denmark

Part I
Understanding Environmental Pollution

Chapter 1
Environmental Pollution: A Global Challenge

Ankesh Kumar, Prasoon Kumar Singh, and Vivek Kumar

Abstract Environmental pollution is a critical global challenge that affects ecosystems, human health, and the economy. This study provides a comprehensive examination of the various forms, causes, impacts, and solutions to environmental pollution. It explores different types of pollution, including air, water, and soil pollution, along with their primary sources, including waste management, transportation, agriculture, and industrial operations. This chapter introduces environmental pollution as a significant global issue, tracing its historical development and highlighting key events that have shaped contemporary understanding. It then delves into the specific types and sources of pollution, providing examples and statistics that illustrate the widespread nature of the problem globally. The discussion extends to the environmental, social, and economic impacts of pollution, including its effects on human health, ecosystems, and climate change, while also presenting examples of successful pollution control initiatives. Furthermore, innovative technologies for pollution reduction and environmental quality improvement are explored, with a focus on technology's role in monitoring and remediation. This study also examines the role of policies and regulations in pollution control, addressing the challenges of implementing and enforcing environmental laws on a global scale. Therefore, this chapter aims to provide a comprehensive understanding of the challenges posed by environmental pollution and to identify effective strategies for addressing this global issue, emphasizing the need for a collaborative and multifaceted approach.

Keywords Environmental pollution · Types of pollution · Sources of pollution · Environmental impacts · Pollution control

A. Kumar · P. K. Singh (✉) · V. Kumar
Department of Environmental Science and Engineering, Indian Institute of Technology, (Indian School of Mines), Dhanbad 826004, India
e-mail: pks0506@iitism.ac.in

A. Kumar
e-mail: ankeshjaiswal1983@gmail.com

V. Kumar
e-mail: vivekpoddar883@gmail.com

S. K. Tiwari et al. (eds.), *AI-Driven Environmental Pollution Management*, Climate Risks and Solutions, https://doi.org/10.1007/978-3-031-96243-1_1

1.1 Introduction

Environmental pollution refers to the contamination of the Earth's ecosystems or natural environment by harmful substances or pollutants, causing negative impact on ecosystems, living organisms, human health, planet, and natural processes as a whole. Human activities such as industrialization, urbanization, deforestation, and unsustainable agriculture methods all directly lead to pollution. It impacts the quality of air we breathe, the water we drink, and the soil that sustains plant and animal life, making it one of the most significant global environmental concerns (Ukaogo et al. 2020). It includes a range of pollutants caused by industrial, agricultural, and urban activities, such as those found in the air, water, soil, and noise. There are many distinct types of environmental pollution, and they all are affecting different aspects of the natural world. The main source of air pollution is the release of dangerous gases and particulate matter from industrial processes, vehicle exhaust, and the burning of any fossil fuels. These pollutants, including carbon monoxide (CO), sulfur dioxide (SO_2), and nitrogen oxides (NO), contribute to global warming of the planet, respiratory diseases, and acid rain (Munsif et al. 2021). Water pollution occurs when harmful substances like waste from industries, agricultural runoff (pesticides and fertilizers), plastics, and untreated sewage are discharged into water bodies, degrading water quality and harming aquatic life (Saw et al. 2021). Soil pollution results from excessive use of chemical fertilizers, pesticides, and improper disposal of industrial waste, which contaminates land, reduces agricultural productivity, and impacts food safety (Hossain et al. 2022). Noise pollution, caused by vehicular traffic, construction, and industrial operations which disrupt human health and wildlife behavior (Patel et al. 2024a). Lastly, thermal pollution arises when industries release heated water or air into the environment, altering natural temperature balances and impacting temperature-sensitive ecosystems (Christianson 1968). These different types of pollution are caused by human activity and have a continuous effect on the ecosystem and public health.

The problem of environmental pollution became more noticeable with the industrial revolution in the eighteenth and nineteenth centuries when the rapid expansion of factories and urban areas led to significant emissions of pollutants. A major turning point in the industrial revolution occurred when coal replaced other energy sources as the main source of energy, severely contaminating the air and water. Significant incidents like the Great Smog of London in 1952, which resulted in thousands of premature deaths due to thick smog, brought attention to the risks of uncontrolled air pollution and sparked public concern (Mitvhell and Sweeney 2018). By the mid-twentieth century, awareness of environmental issues grew, notably with the publication of Rachel Carson's Silent Spring in 1962 (Meiners et al. 2012). The book revealed the detrimental effects of pesticides, especially DDT, on human health and ecosystems, which served as inspiration for the modern environmental movement. The Minamata disease incident in Japan, resulting from mercury poisoning due to industrial waste, further underscored the consequences of pollution, drawing international attention to industrial accountability (Fujigaki 2015). As pollution levels continued to rise, it became clear that this was not just a local or national problem but a global issue, exacerbated by globalization and industrial activities that crossed borders. The 1972 Stockholm Conference marked a significant moment, as it was the first United Nations conference focused on global environmental challenges, leading to increased international collaboration (Paglia 2021). Subsequent milestones, such as the establishment of the U.S. Environmental Protection Agency (EPA) in 1970 and international treaties like the Montreal Protocol (1987) (Gardner 2020) and the Paris Agreement (2015), signified a growing commitment to addressing pollution on a global scale.

Environmental pollution levels vary immensely among countries, depending on factors such as industrial operations, population density, and environmental regulations. Large manufacturing sectors, high energy consumption, and fast urbanization are all associated with higher pollution levels, whereas nations with more stringent environmental laws and a greater uptake of renewable energy typically have lower pollution levels. For example, the Fig. 1.1 shows the air quality trends worldwide over the months of 2022, in which countries like Bangladesh and Kuwait have significantly higher air pollution levels due to rapid industrial growth and coal-based energy, whereas countries like Bahrain and Nepal have lower pollution levels due to stricter environmental policies and investments in clean energy. The Iraq shows moderate pollution levels, largely attributed to transportation emissions and industrial activities, though it has made strides in reducing certain pollutants through regulations like the Clean Air Act.

Nowadays, environmental pollution is not only a national concern but a global challenge due to its far-reaching effects. It is significant due to its vast and linked

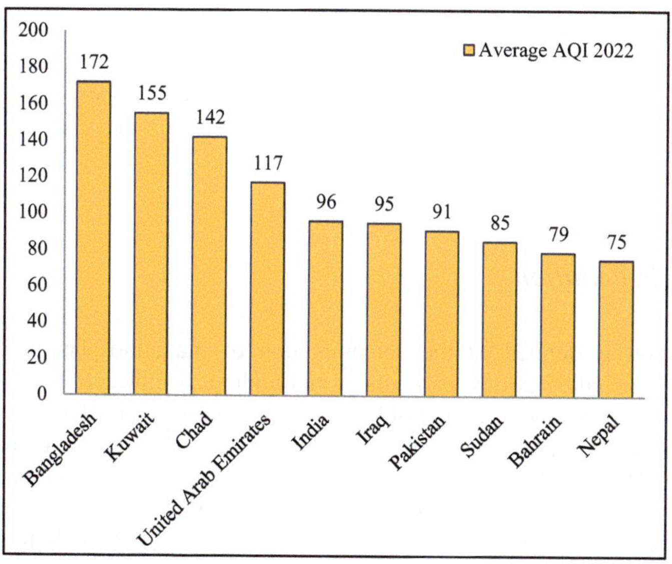

Fig. 1.1 Top 10 most polluted countries in the world in the year 2022 (*Source* www.aqi.in)

impacts on ecosystems, human health, and climate. Pollution damages air, water, and soil resources, disturbs natural ecosystems, and leads to the loss of biodiversity. These effects have a serious negative impact on public health, especially for people with disabilities. Additionally, pollution is a major driver of climate change; intensify global warming as a result of greenhouse gas emissions and carbon degradation sinks like forests and oceans. Due to the global nature of pollution, comprehensive strategies and international collaboration are required as its consequences extend beyond national boundaries. Effective pollution control is essential to ensure the sustainability of natural resources, protect public health, and stabilize the climate for future generations. Key initiatives for reducing these effects include the creation of green technologies, more stringent environmental laws, and global accords like the Paris Agreement.

This study provides an in-depth review of the different types of pollution such as air, water, and soil-along with their major sources, including industrial activities, transportation, agriculture, and waste disposal. It traces the historical development of pollution awareness and highlights key events shaping our current understanding. The chapter highlights effective control measures while discussing the negative effects of pollution on the environment, society, and economy, including the implications on health. Furthermore, the chapter addresses global policies, regulatory challenges, and the importance of collaboration in managing pollution, offering a comprehensive view of the issue and potential strategies for its resolution.

1.2 Types of Pollution and Their Sources

Environmental pollution is the addition of undesirable substances or products into the environment in a manner which leads to adverse impacts to living organisms, ecosystems, and the environment in general. There are many different types of pollution, and each has very different characteristics and origins. Here's a comprehensive overview of the different types of environmental pollution:

1.2.1 Air Pollution

Air pollution is defined as the contamination of the atmosphere by harmful substances, including gasses, particulates and biological molecules. Vehicle emissions, industrial processes, household goods, agriculture, and trash incineration are some of the main sources. Natural events like wildfires and volcanic eruptions also contribute. These pollutants damage ecosystems and contribute to climate change in addition to posing a substantial risk to human health, including respiratory and cardiovascular disorders (Manisalidis et al. 2020). Addressing air pollution requires regulatory measures, technological advancements, and public awareness to reduce emissions and improve air quality, ultimately protecting both human health and the environment.

As shown in Fig. 1.2 and Table 1.1, some of the major sources of air pollutions such as vehicle emissions from cars, trucks, and buses contribute to air pollution by releasing harmful gases like nitrogen oxides (NO_x), carbon monoxide (CO), volatile organic compounds (VOCs), and particulate matter (PM) due to fossil fuel combustion. Industrial activities, such as factories and power plants, emit poisonous pollutants like sulfur dioxide (SO_2), heavy metals, and particulates (Bhanarkar et al. 2005). Agricultural activities also play a role through the use of fertilizers and pesticides, which release ammonia, and livestock farming generates methane (CH_4), a potent greenhouse gas (Burke and Lashof 1990). Household activities, including the use of paints, cleaning products, and air fresheners, release VOCs, while burning wood or coal for heating emits harmful particles and gases. Construction and demolition projects also contribute to air pollution through dust and particulate matter from digging, grinding, and demolition activities (Adepoju et al. 2024). Waste incineration can produce toxic substances like dioxins and furans. Natural events, such as wildfires, volcanic eruptions, and dust storms, also release significant amounts of air pollutants (Hait and Sahu 2024). Additionally, residential heating, especially with solid fuels like wood and coal, can generate PM and CO which further contribute to air pollution.

1.2.2 Water Pollution

The term water pollution refers to the contamination of water bodies, such as rivers, lakes, oceans, and groundwater, due to harmful substances. Figure 1.3 shows the major sources of water pollution including industrial discharges, domestic sewage, agricultural runoff, and plastic waste. Pollutants like heavy metals, pesticides, and pathogens can harm aquatic ecosystems and pose significant risks to human health. Water pollution can lead to problems such as biodiversity loss, contaminated drinking water, and the spread of waterborne diseases like cholera, diarrhea, dysentery, E. coli infection, hepatitis A, salmonellosis, typhoid fever, and traveler's diarrhea (Madhav et al. 2020). Effective management and pollution control measures are essential, such as wastewater treatment, sustainable farming practices, and public awareness campaigns, to protect water quality and preserve aquatic environments for future generations.

Some of the major Sources of water pollutions includes industrial discharges containing wastes from factories, manufacturing plants dump heavy metals and chemical and toxic wastes from industries such as dyeing, metal plating, and oil refining into water sources (Mishra et al. 2019; Saw et al. 2023a, b). Agricultural runoff, caused by the use of fertilizers, pesticides, and herbicides, leads to the introduction of nitrates, phosphates, and other chemicals into nearby rivers and lakes as shown in Table 1.2, resulting in eutrophication and harmful to aquatic life during rainstorms (Saw et al. 2023a, b). Domestic sewage and wastewater from households

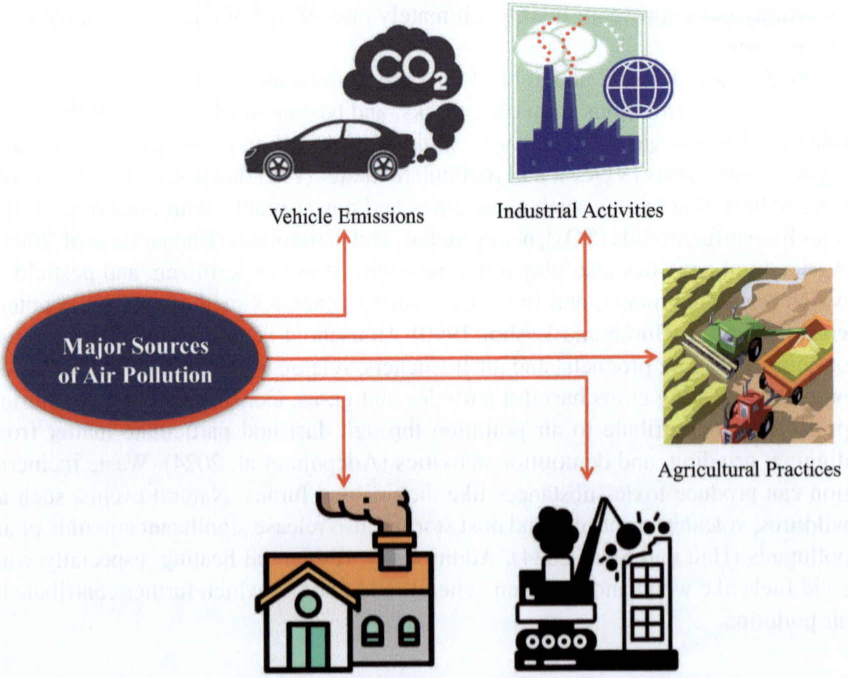

Fig. 1.2 Major sources of air pollution

Table 1.1 Sources of air pollution and the associated pollutants

Source of air pollution	Associated pollutants
Vehicle emissions	Nitrogen oxides (NO_x), Carbon monoxide (CO), VOCs, Particulate matter (PM)
Industrial activities	Sulphur dioxide (SO_2), Heavy metals, Particulates, VOCs
Agricultural practices	Ammonia (NH_3), Methane (CH_4), Pesticides, Nitrous oxide (N_2O)
Household activities	VOCs, Carbon monoxide (CO), Particulate matter (PM)
Construction and demolition	Dust, Particulate matter (PM), Asbestos
Waste incineration	Dioxins, Furans, VOCs, Heavy metals
Natural sources	Particulates, Sulphur dioxide (SO_2), Carbon monoxide (CO)
Residential heating	Particulate matter (PM), Carbon monoxide (CO), Sulphur dioxide (SO_2)

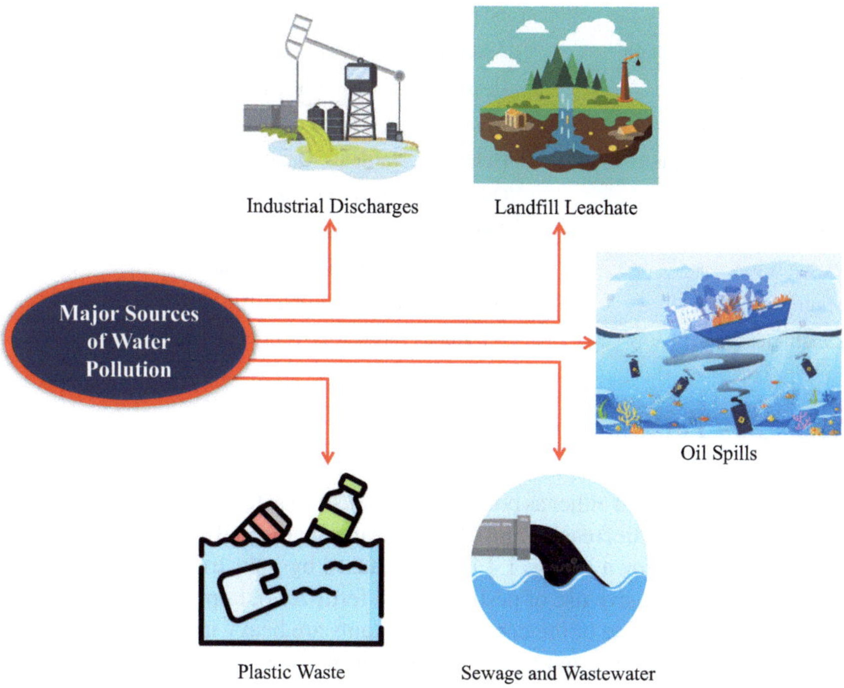

Fig. 1.3 Major sources of water pollution

and industrial facilities can introduce pathogens, nutrients, and organic matter into water bodies when it is inadequately treated which results in posing health risks and degrading water quality (Karri et al. 2021). Also contaminated rainwater picks up a variety of chemicals including oil and grease, heavy metals, debris from buildings all drain into storm water system and finally into water bodies—rivers, lakes etc. Oil spills from tankers, pipelines, or drilling operations can significantly harm marine ecosystems and wildlife. Plastic waste, including microplastics, often enters water bodies through littering, improper disposal, and runoff which results in adversely affecting marine life and infiltrating the food chain (Du et al. 2021). Mining activities can also introduce heavy metals and sediments into nearby water resources that ultimately contaminating drinking water supplies and disrupt ecosystems. Lastly, the landfill leachate, which is a toxic liquid produced from waste buried in landfills, can seep into the ground which leads to contaminating groundwater and nearby water bodies with hazardous substances.

Table 1.2 Sources of water pollution and the associated pollutants

Source of water pollution	Associated pollutants
Industrial discharges	Heavy metals, Chemicals, Toxic substances
Agricultural runoff	Nitrates, Phosphates, Pesticides, Herbicides
Sewage and wastewater	Pathogens, Nutrients, Organic matter
Storm water runoff	Oil, Grease, Heavy metals, Trash
Oil spills	Crude oil, Hydrocarbons, Toxins
Plastic waste	Plastics, Microplastics, Chemicals
Mining activities	Heavy metals, Sediments, Toxic chemicals
Landfill leachate	Hazardous substances, Nutrients, Toxic liquids

1.2.3 Soil Pollution

Soil pollution is defined either as poisonous chemicals waste or toxic substances in the soil that will result in decreasing and deteriorate the quality and health of the soil. This contamination as shown in Fig. 1.4 is often caused by industrial activities, agricultural practices (excessive use of pesticides and fertilizers), improper waste disposal, and deforestation. Some of the common pollutants are heavy metals, chemicals like petroleum hydrocarbons, and plastic waste. Soil pollution can reduce soil fertility, disrupt ecosystems, and harm plants, animals, and humans who depend on contaminated land (Mishra et al. 2016). Over time, it can also affect water quality through leaching, spreading pollutants into groundwater or nearby water bodies which cause further environmental risks.

Some of the major sources of soil pollutions are industrial activities and mining operations which frequently release toxic chemicals including heavy metals like lead and mercury, and hydrocarbons into the soil. It is clear from Table 1.3 also that Agricultural practices are also responsible for soil contamination through over use of pesticide, herbicides and fertilizers which bring about the deleterious effects of harm some chemicals and nutrient imbalance. Further, improper disposal of solid waste: plastics, e-waste and industrial effluents and byproducts create long term soil pollution. Untreated or poorly managed sewage can infiltrate the soil, introducing harmful pathogens and chemicals that degrade the soil quality (Havugimana et al. 2017). Accidental oil spills during extraction or transportation can contaminate vast areas of soil, rendering them unfit for agriculture or habitation. Furthermore, Deforestation increases soil erosion risk, and chemicals used in the process can worsen soil deterioration. Together, these factors contribute significantly to the deterioration of soil health and its ability to sustain life.

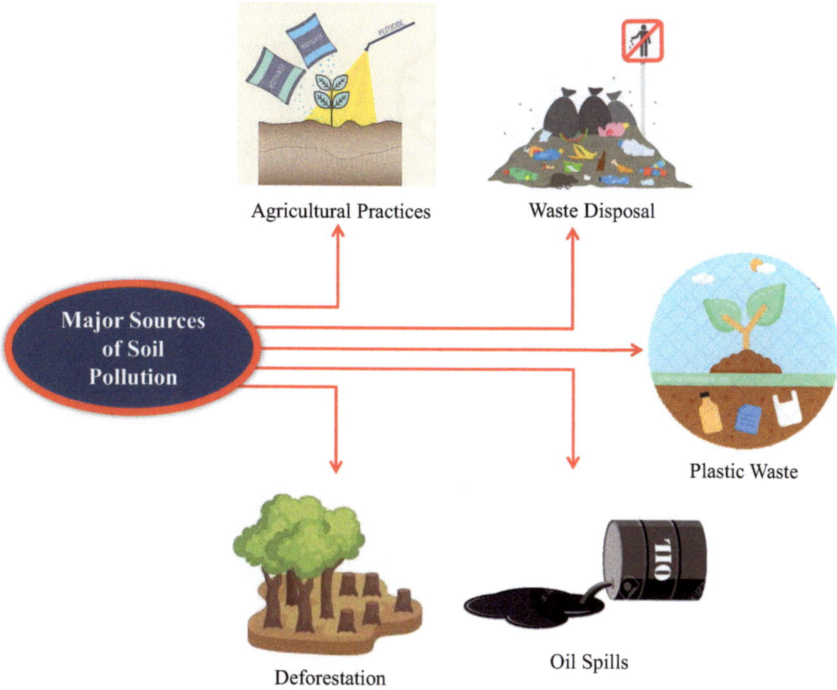

Fig. 1.4 Major sources of soil pollution

Table 1.3 Sources of soil pollution and the associated pollutants

Source of soil pollution	Associated contaminants
Industrial activities	Heavy metals (lead, mercury), Hydrocarbons
Agricultural practices	Pesticides, Herbicides, Fertilizers
Waste disposal	Plastics, E-waste, Industrial waste
Sewage and wastewater	Pathogens, Organic matter, Chemicals
Oil spills	Petroleum hydrocarbons
Deforestation	Soil erosion, Chemicals from land clearing

1.2.4 Noise Pollution

Noise pollution is the excessive or harmful levels of sound in the environment which is often caused by human activities such as traffic, industrial operations, construction, and urbanization as shown in Fig. 1.5. It disrupts daily life, causing stress, hearing loss, and sleep disturbances. Prolonged exposure to high noise levels can also contribute to cardiovascular diseases (Hahad et al. 2019; Tiwari et al. 2023a, b). Wildlife is also affected, as it interferes with animal communication, migration,

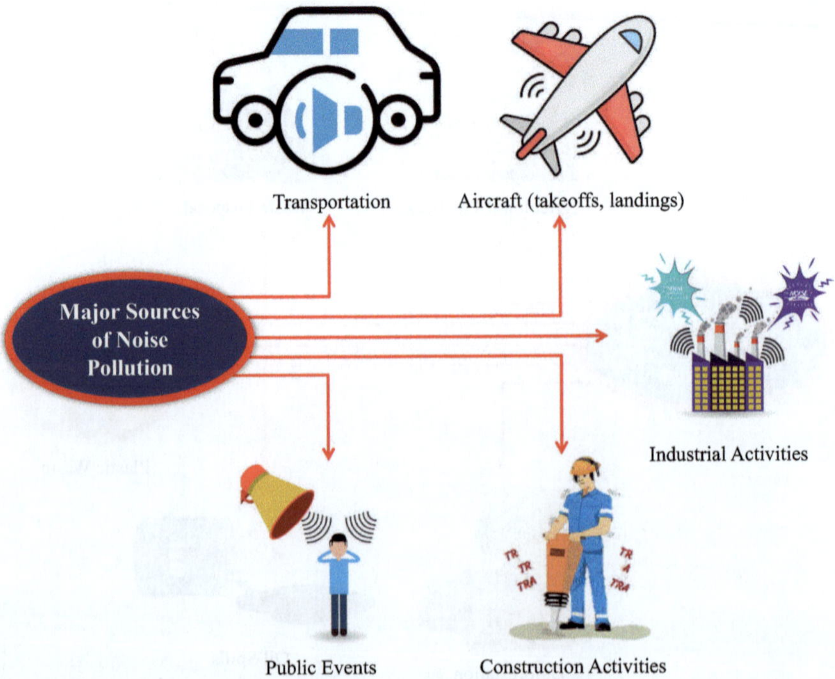

Fig. 1.5 Major sources of noise pollution

and reproduction. In order to safeguard the environment and public health, noise pollution can be reduced in densely populated regions by implementing rules on industrial noise levels, installing noise barriers, and improving urban design (Tiwari et al. 2023a, b, 2024).

Among all the sources of noise pollution, transportation sectors are a major source, including cars, lorries, buses, and motorcyclists generating enormous noise from their engines, horns, and tyre friction on the road (Patel et al. 2024b; Tiwari et al. 2022). Airports have also significant noise levels owing to aircraft takeoffs, landings, and over flights, which damage adjacent neighborhoods. Additionally, trains contribute to noise pollution, particularly during acceleration, braking, and at crossings, impacting areas adjacent to rail lines. Industrial activities, including those at factories, construction regions, and mining operations, generate continual noise from heavy machinery and equipment, such as drilling, welding, and cutting, which can negatively affect workers and neighboring inhabitants. Construction activities, including building construction, roadwork, and infrastructure development, generate loud noises from machinery, demolition, and the transportation of materials. Household and commercial sources, such as loud music, home appliances, air conditioning units, and power tools, also add to noise pollution in residential areas (Patel et al. 2024c; Farooqi et al. 2022). Furthermore, commercial facilities like clubs, restaurants, and entertainment venues contribute significantly to city noise levels. Public events, such as concerts,

Table 1.4 Sources of noise pollution and the associated pollutants

Source of noise pollution	Associated contaminants
Transportation	Engine noise, tire friction, horn noise, exhaust emissions, particulate matter, carbon monoxide, nitrogen oxides
Aircraft (takeoffs, landings)	Jet engine noise, sonic boom, fuel combustion emissions, particulate matter, carbon monoxide, nitrogen oxides
Trains	Train engine noise, brake squeal, wheel-rail friction noise, diesel emissions (if applicable)
Industrial activities	Machinery noise, vibrations, air pollutants (dust, fumes, chemical vapors), particulate matter, sulfur oxides
Construction activities	Machinery noise, drilling, demolition, dust, particulate matter, emissions from construction vehicles
Household and commercial sources	Loud music, air conditioning noise, appliance vibrations, dust, particulate matter, volatile organic compounds (VOCs)
Public events	Loudspeakers, crowd noise, smoke from fireworks or generators, dust, and litter

athletic events, and festivals, frequently include loudspeakers and crowds, resulting in temporary but intense noise pollution (Patel et al. 2022). Together, all these factors shown in Table 1.4 create a complex landscape of noise that affects both urban and rural environments.

1.2.5 Radioactive Pollution

Radioactive pollution takes place when radioactive substances are released into the environment, contaminating the air, water, and soil. As shown in Fig. 1.6, this can be caused by nuclear power plant accidents, inappropriate disposal of nuclear waste, mining of radioactive materials, or nuclear weapons testing. Radioactive particles, such as uranium, plutonium, and cesium, emit harmful ionizing radiation, which can cause serious health issues like cancer, genetic alterations, and radiation sickness (Vaiserman et al. 2018). These pollutants remain dangerous for long periods due to their long half-lives, making decontamination difficult. It also harms ecosystems by contaminating plants and animals, resulting in bioaccumulation and long-term environmental damage.

One of the major sources of radioactive pollutions is nuclear power plants which pose significant environmental risks, as accidents or leaks can release radioactive materials, including uranium, plutonium, and cesium, into the air, water, or soil. Similarly, Table 1.5 shows that atmospheric testing of nuclear weapons generates radioactive fallout that contaminates extensive areas, impacting air, water, and soil quality (Adeola et al. 2023). Medical waste also contributes to pollution when radioactive isotopes used in treatments and diagnostic procedures are improperly disposed of.

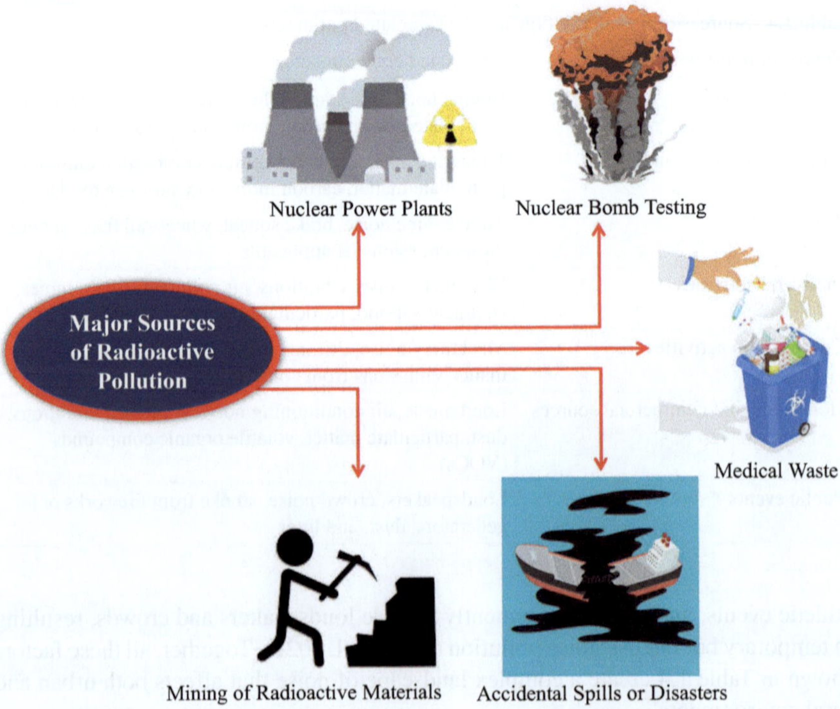

Fig. 1.6 Major sources of radioactive pollution

Additionally, the mining and processing of radioactive materials, such as uranium, can release radiation and toxic byproducts that contaminate nearby environments. Improper disposal of radioactive waste from nuclear reactors, hospitals, and industrial activity worsens the problem, as seeping radiation can impair both groundwater and soil health (Rajkhowa et al. 2021). Notably, catastrophic tragedies like the Chernobyl and Fukushima disasters have resulted in huge radiation releases, leading to widespread contamination and long-term environmental implications. Collectively, these factors underscore the urgent need for stringent safety measures and responsible management of radioactive materials to protect the environment and public health.

1.3 Impacts of Pollution

The environment is greatly impacted by pollution, which has detrimental effects on ecosystems and human health in addition to lowering the quality of the air, water, and soil. Air pollution, primarily from vehicle emissions and emissions from industries and burning of fossil fuel leads to air pollution which in turn leads to global

Table 1.5 Sources of radioactive pollution and the associated pollutants

Source of radioactive pollution	Associated contaminants
Nuclear power plants	Uranium, Plutonium, cesium
Nuclear weapons testing	Radioactive fallout
Medical waste	Radioactive isotopes
Mining and processing of Radioactive materials	Radiation, toxic byproducts
Improper disposal of radioactive waste	Radiation
Accidental spills or disasters	Large amounts of radiation

warming, respiratory diseases, and effects such as the formation of acid rain, which is destructive to plant life, aquatic creatures and to civil structures. (Manisalidis et al. 2020). Water pollution is caused by the industrial waste, agricultural runoff, and sewage discharge, leads to the contamination of freshwater sources, threatening marine ecosystems and human access to clean water. Soil pollution, often due to chemicals, pesticides, and waste, reduces agricultural productivity and threatens food security. All together, these pollution types of damages biodiversity, worsen climate change, and lead to the destruction of wildlife habitats, which eventually affects the delicate balance of natural systems that support life.

1.3.1 Environmental Impacts of Pollution

Environmental pollution has far-reaching impacts on air quality, water sources, soil health, biodiversity, and climate stability. Air quality degradation causes smog and acid rain, which harms flora, aquatic ecosystems, and human health. Greenhouse gases like carbon dioxide and methane when released into the atmosphere causes climate change, global warming and changes weather patterns that cause phenomena like droughts, tsunamis (Manisalidis et al. 2020). Water contamination poses another major challenge, with polluted water bodies harming aquatic life and compromising access to clean drinking water for humans. Contaminants disturb the food chain and can cause the deaths of marine species, further diminishing biodiversity. Similarly, Table 1.6 shows that soil degradation happens when contaminants impair soil fertility, thereby impacting plant development and disrupting entire ecosystems. Contaminated soil can hinder agricultural productivity, leading to food insecurity, as harmful substances absorbed by plants can also impact human health (Edo et al. 2024). The loss of biodiversity is driven by habitat destruction, pollution, and the introduction of toxic chemicals, contributing to species extinction, and disrupting ecological balance. As species decline or disappear, ecosystems become less resilient, increasing overall environmental instability. Additionally, climate change stemming from heightened greenhouse gas emissions alters ecosystems and poses significant threats to various species and habitats. This shift can lead to altered migration patterns, disrupted breeding seasons, and greater vulnerability to diseases (Kovaleva

Table 1.6 Summarizing the types of pollution and their associated impacts

Type of pollution	Impact
Air quality degradation	Leads to smog, acid rain, and contributes to climate change through greenhouse gas emissions.
Water contamination	Polluted water bodies affect the life of aquatic plants or animals and reduce access to clean water for human consumption
Soil degradation	Reduces fertility, affects plant growth, and disrupts ecosystems
Loss of biodiversity	Habitat destruction and toxic chemicals cause species extinction and disrupt ecological balance
Climate change	Increased emissions of CO_2 and methane contribute to global warming, altering weather patterns
Ozone layer depletion	Release of harmful chemicals, such as CFCs, depletes the ozone layer, increasing UV radiation

et al. 2019). Lastly, the depletion of the ozone layer, primarily due to the release of hazardous substance like chlorofluorocarbons (CFCs), increases ultraviolet (UV) radiation reaching the Earth's surface. This increase in UV radiation intensifies risks to human health, including enhanced rates of skin cancer and eye cataracts, as well as adverse impacts on plant development and marine ecosystems. Together, these impacts illustrate the interconnectedness of environmental health, human well-being, and the intricate balance of ecosystems, emphasizing the urgent need for comprehensive environmental management and policy initiatives.

1.3.2 Social Impacts of Pollution

Pollution significantly impacts human health, leading to a range of serious issues. Table 1.7 represents the various forms of pollution and their associated impacts on health, quality of life, displacement, social justice, and cultural integrity. Poor air quality is linked to respiratory diseases, heart problems, and increased cancer rates, as harmful pollutants enter the body and aggravate existing health conditions. Similarly, contaminated water sources pose severe effect, causing gastrointestinal diseases and other health problems when consumed. Beyond direct health effects, pollution reduces the overall quality of life for affected populations. Noise, air, and water pollution create discomfort and stress, leading to diminished well-being. The long-term presence of these contaminants can make routine activities difficult, contributing to a general decrease in life satisfaction and mental health (Tzivian et al. 2015). Furthermore, environmental degradation caused by pollution can result in relocation, as pollution-induced natural disasters such as floods or chemical spills compel communities to flee from their homes. This relocation not only disrupts lives but also creates economic challenges for those who must start anew in unfamiliar areas. The burden of pollution is often inequitably distributed, with vulnerable communities

Table 1.7 Various types of pollution and their associated impacts on health, quality of life, displacement, social justice, and cultural integrity

Type of pollution	Impact
Air pollution	Increases respiratory diseases, heart problems, and cancer due to poor air quality
Water pollution	Leads to health issues from contaminated drinking water, contributing to diseases and overall health risks
Noise pollution	Causes discomfort, stress, and reduces well-being in affected populations
Environmental degradation	Pollution-induced natural disasters force people to migrate and displace communities
Social inequities	Vulnerable communities disproportionately bear the pollution burden, leading to social injustice
Cultural loss	Damages natural heritage sites and disrupts traditional livelihoods, leading to cultural erosion

bearing a disproportionate share of its consequences. These populations frequently experience higher exposure to pollutants and fewer resources for mitigation, leading to social injustice and aggravating existing inequalities. Finally, pollution can lead to cultural loss as environmental degradation damages natural heritage sites and disrupts traditional livelihoods (Mutugi and Kiiru 2015). Communities that rely on their environment for cultural practices, nourishment, and identity may face threats to their ways of life, culminating in the loss of cultural heritage and social cohesiveness. This diverse impact of pollution emphasizes the critical need for comprehensive solutions to reduce environmental damage and preserve public health.

1.3.3 Economic Impacts of Pollution

Beyond just degrading the environment, pollution has a substantial negative economic impact on a number of industries and towns. Table 1.8 represents the various economic impacts of pollution along with their associated consequences. Some of these are increased healthcare costs which are a primary concern, as the rise in health issues linked to pollution-such as respiratory diseases, heart problems, and cancer-results in higher medical expenses for individuals and governments alike. This financial burden can strain healthcare systems and divert resources from other essential services. Additionally, pollution-induced illnesses lead to a noticeable loss of productivity in the workforce. When employees are unwell, their ability to work effectively diminishes, ultimately reducing overall economic output. This decline in productivity can adversely affect businesses and economies as a whole (Zhang et al. 2011). Pollution is also a threat since it affects structures within the natural environment. For example, acid rain can erode roads, buildings, and bridges, necessitating expensive upkeep and repairs. The long-term deterioration of infrastructure not

Table 1.8 Various economic impacts of pollution along with their associated consequences

Type of impact	Description
Healthcare costs	Increased health issues lead to higher medical expenses for both individuals and governments
Loss of productivity	Illnesses caused by pollution reduce workforce productivity and economic output
Damage to infrastructure	Acid rain and pollution corrode buildings, roads, and other infrastructure, leading to costly repairs
Agricultural losses	Soil pollution and water contamination reduce crop yields, affecting food supply and farmer incomes
Tourism decline	Pollution of natural areas can deter tourists, impacting local economies dependent on tourism
Environmental remediation costs	Governments and businesses face high expenses for cleaning up polluted areas

only poses safety risks but also requires significant financial investments to address (Kumar and Imam 2013). Agricultural losses represent another major impact, as soil pollution and water contamination can severely reduce crop yields. This decline affects the food supply chain and farmer incomes, threatening food security and the livelihoods of those dependent on agriculture. Moreover, pollution in natural areas can deter tourists, leading to a decline in tourism revenue. Regions that rely heavily on tourism may experience economic downturns as visitors seek cleaner, more pristine environments. Lastly, environmental remediation costs pose a significant financial challenge. The cleanup of polluted areas requires significant resources from both corporations and governments, which can put a strain on budgets and postpone investments in other important projects. Collectively, these economic impacts underscore the urgent need for effective pollution control measures and sustainable practices to protect both public health and economic stability.

1.3.4 Impacts of Pollution on Climate Change

Greenhouse gas emissions, primarily in the form of carbon dioxide (CO_2), methane, and nitrous oxide, play a significant role in trapping heat in the Earth's atmosphere, resulting in global warming as shown in Table 1.9. This rise in temperatures disturbs ecosystems and contributes to a host of environmental problems. One major consequence of global warming is ozone depletion, as chemicals like chlorofluorocarbons (CFCs) damage the ozone layer, allowing increased ultraviolet (UV) radiation to penetrate the atmosphere. There is major health hazards associated with this increased UV exposure for both individuals and ecosystems (Saxena 2018). Furthermore, pollution compels climate change, wheeling about weather which causes hurricanes, floods, and droughts through change of weather conditions. These events not only threaten human life and infrastructure but also disrupt agricultural productivity

Table 1.9 Impacts of various forms of pollution on the environment

Type of pollution	Impact
Greenhouse gas emissions	Air pollutants like CO_2, methane, and nitrous oxide trap heat, leading to global warming.
Ozone depletion	Chlorofluorocarbons (CFCs) and other chemicals damage the ozone layer, increasing UV radiation exposure
Altered weather patterns	Climate change exacerbated by pollution results in extreme weather events such as hurricanes, floods, and droughts.
Rising sea levels	Melting ice caps due to warming oceans threaten coastal ecosystems and human settlements
Increased wildfires	Pollution-induced climate change intensifies wildfires, destroying habitats and releasing more pollutants into the atmosphere.

and water supply, impacting food security. Rising sea levels, driven by the melting of ice caps and glaciers due to warming oceans, further threaten coastal ecosystems and human settlements, leading to increased flooding and habitat loss (Griggs and Reguero 2021). Moreover, the interplay between pollution and climate change results in more frequent and intense wildfires. These wildfires not only destroy habitats but also release additional pollutants into the atmosphere, creating a vicious cycle of environmental degradation. Collectively, these impacts illustrate the complex and interconnected challenges posed by pollution and climate change, emphasizing the urgent need for comprehensive strategies to mitigate their effects.

1.4 Pollution Control and Mitigation

Pollution control and mitigation strategies are essential for minimizing the harmful effects of various types of pollution, including air, water, soil, and noise pollution. Effective management requires tailored approaches for each category, supported by regulatory frameworks, technological solutions, and sustainable practices.

To control and mitigate air pollution, governments and industries implement regulatory measures such as air quality standards (e.g., WHO guidelines and national ambient air quality standards). Technological solutions like scrubbers, electrostatic precipitators, and catalytic converters help reduce emissions of harmful particulates and gases from industrial processes and vehicles (Asif et al. 2022). Transitioning to renewable energy sources such as wind, solar, and hydropower reduces the reliance on fossil fuels, thus lowering greenhouse gas emissions. Promoting public transportation, cycling, walking, and the use of electric vehicles further minimizes vehicular emissions. Additionally, afforestation plays a critical role in absorbing carbon dioxide and improving overall air quality. In tackling water pollution, wastewater treatment processes-primary, secondary, and tertiary-are employed to remove physical, biological, and chemical contaminants. Effluent standards ensure that pollutants discharged into water bodies are within legal limits. Techniques like rainwater harvesting help reduce surface runoff and pollution. Bioremediation, which uses microorganisms to break down contaminants, and wetland restoration, where natural filters remove pollutants before they reach water bodies, are also key strategies (Saeed et al. 2022). Regulating industrial discharges, particularly from sectors like mining and agriculture, helps control effluent pollution, while international agreements like MARPOL address marine pollution from ships. For soil pollution, techniques such as phytoremediation, where plants absorb contaminants like heavy metals and soil treatment technologies like thermal desorption and chemical remediation are effective. Bioremediation, introducing microorganisms to degrade organic contaminants, is also employed. Sustainable agricultural practices, such as organic farming, reduce the use of synthetic pesticides and fertilizers, while crop rotation and conservation tillage improve soil health and reduce erosion (Tahat et al. 2020). Proper management of landfills and hazardous waste is critical to preventing soil contamination from leachates and other pollutants. Noise pollution can be mitigated through the construction of noise barriers along highways and railways to protect residential areas from excessive traffic noise. Zoning laws are used to restrict noisy industrial activities to designated areas far from residential zones, and soundproofing materials are incorporated in buildings near high-noise areas like airports and busy roads. Vehicle noise standards are also imposed, limiting the maximum allowable noise levels from vehicles, with mandatory use of mufflers. Urban green spaces, like trees and green belts, help reduce noise levels, and regulations can restrict the use of loud machinery and equipment during specific hours (Baydere 2000). Finally, integrated pollution control (IPC) adopts a cross-media approach, considering air, water, and soil pollution together to avoid shifting pollutants from one medium to another. Sustainability practices, such as circular economy models, ensure that industries design products and processes with minimal pollution and waste generation, fostering a healthier and more sustainable environment.

Some of the successful pollution control initiatives around the world have demonstrated the importance of coordinated efforts in addressing environmental challenges. One prominent example is the drastic reduction of air pollution in Los Angeles, California, through the implementation of stringent vehicle emissions standards, such as the introduction of catalytic converters in cars and the enforcement of the Clean Air Act (Collantes et al. 2008). This initiative helped to significantly reduce smog levels and improve public health. Another success story is the rehabilitation of the Rhine River in Europe. Once heavily polluted by industrial discharges, the river was restored through the coordinated efforts of neighboring countries. Strict regulations on industrial waste and large-scale water treatment programs transformed the Rhine into a cleaner, healthier body, benefiting biodiversity and communities reliant on its resources. In India, the Namami Gange project is a major pollution control initiative aimed at cleaning the Ganga River. The program focuses on sewage treatment, riverfront development, and rural sanitation to address the river's pollution caused by urban waste and industrial effluents (Mathur 2020). With on-going efforts, the project is showing promising results in reducing contamination levels. Innovative technologies are playing a crucial role in pollution reduction and environmental monitoring. For instance, real-time air quality monitoring systems, using sensor networks and data analytics, help cities like Beijing and New Delhi track and manage pollution levels. Technologies like satellite imaging and drones have enhanced environmental surveillance, allowing authorities to detect illegal activities such as deforestation and unauthorized dumping of waste (Bakirci 2024).

1.5 Policies and Regulations

India has established a robust legal framework to control environmental pollution, addressing issues related to air, water, and soil contamination. Table 1.10 provides an overview of significant policies and regulations aimed at controlling environmental pollution in India including the Air (Prevention and Control of Pollution) Act, 1981, which regulates air pollution from industrial and vehicular emissions, and the Water (Prevention and Control of Pollution) Act, 1974, aimed at preventing water contamination by industries and municipalities. The Environment Protection Act, 1986 serves as a comprehensive law covering various aspects of environmental pollution control, including hazardous waste management, solid waste handling, and setting environmental standards. Other meaningful policies are the National Green Tribunal Act, 2010 to make provision for the establishment of National Green Tribunal for the purposes of hear and dispose of cases involving questions relating to environment; and the National Clean Air Programme (NCAP), launched in the year 2019 to have an emission standards target of 20–30% reduction of particulate matter concentration in the ozone by 2024. These regulations, along with state-level initiatives, form the backbone of India's efforts to mitigate environmental pollution.

The role of policies and regulations in pollution control is multifaceted, providing guidelines for pollution standards, penalties for violations, and frameworks for sustainable development. They set specific emission limits for industries, vehicles, and waste disposal, and enforce the adoption of cleaner technologies, such as the Bharat Stage VI (BS-VI) emission standards for vehicles. Furthermore, policies encourage renewable energy use and regulate industrial waste discharge, limiting pollutants entering water bodies and soil. Regulations also promote environmental impact assessments (EIAs) for large projects to assess their ecological consequences, ensuring that industries operate in an environmentally responsible manner (Glasson and Therivel 2013). In addition, the role of judicial bodies like the NGT has strengthened enforcement by holding industries and governments accountable for non-compliance, offering a legal avenue for citizens to seek redress for environmental degradation.

Table 1.10 Overview of significant policies and regulations aimed at controlling environmental pollution in India

Policy/regulation	Year of enactment	Description
Water (Prevention and control of pollution) act	1974	Regulates water pollution by controlling the discharge of pollutants into water bodies
Air (Prevention and control of pollution) act	1981	Addresses air pollution from industries and vehicles, setting standards for air quality
Environment protection act	1986	Provides a framework for environmental protection, including hazardous waste management and pollution control.
Public liability insurance act	1991	Mandates insurance for industries handling hazardous substances to provide compensation in case of accidents
National forest policy	1988	Aims to maintain environmental stability and protect forests as vital resources for sustainable development
Hazardous waste (Management and handling) rules	1989	Regulates the management and handling of hazardous waste to minimize pollution risks
Municipal solid waste (Management and handling) rules	2000	Establishes guidelines for the management of municipal solid waste, including segregation and treatment
National biodiversity act	2002	Concerns the preservation of life diversity, utilization of the components, including their equitable utilization
National environmental policy	2006	Sets out a framework for sustainable development and environmental protection across sectors
National green tribunal act	2010	Builds the National Green Tribunal to deal with issues of environmental conservation and compliance with the law
National clean air programme (NCAP)	2019	Target to reduce particulate matter pollution in cities by 20–30% by 2024 through various initiatives
Plastic waste management rules	2016	Regulates the management of plastic waste and promotes recycling and reuse of plastics
Environmental impact assessment (EIA) Notification	2006 (revised 2020)	Requires an environmental impact assessment for projects to evaluate their effects on the environment

1.6 Conclusion

In conclusion, this comprehensive exploration of environmental pollution reveals it as a significant global challenge that impacts human health, ecosystems, and economies. Various types of pollution that is air, water, soil, and noise-emanate from diverse sources such as industrial activities, agriculture, and transportation. Understanding the historical evolution of pollution in environmental awareness has been crucial in shaping our current strategies for mitigation. The severe impacts of pollution underscore the urgency of addressing this issue, with far-reaching consequences for climate change and public health. Successful pollution control and mitigation efforts worldwide demonstrate that effective strategies are attainable through innovative technologies and collaborative approaches. Moreover, the role of policies and regulations is pivotal in ensuring compliance and accountability among industries and communities. Addressing pollution requires a united effort from governments, industries, and individuals, emphasizing the need for continued education and awareness. By fostering sustainable practices and leveraging technology for monitoring and remediation, we can then build a healthier environment and develop a better future for the next generation.

References

Adeola AO, Iwuozor KO, Akpomie KG, Adegoke KA, Oyedotun KO, Ighalo JO, Conradie J (2023) Advances in the management of radioactive wastes and radionuclide contamination in environmental compartments: a review. Environ Geochem Health 45(6):2663–2689

Adepoju AO, Omotoso IO, Tiamiyu OG (2024) Air pollution: prevention and control strategies. In: Environmental pollution and public health. Elsevier, pp 49–62

Asif M, Haq R, Gulfreen E, Arshad S, Tasleem MW, Rajpoot SR, Maqbool M (2022) Particulate matter emission sources and their control technologies. Pollut Res 30:696–706

Bakirci M (2024) Smart city air quality management through leveraging drones for precision monitoring. Sustain Cities Soc 106:105390

Baydere S (2000) Noise analysis in urban areas and alternative solutions. Izmir Institute of Technology (Turkey)

Bhanarkar AD, Rao PS, Gajghate DG, Nema P (2005) Inventory of SO2, PM and toxic metals emissions from industrial sources in Greater Mumbai, India. Atmospher Environ 39(21):3851–3864

Burke LM, Lashof DA (1990) Greenhouse gas emissions related to agriculture and land-use practices. In: Impact of carbon dioxide, trace gases, and climate change on global agriculture, vol 53. pp 27–43

Christianson AG (1968) Industrial waste guide on thermal pollution (Vol. 2). US Department of the Interior, Federal Water Pollution Control Administration, Northwest Region, Pacific Northwest Water Laboratory

Collantes G, Sperling D (2008) The origin of California's zero emission vehicle mandate. Transport Res Part a: Policy Pract 42(10):1302–1313

Du S, Zhu R, Cai Y, Xu N, Yap PS, Zhang Y, Zhang Y (2021) Environmental fate and impacts of microplastics in aquatic ecosystems: a review. RSC Adv 11(26):15762–15784

Edo GI, Itoje-akpokiniovo LO, Obasohan P, Ikpekoro VO, Samuel PO, Jikah AN, Agbo JJ (2024) Impact of environmental pollution from human activities on water, air quality and climate change. Ecol Front

Farooqi ZUR, Ahmad I, Ditta A, Ilic P, Amin M, Naveed AB, Gulzar A (2022) Types, sources, socioeconomic impacts, and control strategies of environmental noise: a review. Environ Sci Pollut Res 29(54):81087–81111

Fujigaki Y (Ed) (2015) In: Lessons from Fukushima: Japanese case studies on science, technology and society. Springer

Gardner AL, Gardner AL (2020) Climate change and the environment. In: Stars with stripes: the essential partnership between the European Union and the United States. pp 379–421

Glasson J, Therivel R (2013) Introduction to environmental impact assessment. Routledge
Griggs G, Reguero BG (2021) Coastal adaptation to climate change and sea-level rise. Water 13(16):2151
Hahad O, Prochaska JH, Daiber A, Muenzel T (2019) Environmental noise-induced effects on stress hormones, oxidative stress, and vascular dysfunction: key factors in the relationship between cerebrocardiovascular and psychological disorders. Oxid Med Cell Longev 2019(1):4623109
Hait M, Sahu P (2024) Natural hazards as a source of environmental pollution in the global south. In: Biomonitoring of pollutants in the global south. Singapore, Springer Nature, pp 153–190
Havugimana E, Bhople BS, Kumar A, Byiringiro E, Mugabo JP, Kumar A (2017) Soil pollution–major sources and types of soil pollutants. Environ Sci Eng 11:53–86
Hossain ME, Shahrukh S, Hossain SA (2022) Chemical fertilizers and pesticides: impacts on soil degradation, groundwater, and human health in Bangladesh. Environmental degradation: challenges and strategies for mitigation. Springer International Publishing, Cham, pp 63–92
Karri RR, Ravindran G, Dehghani MH (2021) Wastewater—sources, toxicity, and their consequences to human health. In: Soft computing techniques in solid waste and wastewater management. Elsevier, pp 3–33
Kovaleva TN, Shtofer LL, Stukalov SY, Spektor LA, Chernov IV (2019) Ecological safety and hazards to public health: the search for balance within the" human-biosphere" system. Ekoloji 28(107):5039–5043
Kumar P, Imam B (2013) Footprints of air pollution and changing environment on the sustainability of built infrastructure. Sci Total Environ 444:85–101
Madhav S, Ahamad A, Singh AK, Kushawaha J, Chauhan JS, Sharma S, Singh P (2020) Water pollutants: sources and impact on the environment and human health. In: Sensors in water pollutants monitoring: role of material. pp 43–62
Manisalidis I, Stavropoulou E, Stavropoulos A, Bezirtzoglou E (2020) Environmental and health impacts of air pollution: a review. Front Public Health 8:14
Mathur A (2020) Namami gange scheme-a success or mere propaganda? GLS Law J 2(2):54–64
Meiners R, Desrochers P, Morriss A (Eds.) (2012) Silent spring at 50: the false crises of Rachel Carson. Cato Institute
Mishra RK, Mohammad N, Roychoudhury N (2016) Soil pollution: causes, effects and control. Van Sangyan 3(1):1–14
Mishra S, Bharagava RN, More N, Yadav A, Zainith S, Mani S, Chowdhary P (2019) Heavy metal contamination: an alarming threat to environment and human health. Environ Biotechnol: For Sustain Future 103–125
Mitchell R, Sweeney F (2018) Air pollution. Scientific e-Resources
Munsif R, Zubair M, Aziz A, Zafar MN (2021) Industrial air emission pollution: potential sources and sustainable mitigation. In: Environmental emissions. IntechOpen
Mutugi M, Kiiru W (2015) Biodiversity, local resource, national heritage, regional concern and global impact: the case of Mau forest, Kenya. Europ Scient J 1:681–691
Paglia E (2021) The Swedish initiative and the 1972 stockholm conference: the decisive role of science diplomacy in the emergence of global environmental governance. Humanit Soc Sci Commun 8(1):1–10
Patel R, Singh PK, Saw S (2022) Recent advancements in the challenges and strategies of globally used traffic noise prediction models. Environ Sci Pollut Res 29(32):48168–48184
Patel R, Singh PK, Saw S (2024a) A modeling approach for suitability evaluation of traffic noise prediction under mixed traffic situation in mid-sized Indian cities. Innov Infrastruct Solut 9(6):183
Patel R, Singh PK, Saw S (2024) Traffic noise modeling under mixed traffic condition in mid-sized indian city: a linear regression and neural network-based approach. Int J Mathemat Eng Managem Sci 9(3)
Patel R, Singh PK, Saw S, Tiwari SK, Kumar R (2024) Assessment and mapping of traffic noise levels in coal capital of India, Dhanbad: implications for noise pollution management. In:

Dynamics of transportation ecosystem, modeling, and control. Singapore, Springer Nature, pp 317–332

Rajkhowa S, Sarma J, Das AR (2021) Radiological contaminants in water: pollution, health risk, and treatment. In: Contamination of water. Academic Press, pp 217–236

Saeed MU, Hussain N, Sumrin A, Shahbaz A, Noor S, Bilal M, Iqbal HM (2022) Microbial bioremediation strategies with wastewater treatment potentialities–a review. Sci Total Environ 818:151754

Saw S, Singh PK, Mahato JK, Patel R, Deoli V, Singh L (2023b) Application of GIS-based DRASTIC model and its validation with solute transport model in the assessment of groundwater vulnerability index: a case study from coal mining region of India. Environ Qual Manage 33(2):265–276

Saw S, Mahato JK, Singh PK (2021) Suitability evaluation of CCME-WQI and GWQI for the modeling of groundwater and human health risk assessment of heavy metals-Eastern India

Saw S, Singh PK, Mahato JK, Patel R, Shikha D (2023) A modelling approach for the suitability evaluation and human health risk assessment of heavy metals dispersion in groundwater resources. Environ Developm Sustain 1–25

Saxena AK (2018) Environmental pollution and effect on ozone layer depletion: systematic approach towards remedial measures. Environ Pollut 5(6)

Tahat MM, Alananbeh K, Othman YA, Leskovar DI (2020) Soil health and sustainable agriculture. Sustainability 12(12):4859

Tiwari SK, Kumaraswamidhas LA, Gautam C, Garg N (2022) An auto-encoder based LSTM model for prediction of ambient noise levels. Appl Acoust 195:108849

Tiwari SK, Kumaraswamidhas LA, Garg N (2023b) Assessment of noise pollution and associated subjective health complaints in Jharia Coalfield, India: a structural equation model analysis. Noise Mapping 10(1):20220172

Tiwari SK, Kumaraswamidhas LA, Garg N (2023) Strategies and implications of noise pollution monitoring, modelling, and mitigation in urban cities. In: Handbook of metrology and applications. Singapore, Springer Nature, pp 1571–1593

Tiwari SK, Kumaraswamidhas LA, Patel R, Garg N, Vallisree S (2024) Traffic noise measurement, mapping, and modelling using soft computing techniques for mid-sized smart Indian city. Measurem Sens 33:101203

Tzivian L, Winkler A, Dlugaj M, Schikowski T, Vossoughi M, Fuks K, Hoffmann B (2015) Effect of long-term outdoor air pollution and noise on cognitive and psychological functions in adults. Int J Hyg Environ Health 218(1):1–11

Ukaogo PO, Ewuzie U, Onwuka CV (2020) Environmental pollution: causes, effects, and the remedies. In: Microorganisms for sustainable environment and health. Elsevier, pp 419–429

Vaiserman A, Koliada A, Zabuga O, Socol Y (2018) Health impacts of low-dose ionizing radiation: current scientific debates and regulatory issues. Dose-Response 16(3):1559325818796331

Zhang W, Bansback N, Anis AH (2011) Measuring and valuing productivity loss due to poor health: a critical review. Soc Sci Med 72(2):185–192

Chapter 2
Trends and Health Impacts of Nitrogen Dioxide Pollution in China

R. Rahul and Aman Kumar

Abstract In China, nitrogen dioxide (NO_2) has emerged as a major pollutant in the atmosphere due to rising urbanization and industrialization. The present study (1) from 2015 to 2020 in China distinguished long-term NO_2 ascribed health metric; (2) evaluate NO_2 concentration trends; (3) comparison of NO_2 levels before and at the time of COVID-19 pandemic. The research utilizes data from national air quality monitoring networks and health impact assessments to elucidate the trends and impacts of NO_2 pollution. Results indicate a general decline in NO_2 concentrations across the majority of Chinese cities, reflecting the effects of governmental pollution control measures. Notably, a substantial reduction in NO_2 levels was observed at the time of COVID-19 situation, ascribed to the strict lockdowns and significant depletion in vehicular and industrial activities imposed by the Chinese government. This temporary improvement in air quality during the pandemic underscores the potential effectiveness of stringent pollution control measures and provides a unique natural experiment for evaluating such interventions. Additionally, the study reveals spatial disparities in NO_2 concentration reductions, with some regions experiencing more pronounced declines than others, highlighting the need for region-specific strategies. By providing a detailed analysis of NO_2 pollution trends and health impacts over a significant period, the research glimpses to the understanding of urban air quality dynamics in China and informs the development of more effective environmental and public health policies. Furthermore, advancements in AI and ML techniques can be instrumental in enhancing the monitoring and analysis of NO_2 pollution. These technologies enable accurate predictions of pollution trends, real-time detection of emission hotspots, and effective modelling of health impacts, aiding in the formulation of targeted and data-driven environmental policies.

Keywords Nitrogen dioxide (NO_2) pollution · Long-term NO_2 · Air quality trends · COVID-19 pandemic · Health metrics

---

R. Rahul (✉)
Department of Mechanical Engineering, Birla Institute of Technology and Science, Pilani, India
e-mail: p20240900@pilani.bits-pilani.ac.in

A. Kumar
Indian Institute of Management–Ranchi (IIM–Ranchi), Ranchi, India
e-mail: aman.kumar@iimranchi.ac.in

S. K. Tiwari et al. (eds.), *AI-Driven Environmental Pollution Management*, Climate Risks and Solutions, https://doi.org/10.1007/978-3-031-96243-1_2

2.1 Introduction

Urbanization and industrialization have significantly altered atmospheric compositions, leading to elevated levels of various pollutants. Among these, nitrogen dioxide (NO_2) stands out as a critical pollutant with substantial adverse effects on the environment and human health. NO_2, primarily produced by vehicle emissions and industrial activities, contributes to respiratory problems, cardiovascular diseases, and deteriorates overall air quality.

In China, one of the world's most rapidly developing nations, the impact of NO_2 pollution has become increasingly evident. To control the air pollution the government of China have administered numerous strategy and measures, yet the effectiveness of these interventions remains a topic of significant interest. The pandemic in 2020, which led to unprecedented lockdowns and a halt in many industrial activities, provides a unique opportunity to analyze NO_2 pollution dynamics under drastically reduced human activities. This research aims to categorize the long-term health impacts of NO_2 pollution in China from 2015 to 2020, estimate trends in NO_2 concentrations, and compare these levels before and during the COVID-19 pandemic. The findings offer valuable insights into the effectiveness of pollution control measures and highlight areas requiring further attention to improve air quality and public health. Furthermore, research also shows poor air quality is a severe environmental issue that harms human health (Pope III and Dockery 2006). Air pollutants generated by many sources in metropolitan areas can result in serious health consequences (World Health Organization 2021). The recent urbanization and industrialization in China have led it to surpass other developed nations in terms of nitrogen dioxide (NO_2) concentration (Zhu et al. 2019). Cui et al. (2021) mention the economic expansion has resulted in the utilization of a considerable quantity of fossil fuels, which has resulted in a significant rise in NO_2 concentrations in the atmosphere. NO_2 contributes significantly to tropospheric pollution, which includes ozone depletion and occurrence of acid rain (Zhang et al. 2017). The world's most populated and industrialized regions include China, Western Europe, the United States, and India, have the highest NO_2 levels (Schneider et al. 2015). Carmona-Cabezas et al. (2020) have explained the NO_2 is a significant precursor to $PM_{2.5}$ and ozone formation and is associated with several environmental concerns. Moreover, Mao et al. (2020) explained NO_2 leads to several disorders and relates to increased mortality from pulmonary heart disease, impaired lung function, depression. By Silvern et al. (2019) and Xie et al. (2018) the spatiotemporal variations of NO_2 emission have been extensively studied using satellite NO_2 retrievals from the Ozone Monitoring Instrument (OMI) for large-scale, global coverage and high-resolution data. OMI retrievals were used to examine the pattern and geographic variation of atmospheric NO_2 levels in China, including the Sichuan Basin, the Inner Mongolia region, and Henan Province (Zhang et al. 2017; Ai et al. 2018). Prior literature states that different studies produced inconclusive results on the quantities of NO_2 that potentially harm people (World Health Organization 2021). In this context, the present study will have significant contribution to the literature on NO_2 pollution with context to China. Figure 2.1a, b 1623 Monitoring Stations and lists the major cities in China in a geospatial map.

Both artificial and natural sources contribute to the production of NO_2 in the troposphere. Common reasons for NO_2 generation include the usage of fossil fuels, biomass burning, lightning, and soil release (Boersma et al. 2004). Furthermore, the presence of automobiles, power stations, and certain excessive emission industries play a role in NO_2 output (Heue et al. 2005). Significant attention has been paid to China's severe air pollution caused by rapid urbanization (Zhang et al. 2019, 2020). Ma et al. (2019) have mentioned that the China Communist Party has recently created regulations and strategies to mitigate the negative consequences of NO_2 emissions. The 12th Five-year plan, for example, established an aim of reducing NO_x during 2011–2015 by 10% (12th Five-year Plan 2010). In the 13th Five-year plan, the objective of pollution reduction and energy-saving remains steadfast (13th Five-year Plan 2015). In 2013, the control action plan and air pollution prevention to enhance national air quality in several regions of China announced by the Clean Air Alliance of China (CAAC). However, very limited research has been conducted to investigate the variations in NO_2 around China following the execution of these preventive measures. Previous literature suggests a high correlation between the respiratory diseases and long-term exposure to NO_2 (Costa et al. 2014). Few studies provide fragmented results while evaluating the respective contribution of NO_2 on health outcomes (Pope et al. 1995; Gustafsson et al. 2018). NO_2 is a significant contributor to the growth of harmful pollutants concentration in the atmosphere, which impacts human health; therefore, the need of the hour is to call for a policy that restricts the associated health consequences (Khaniabadi et al. 2017; Samet et al. 2000). Moreover, some supplementary importance should be given to NO_2 due to its indication of traffic-related air quality and a key contributor of a variety of harmful gases that, when combined, are likely to cause serious health consequences (Hesterberg et al. 2009; Touloumi et al. 1997). Several scholars examined the consequences of various air pollutants (Katsouyanni et al. 2001). Typically, atmospheric particulates were utilized as pollution indicators (Katsouyanni et al. 1997); however, atmospheric toxins, including carbon monoxide (CO), ozone, and nitrogen dioxide (NO_2), negatively impact mortality (Gryparis et al. 2004; Schwartz 2005). Strong respiratory irritating gas NO_2 is produced by high-temperature combustion. The significant producers of NO_2 are cookers, heaters, and combustion appliances, whereas the main outside sources are automobiles (especially those powered by diesel) and power plants (Kraft et al. 2005). Several regions have implemented quarantine measures to restrict virus transmission since January 2020 (Filonchyk et al. 2020). Global industrial activity has stopped as countries have implemented the lockdown. The effect of the lockdown during the COVID-19 pandemic was most significant for the transportation sector. Many countries have suspended road and air transport, significantly restricting human movement (Muhammad et al. 2020). Besides transportation, the manufacturing industry also experienced a slowdown during the pandemic. As the world's industrial and transportation sectors came to a halt, so did oil consumption and prices. The global spreading and public health threats posed by COVID-19 were severe enough to have a significant impact on global GDP (Wu et al. 2020).

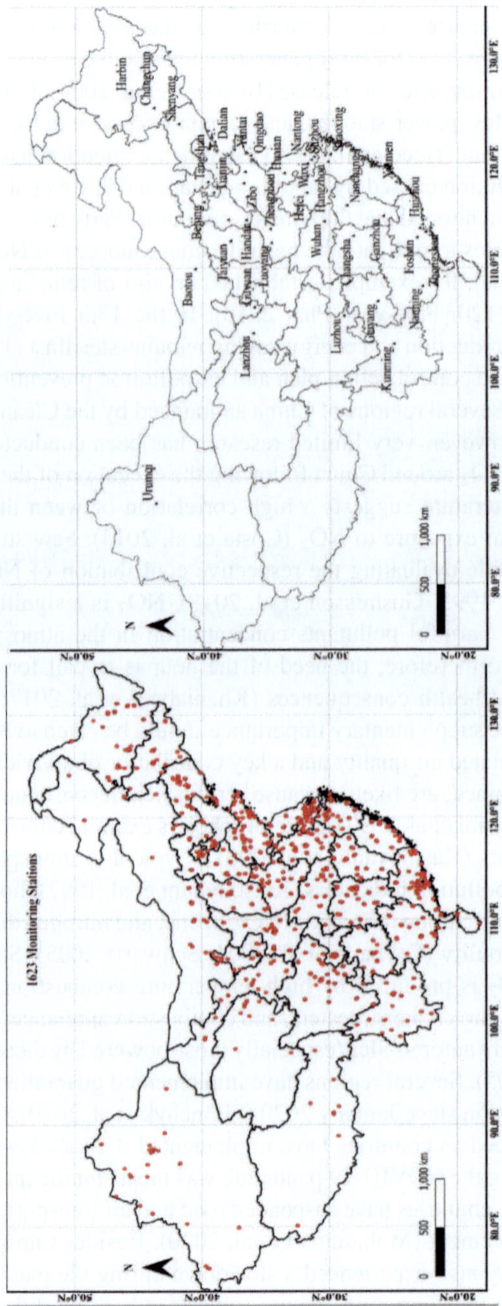

Fig. 2.1 a 1623 monitoring stations and b lists the major cities in China a geospatial map

The effects of O_3, $PM_{2.5}$, and PM_{10}, have been extensively examined in the context of China, however, quantitative research on the consequences of NO_2 pollution is limited NO_2. The real-time pollution data was retrieved from China National Environmental Monitoring Centre (CNEMC), which provided a new avenue for examining the relationship between variations in spatio-temporal NO_2 concentrations and various health effects due to NO_2 in China. The present research focuses on the ground NO_2 concentration from 2015 to 2020. First the non-parametric linear trend for NO_2 concentration during 2015–2020 are examined.

Furthermore, Advancements in artificial intelligence (AI) and machine learning (ML) have opened new frontiers in addressing complex environmental challenges, including nitrogen dioxide (NO_2) pollution. As one of the primary pollutants contributing to urban air quality degradation and adverse health effects, NO_2 demands robust monitoring and analytical frameworks. AI and ML techniques such as SVM, LSTM, MLR, and ANN have emerged as transformative tools in this domain, enabling high-precision prediction of pollution trends, real-time identification of pollution emission hotspots, and comprehensive modelling of exposure-related health impacts (Tiwari et al. 2022, 2024; Rehman et al. 2025). These technologies facilitate the integration of large-scale, heterogeneous datasets—such as satellite imagery, meteorological parameters, and ground-based sensor readings—into sophisticated analytical models that enhance both accuracy and efficiency. Moreover, AI-driven solutions empower policymakers with actionable insights, supporting the design and implementation of targeted interventions and evidence-based regulations. Beyond immediate monitoring, the predictive capabilities of ML algorithms enable proactive planning to mitigate the long-term impacts of NO_2 pollution on public health and ecosystems. By bridging the gap between traditional environmental monitoring methods and modern computational advancements, AI and ML hold immense potential to redefine how societies understand, manage, and reduce air pollution, paving the way for sustainable development and improved quality of life.

Atmospheric pollution is increasingly seen as a persistent worry and a major environmental issue on a global scale, with adverse health consequences. Because air pollution can impact on human health, quality of life, and the environment, predicting air pollution values has piqued the interest of many researchers due to the strong relationship between people's health and the environment. The findings indicated that an accurate prediction approach based on data processing techniques and an artificial neural network model could be a powerful, effective, and appropriate tool for analyzing and modelling complex and non-linear relationships of environmental variables, such as the ability to forecast air pollution. (Shams et al. 2021) AI models have been utilized in various engineering domains, including air pollution modelling concentrations of PM_{10}, CO, SO_2, and NO_2 atmospheric pollutants. (e.g. Agirre-Basurko et al. 2006; Azid et al. 2014; He et al. 2015; Feng et al. 2015; Perez and Gramsch 2016; Cabaneros et al. 2017; Murillo-Escobar et al. 2019, Baghanam et al. 2024) as well as its machine learning (ML) component including ANN could be a powerful, effective and suitable tool for analysis and modelling complex and non-linear relation of environmental variables such as ability in forecasting air pollution (Mishra and Goyal 2015; Bai et al. 2016; Shams et al. 2021) and predicts environmental contamination and its detrimental effects on humans. (Garlík et al. 2024). Moreover, SVR (Osowski and Granty 2007; Moazami et al. 2016) have recently shown reliable abilities in air quality modelling. (Alimissis et al. 2018) ANN was applied to estimate NO_2 concentrations at each monitoring site considering a specific site as the target and using concentrations at remaining monitoring sites as independent variables. Radojevic et al. (2019) tested the ability of an artificial neural network (ANN) to forecast the daily average of NO_x concentrations using periodic parameters and metrological data. Linear models, illustrated by Kamińska (2019) and Shang et al. (2019), including decision trees and random forests, have successfully forecasted pollutant concentrations with pre- and post-processing approaches. For future research, traffic-related pollution should be used as an input to explore its sensitivity to NO_2 fluctuations. It is also recommended that apply various machine learning models be applied for sensitivity analysis (Nourani et al. 2021).

2.2 Methodology

2.2.1 *Trend of Ground-Level NO_2 Concentration*

The time series data of NO_2 between January 2015 and December 2020 has been recorded in China's National air quality (NAQSs) on hourly basis. The information is retrieved from http://beijingair.sinaapp.com/, from National Environmental Monitoring Centre's authorized download site (http://www.cnemc.cn). The dataset had NO_2 concentration data at monitoring station level measured in $\mu g/m^3$. We have screened the data using three filters as suggested by Maji (2020); firstly, we have removed datasets which had repeated values for more than 5 times simultaneously; secondly, we have removed columns (monitoring stations) which had more than 10% missing data; thirdly, we have removed the outliers i.e. greater than 500 and less than 1 values from the complete dataset. The initial data included an aggregate of 1633 monitoring sites, but after putting all the filter we were left with just 1302 monitoring stations. The annual mean NO_2 levels during 2015–20 was derived by averaging the hourly data from each city's monitoring stations.

We analyzed a five-year time series in order to discover monotonic linear trends. As suggested by Maji (2020) Mann–Kendall test and Theil-Sen estimation was employed to determine the statistical significance of trends and the degree of annual trends using de-seasonalized data respectively. The trends are interpreted as yearly increases or decreases in NO_2 concentrations, expressed in $\mu g/m^3$/year. Moreover, for seasonal analysis, we have considered the month of December, January and February as Winter season, whereas we considered the month of June, July and August as Summer season as suggested by Maji and Sarkar (2020). Statistical analyses were performed in RStudio 3.6.0.

2.3 Past Studies and Comparison

A myriad of studies which discussed the spatio-temporal aspects of the NO_2 pollution in China (Cui et al. 2021). However, there exists no study in NO_2 context which uses the most recent data of COVID-19 era to explain the NO_2 distribution and trends across China with annual and seasonal focus as well and also provides a pre-pandemic and post-pandemic comparison. Moreover, NO_2 concentrations in China were measured at various temporal and geographical resolutions using various methods. Therefore, the present study gives a comprehensive understanding of long term NO_2 related health risks across China and also provides policy implications on the basis of the results found through this research.

2.4 Results and Discussion

2.4.1 Distribution and Trend of NO_2 from 2015 to 2019

The nationwide NO_2 concentration at 1302 stations for 2015 to 2019 are shown in Fig. 2.2. In 2015, the average of NO_2 concentration at 1302 stations ranged from 9.8 μg/m^3 (Yuxi city) to 61.3 μg/m^3 (Zibo city), with nation mean 30.2 ± 10.4 μg/m^3 (mean ± standard deviation) (Fig. 2.2a). Nationwide NO_2 concentration increased in 2016 (30.3 ± 10.2 μg/m^3) and 2017 (31.4 ± 10.2 μg/m^3) (Fig. 2.2b, c). In 2018 and 2019 nationwide NO_2 concentration value slightly decreased to 28.7 ± 9.4 μg/m^3 (Fig. 2.2d) and 27.8 ± 8.9 μg/m^3 (Fig. 2.2e) respectively. According to the World Health Organization 2021 interim targets among 1302 monitoring sites, 1064 sites in 2015, 1071 in 2016, 1089 in 2017, 1050 in 2018 and 1041 in 2019 surpasses the interim target-3 of 20 μg/m^3 recommended by the air quality guideline (AQG) level. The number of cities exceeding Interim target-1 i.e., 40 μg/m^3 decreased from 67 in 2015 to 60 in 2017, then reduced to 50 and 34 in 2018 and 2019, respectively.

Figure 2.3a shows from 2015 to 2020, the cities with high NO_2 values were Tangshan (56.6 μg/m^3), Xingtai (54.2 μg/m^3), Urumqi (51 μg/m^3), Xi'an (50.8 μg/m^3) and Zhengzhou (50.4 μg/m^3). Within five years, NO_2 concentration rises by 40 μg/m^3 in 53 cities and by 20 μg/m^3 in 285 cities. In the megacity-cluster, Yangtze River Delta (YRD) and Beijing- Tianjin-Hebei Urban Agglomerations (BTH) region, the NO_2 concentration values decreased from 46 ± 9.4 and 37.1 ± 8.8 μg/m^3 in 2015 to 38.3 ± 7.1 and 34.9 ± 6.2 μg/m^3 in 2019. In PRD region NO_2 concentration values slightly increased from 32.6 ± 6.6 μg/m^3 in 2015 to 32.9 ± 6.4 μg/m^3 in 2019. Around 2 μg/m^3 from 2015 to 2019 nationwide annual average of NO_2 concentration has constantly reduced. The NO_2 concentration in the megacities Beijing, Chengdu, Chongqing, Harbin, Shanghai, Shenzhen, Tianjin, and Wuhan is reduced from 48.6, 49.3, 41.0, 46.1, 45.6, 33.4, 42.3, and 52.3 μg/m^3 to 35.7, 39.8, 37.8, 32.4, 41.4, 25.8, 42.3, and 45.0 μg/m^3 respectively. The most significant decrease in NO_2

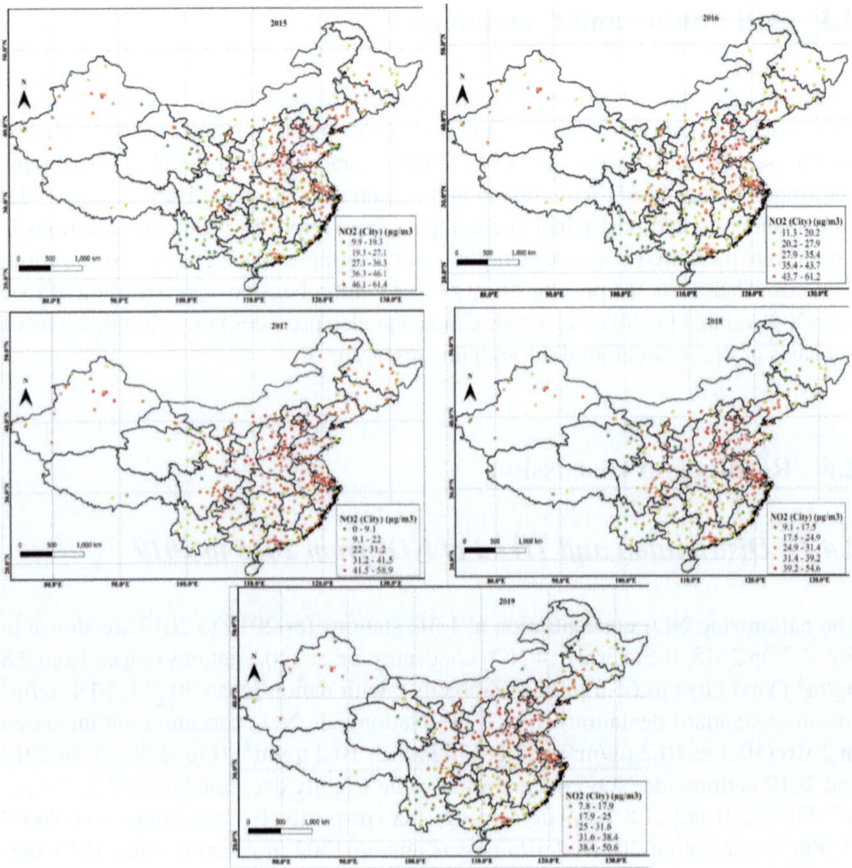

Fig. 2.2 **a–e** Monitoring sites-specific annual averaged distribution of NO_2 (µg/m^3) in China 2015–2019

concentration was observed in Linxia, Zibo, Harbin, Xuchang, Xingtai city by 31.9, 19.3, 16.8, 14.2 and 14.1 µg/m^3, respectively.

Effective control methods including the utilization of hydroelectric and nuclear power, the removal of small coal-fired energy plants, and the adoption of rigorous emission rules for industrial zones are responsible for the decrease in NO_2 concentrations (Tian et al. 2022). However, meteorological conditions also play a role in NO_2 reduction (Xue et al. 2019). Similarly, the most significant increase in NO_2 concentration was observed in Luliang, Hami, Yangzhou, Lu'an, Jinzhong city by 21.1, 12.9, 11.3, 11.0 and 10.5 µg/m^3, respectively. This study shows consistent results with previous research which shows significant increase in NO_2 concentrations in few regions of China like central east China (He et al. 2007). Figure 2.3b shows the absolute trends (µg/m^3/yr) of NO_2. The overall average annual trend is -0.7 ± 1.5 µg/m^3/yr across China, with reductions in some cities, including Linxia (-5.0 ± 2.3

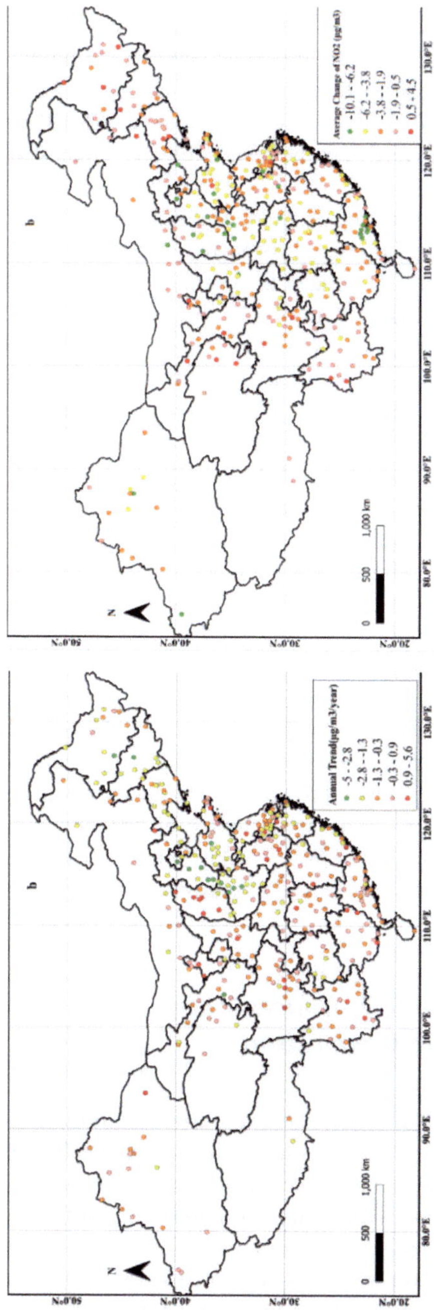

Fig. 2.3 City specific **a** annual averaged NO_2 (μg/m^3) change in China 2015–2019 and **b** absolute annual trend

μg/m^3/yr), Zibo (−4.9 ± 1.9 μg/m^3/yr), Hebi (−4.2 ± 0.6 μg/m^3/yr), Xingtai (−3.9 ± 0.8 μg/m^3/yr) and Handan (−3.8 ± 1.1 μg/m^3/yr) city. 94 out of 332 cities tend to have a positive trend; however, most cities, i.e., 24 cities, have a negative trend. The cities with the highest positive trend are Luliang (5.6 ± 0.3 μg/m^3/yr), Hami (3.2 ± 0.8 μg/m^3/yr), Yangzhou (3.0 ± 0.5 μg/m^3/yr), Jinzhong (2.4 ± 0.4 μg/m^3/yr) and Xinzhou (2.3 ± 0.1 μg/m^3/yr) city. However, Silver et al. (2018) reported an average annual trend that remains more than the reported median annual trends of −1.9 μg/m^3/yr from 2015 to 2017.

From the seasonal analysis, the NO_2 concentration tend to rise in winter significantly, when compared to summers irrespective of year considered. Winter NO_2 emissions were discovered to be greater than summer levels, and Wang et al. (2019) earlier reported a similar finding. Out of 332 cities, 92 cities have seen an increase in NO_2 concentration in summers from 2015 to 2019, whereas 240 cites has seen a decrease in NO_2 concentration. Similarly, 86 cities have seen an increase in NO_2 concentration in winters from 2015 to 2019, whereas 246 cites has seen a decrease in NO_2 concentration. With context to winter season, the results show that the cities with most increase in winter NO_2 levels between 2019 and 2015 are Luliang (24.7 μg/m^3), Yangzhou (23.8 μg/m^3) and Taiyuan (18.5 μg/m^3); whereas the cities with most decrease in summer NO_2 levels are Linxia (−54.9 μg/m^3), Harbin (−31.0 μg/m^3) and Loudi (−24.1 μg/m^3); (Fig. 2.4a). With context to summer season, the results show that the cities with most increase in summer NO_2 levels between 2019 and 2015 are Luliang (15.7 μg/m^3), Datong (9.9 μg/m^3) and Hami (8.8 μg/m^3); whereas the cities with most decrease in summer NO_2 levels are Zibo (−24.6 μg/m^3), Linxia (−22.0 μg/m^3) and Xigaze (−21.8 μg/m^3) (Fig. 2.4b).

This disparity is most likely related to variations in NO_2 production in various areas. According to Van Der A et al. (2006), the major contributors to NO_2 in China (western) were soil, temperature, and precipitation-related natural emissions, whereas industrial emission are the main source of NO_2 in China (eastern). Coal-fired heating, in particular, is a major source of NO_2 emissions throughout the winter. Winter has a longer NO_2 lifespan than other seasons, resulting in severe air pollution.

2.4.2 Distribution of NO_2 Concentration in 2020: The COVID-19 Era Comparison

The concentration of NO_2 in 2020 was much lower than the distribution levels of NO_2 in 2019. Air pollution levels were high approximately a week prior to the Chinese government implemented severe preventative and control measures across the country. One of China's busiest industrial areas is East China, where intense industrial activity has resulted in large atmospheric releases of NO_2 and other aerosols. Furthermore, once the government implemented stringent regulatory measures to limit COVID-19, which resulted in the interruption of several facilities and companies, emission of NO_2 is considerably decreased throughout the nation, resulting in

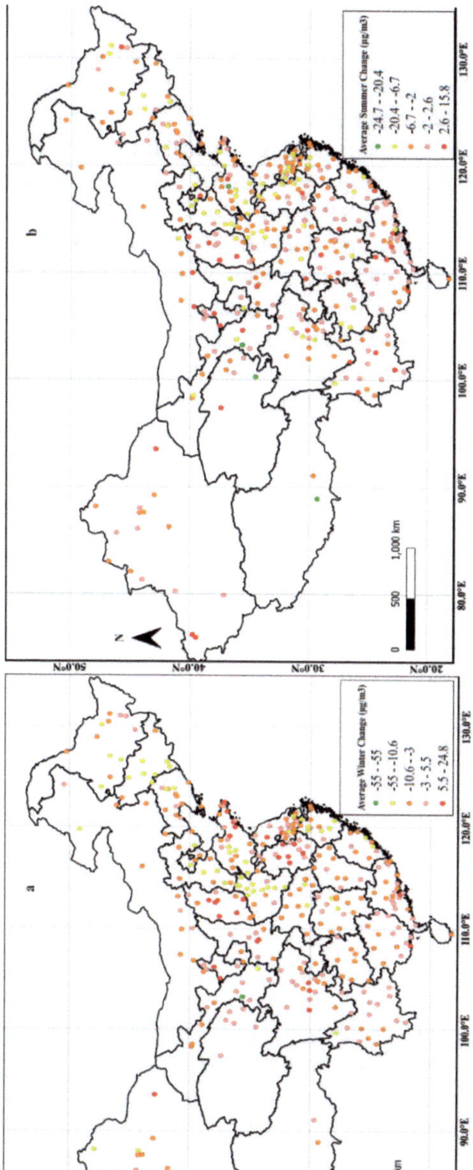

Fig. 2.4 City specific **a** winter average change and **b** summer average change of NO_2 (µg/m^3) in China 2015–2019

better air quality (Huang and Sun 2020). Our results are consistent with several recent studies, where significant declines in NO_2 concentration were found in the United Kingdom, Germany and Switzerland. The main reason attributed to this decline is technological advancements in vehicle and industry vents which purify the NO_2 before discharge, and also awareness of compliance with government regulations regarding air quality standards (Kurtenbach et al. 2012).

Figure 2.5a illustrate the comparison of NO_2 distribution of 2019 and 2020. Similarly, Fig. 2.5b shows annual average NO_2 (μg/m^3) change in China from 2019 to 2020. The average NO_2 of 27.8 ± 8.9 μg/m^3 (2019) is decreased to 24.9 ± 7.9 μg/m^3 (2020) due to lockdown measures to control COVID-19 over entire China. The average NO_2 concentration at 332 Cities stations ranged from 8.3 μg/m^3 (Diqing City) to 46.2 μg/m^3 (Tangshan City), with a national mean of 24.9 ± 7.9 μg/m^3. In 2019, cities with highest average of NO_2 concentrations are Tangshan 50.6 μg/m^3, Urumqi 48.9 μg/m^3, Xi'an 48 μg/m^3, Jinan 47 μg/m^3, Taiyuan 46.2 μg/m^3 is decreased to 46.2, 42.2, 41.3, 36.2 and 42 μg/m^3 respectively in 2020. The five cities with the highest average of NO_2 concentrations in 2020 are Tangshan 46.2 μg/m^3, Urumqi 42.2 μg/m^3, Taiyuan 42.0 μg/m^3, Huzhou 41.8 μg/m^3 and Xi'an 41.3 μg/m^3.

In the megacity-cluster, YRD and BTH region NO_2 concentration values decreased from 38.2 34.7 ± 6.1 μg/m^3 and ± 7.7 μg/m^3 in 2019 to 31 ± 5.9 μg/m^3 and 34.1 ± 7.0 μg/m^3. In the PRD region, NO_2 concentration values also decreased

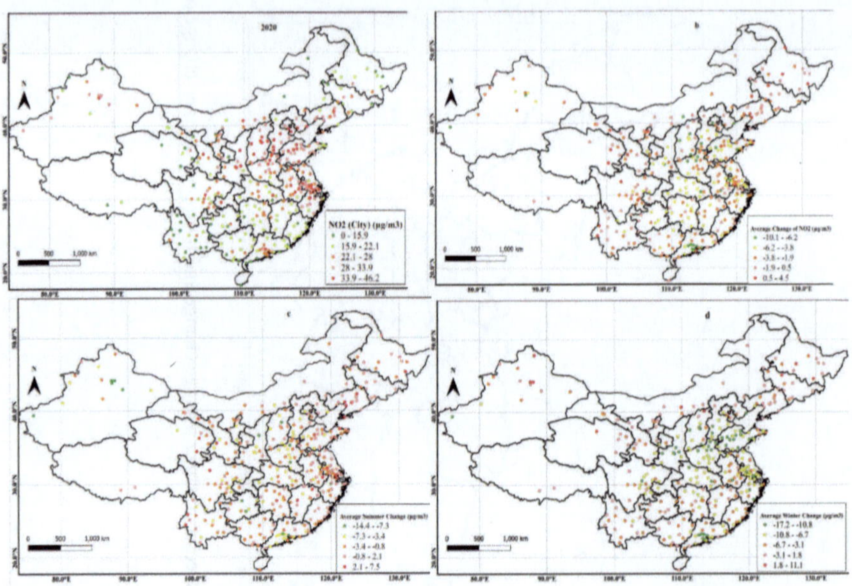

Fig. 2.5 City specific **a** annual averaged distribution of NO_2 (μg/m^3) in China 2020, **b** annual averaged NO_2 (μg/m^3) change in China 2019–2020, **c** summer average change and **d** winter average change of NO_2 (μg/m^3) in China 2020

from 30.0 ± 6.2 μg/m^3 in 2019 to 23.7 ± 4.3 μg/m^3 in 2020. NO_2 concentration in the megacity like Beijing, Chongqing, Chengdu, Shanghai, Shenzhen, Tianjin, Wuhan, Guangzhou, Dongguan, and Shenyang is decreased from 35.7, 37.8, 39.8, 41.4, 25.8, 42.3, 45.0, 22.1, 36.8, and 35.7 μg/m^3 to 28.4, 36.5, 34.7, 36.8, 22.9, 39.1, 36.7, 17.0, 26.9 and 34.3 μg/m^3 respectively.

Out of 332 cities, 98 cities have seen an increase in NO_2 concentration in summers from 2019 to 2020, whereas 234 cities have seen a decrease in NO_2 concentration. Similarly, only 25 cities have seen an increase in NO_2 concentration in winters from 2019 to 2020, whereas 307 cites has seen a decrease in NO_2 concentration. With context to winter season, the results show that the cities with most increase in winter NO_2 levels between 2019 and 2020 are Matsubara (11.0 μg/m^3), Changji (7.2 μg/m^3) and Baotou (5.7 μg/m^3); whereas the cities with most decrease in winter NO_2 levels are Jimo (−17.2 μg/m^3), Xi'an (−16.2 μg/m^3) and Kashi Prefecture (−15.6 μg/m^3); (Fig. 2.5c). With context to summer season, the results show that the cities with most increase in summer NO_2 levels between 2019 and 2020 are Zhangqiu (7.5 μg/m^3), Dahong (7.2 μg/m^3) and Zhangye (6.6 μg/m^3); whereas the cities with most decrease in summer NO_2 levels in 2019–2020 are Changji (−14.4 μg/m^3), Urumqi (−11.6 μg/m^3) and Kashi Prefecture (−11.5 μg/m^3); (Fig. 2.5d). The reduction in the NO_2 concentration during pandemic is in line with previous literature which revealed that air quality in China could be improved with implementing strict control measures to reduce pollution from vehicles and industries (Chen et al. 2020).

2.5 Policy Implication

There should be a close look at China's current and projected levels of NO_2 air pollution and its consequences on human health, as well as an assessment of the related accountability that will have important policy implications for reducing pollution. We recommend intensifying the emphasis on the regulation of NO_2 emissions from the automotive sector in order to significantly enhance the quality of air in China in the subsequent years. To safeguard China's ecological sustainability and long-term growth, persistent air pollution reduction policies will be required (Huang and Sun 2020). To more precisely measure the effects on health, additional city-specific, long-term epidemiological studies using demographic data are being developed. To mitigate the negative consequences, public education and the use of technical measures for reducing sulfur emissions from various sources like the petrochemical and oil sectors are being implemented. As control techniques, reducing fossil fuel and diesel usage, as well as thorough air pollution monitoring, will play a significant role (Li et al. 2021). To mitigate the negative health effects of air pollution, public health officials should provide health advice to the whole community, particularly vulnerable people with chronic heart and lung diseases, aging, and kids. Further, precautionary risk reduction actions at the political level should be implemented to minimize dust infiltrating the nation, such as laying mulch, cleaning roadways, managing water sources, and introducing new plant species to absorb airborne contaminants (Zhang et al. 2013). Additional initiatives should concentrate on ecologically friendly transportation networks, stringent actions required to eliminate excessive traffic density, and proper traffic management systems (Zhou et al. 2010). For future studies, traffic-related pollution should be used to investigate its susceptibility to NO_2 changes. It is also recommended that different machine learning models be used for sensitivity analysis.

2.6 Conclusions

The current research highlighted that NO_2 levels have reduced at most of the places in China due to stringent rules and regulations implemented by China. However, due to economic growth which involves industrialization NO_2 levels have increased at some places as well. Between 2015 and 2020, the mean NO_2 concentrations changed dramatically. However, NO_2 concentrations in specific areas continued to rise. NO_2 concentration increases in the mentioned provinces may be accounted for mostly by rising population, increase in vehicles, and recently built energy stations in under-developed regions. In contrast, APPC-AP majorly concentrated on significant businesses (Maji and Sarkar 2020). The main source of NO_x emissions in Chinese cities has been the transportation sector (Seltzer et al. 2018). The Ningxia province's increased NO_x concentrations are probably the result of higher NO_x emissions from recently built power units that began operating in 2010 (Lyu et al. 2016). The present study is unique in nature as it is the first to assess China's health risk attributed to NO_2 during 2015–20. This study found that for 2015–2019, average NO_2 concentration continually decreased across China at a rate of 0.8 ± 2.6 μg/m^3/yr. The study also highlights that densely populated cities with high number of transport vehicles showed a significant decrease in NO_2 levels, this shows that rigorous steps to minimize air pollution have been adopted. Moreover, our results clearly show a significant drop in NO_2 concentration during 2020 in most parts of China. The reason for such a significant drop was that China's authorities had promptly instituted lockdown measures to halt the virus's spread. These policies had a substantial influence on industrial production and transportation throughout the nation, which resulted in significant reductions in NO_2 pollution levels. (Dong et al. 2021). In conclusion, we offer strong epidemiological proof that long exposure to NO_2, a stand-in for air pollutants originating from traffic, is linked to a higher likelihood of mortality from all causes, respiratory disease, and cardiovascular disease. This relationship may exist independently of other contaminants in the air. This study can help educate public health policy on the health impacts of traffic pollution and how to take appropriate measures to decrease exposure to vehicular emissions, particularly in vulnerable groups.

References

12th Five-year Plan (2010) Notice of the state council on issuing the comprehensive work plan for energy conservation and emission reduction during the twelfth five-year plan. http://www.gov.cn/zwgk/2011-09/07/content_1941731.htm

13th Five-year Plan (2015) The 13th five-year plan for economic and social development of the People's Republic of China (2016–2020). https://en.ndrc.gov.cn/policies/202105/P020210527785800103339.pdf

Agirre-Basurko E, Ibarra-Berastegi G, Madariaga I (2006) Regression and multilayer perceptron-based models to forecast hourly O_3 and NO_2 levels in the Bilbao area. Environ Model Softw 21(4):430–446

Ai J, Sun Y, Zheng F, Ni C, Gui K, Zhang X, Jiang W, Liao T (2018) The spatial temporal variation and factor analysis of the tropospheric NO_2 columns in the sichuan basin from 2005 to 2016. Atmos Pollut Res 9:1157–1166

Alimissis A, Philippopoulos K, Tzanis CG, Deligiorgi D (2018) Spatial estimation of urban air pollution with the use of artificial neural network models. Atmos Environ 191:205–213

Azid A, Juahir H, Toriman ME, Kamarudin MKA, Saudi ASM, Hasnam CNC, Aziz NAA, Azaman F, Latif MT, Zainuddin SFM, Osman MR, Yamin M (2014) Prediction of the level of air pollution using principal component analysis and artificial neural network techniques: a case study in Malaysia. Water Air Soil Pollut 225(8):1–14

Baghanam AH, Nourani V, Karimzadeh H (2024) Improving artificial intelligence-based air pollution modelling with the application of meteorological data. Int J Environ Sci Technol 21:431–446

Bai Y, Li Y, Wang X, Xie J, Li C (2016) Air pollutants concentrations forecasting using back propagation neural network based on wavelet decomposition with meteorological conditions. Atmos Pollut Res 7(3):557–566

Boersma K, Eskes H, Brinksma E (2004) Error analysis for tropospheric NO_2 retrieval from space. J Geophys Res Atmos 109

Cabaneros SMS, Calautit JKS, Hughes BR (2017) Hybrid artificial neural network models for effective prediction and mitigation of urban roadside NO_2 pollution. Energy Procedia 142:3524–3530

Carmona-Cabezas R, Gómez-Gómez J, de Ravé EG, Jiménez-Hornero FJ (2020) Checking complex networks indicators in search of singular episodes of the photochemical smog. Chemosphere 241:125085

Chen K, Wang M, Huang C, Kinney PL, Anastas PT (2020) Air pollution reduction and mortality benefit during the COVID-19 outbreak in China. Lancet Planet Health 4:e210–e212

Costa S, Ferreira J, Silveira C, Costa C, Lopes D, Relvas H, Borrego C, Roebeling P, Miranda AI, Paulo Teixeira J (2014) Integrating health on air quality assessment—review report on health risks of two major European outdoor air pollutants: PM and NO_2. J Toxicol Environ Health Part B 17:307–340

Cui Y, Wang L, Jiang L, Liu M, Wang J, Shi K, Duan X (2021) Dynamic spatial analysis of NO_2 pollution over China: satellite observations and spatial convergence models. Atmos Pollut Res 12:89–99

Dong L, Chen B, Huang Y, Song Z, Yang T (2021) Analysis on the characteristics of air pollution in China during the COVID-19 outbreak. Atmosphere 12:205

Feng X, Li Q, Zhu Y, Hou J, Jin L, Wang J (2015) Artificial neural networks forecasting of $PM_{2.5}$ pollution using air mass trajectory based geographic model and wavelet transformation. Atmos Environ 107:118–128

Filonchyk M, Hurynovich V, Yan H, Gusev A, Shpilevskaya N (2020) Impact assessment of COVID-19 on variations of SO_2, NO_2, CO and AOD over East China. Aerosol Air Qual Res 20(7):1530–1540

Garlík B, Přívětivý J (2024) Artificial intelligence algorithms for prediction and diagnosis of air pollution affecting human health. In: Journal of physics: conference series, vol 2701, no. 1. IOP Publishing, p 012072

Gryparis A, Forsberg B, Katsouyanni K, Analitis A, Touloumi G, Schwartz J, Samoli E, Medina S, Ross Anderson H, Niciu EM, Wichmann H-E, Kriz B, Kosnik M, Skorkovsky J, Vonk JM, Dörtbudak Z (2004) Acute effects of ozone on mortality from the "air pollution and health: a European approach" project. Am J Respir Crit Care Med 170(10):1080–1087

Gustafsson M, Lindén J, Åström S, Sjöberg K, Tang L (2018) Quantification of population exposure to NO_2, $PM_{2.5}$ and PM_{10} and estimated health impacts

He HD, Lu WZ, Xue Y (2015) Prediction of particulate matters at urban intersection by using multilayer perceptron model based on principal components. Stoch Env Res Risk Assess 29:2107–2114

He Y, Uno I, Wang Z, Ohara T, Sugimoto N, Shimizu A, Richter A, Burrows JP (2007) Variations of the increasing trend of tropospheric NO_2 over central east China during the past decade. Atmos Environ 41(23):4865–4876

Hesterberg TW, Bunn WB, McClellan RO, Hamade AK, Long CM, Valberg PA (2009) Critical review of the human data on short-term nitrogen dioxide (NO_2) exposures: evidence for NO_2 no-effect levels. Crit Rev Toxicol 39(9):743–781

Heue KP, Richter A, Bruns M, Burrows JP, Platt U, Pundt I, Wang P, Wagner T (2005) Validation of SCIAMACHY tropospheric NO_2-columns with AMAXDOAS measurements. Atmos Chem Phys 5(4):1039–1051

Huang G, Sun K (2020) Non-negligible impacts of clean air regulations on the reduction of tropospheric NO_2 over East China during the COVID-19 pandemic observed by OMI and TROPOMI. Sci Total Environ 745:141023

Kamińska JA (2019) A random forest partition model for predicting NO_2 concentrations from traffic flow and meteorological conditions. Sci Total Environ 651:475–483

Katsouyanni K, Touloumi G, Samoli E, Gryparis A, Le Tertre A, Monopolis Y, Rossi G, Zmirou D, Ballester F, Boumghar A, Anderson HR, Wojtyniak B, Paldy A, Braunstein R, Pekkanen J, Schindler C, Schwartz J (2001) Confounding and effect modification in the short-term effects of ambient particles on total mortality: results from 29 European cities within the APHEA2 project. Epidemiology 12(5):521–531

Katsouyanni K, Touloumi G, Spix C, Schwartz J, Balducci F, Medina S, Rossi G, Wojtyniak B, Sunyer J, Bacharova L, Schouten JP, Ponka A, Anderson HR (1997) Short term effects of ambient sulfur dioxide and particulate matter on mortality in 12 European cities: results from time series data from the APHEA project. BMJ 314(7095):1658

Khaniabadi YO, Goudarzi G, Daryanoosh SM, Borgini A, Tittarelli A, De Marco A (2017) Exposure to PM_{10}, NO_2, and O_3 and impacts on human health. Environ Sci Pollut Res 24:2781–2789

Kraft M, Eikmann T, Kappos A, Künzli N, Rapp R, Schneider K, Seitz H, Voss JW, Wichmann HE (2005) The German view: effects of nitrogen dioxide on human health–derivation of health-related short-term and long-term values. Int J Hyg Environ Health 208(4):305–318

Kurtenbach R, Kleffmann J, Niedojadlo A, Wiesen P (2012) Primary NO_2 emissions and their impact on air quality in traffic environments in Germany. Environ Sci Eur 24:1–8

Li P, Lin Z, Du H, Feng T, Zuo J (2021) Do environmental taxes reduce air pollution? Evidence from fossil-fuel power plants in China. J Environ Manage 295:113112

Lyu W, Li Y, Guan D, Zhao H, Zhang Q, Liu Z (2016) Driving forces of Chinese primary air pollution emissions: an index decomposition analysis. J Clean Prod 133:136–144

Ma Z, Liu R, Liu Y, Bi J (2019) Effects of air pollution control policies on $PM_{2.5}$ pollution improvement in China from 2005 to 2017: a satellite-based perspective. Atmos Chem Phys 19:6861–6877

Maji KJ (2020). Substantial changes in $PM_{2.5}$ pollution and corresponding premature deaths across China during 2015–2019: a model prospective. Sci Total Environ 729:138838

Maji KJ, Sarkar C (2020) Spatio-temporal variations and trends of major air pollutants in China during 2015–2018. Environ Sci Pollut Res 27:33792–33808

Mao S, Chen G, Liu F, Li N, Wang C, Liu Y, Liu S, Lu Y, Xiang H, Guo Y, Li S (2020) Long-term effects of ambient air pollutants to blood lipids and dyslipidemias in a Chinese rural population. Environ Pollut 256:113403

Mishra D, Goyal P (2015) Development of artificial intelligence based NO_2 forecasting models at Taj Mahal, Agra. Atmos Pollut Res 6(1):99–106

Moazami S, Noori R, Amiri BJ, Yeganeh B, Partani S, Safavi S (2016) Reliable prediction of carbon monoxide using developed support vector machine. Atmos Pollut Res 7(3):412–418

Muhammad S, Long X, Salman M (2020) COVID-19 pandemic and environmental pollution: a blessing in disguise? Sci Total Environ 728:138820

Murillo-Escobar J, Sepulveda-Suescun JP, Correa MA, Orrego-Metaute D (2019) Forecasting concentrations of air pollutants using support vector regression improved with particle swarm optimization: case study in Aburrá Valley, Colombia. Urban Clim 29:100473

Nourani V, Abdollahi Z, Sharghi E (2021) Sensitivity analysis and ensemble artificial intelligence-based model for short-term prediction of NO_2 concentration. Int J Environ Sci Technol 18(9):2703–2722

Osowski S, Garanty K (2007) Forecasting of the daily meteorological pollution using wavelets and support vector machine. Eng Appl Artif Intell 20(6):745–755

Perez P, Gramsch E (2016) Forecasting hourly $PM_{2.5}$ in Santiago de Chile with emphasis on night episodes. Atmos Environ 124:22–27

Pope CA III, Dockery DW (2006) Health effects of fine particulate air pollution: lines that connect. J Air Waste Manag Assoc 56(6):709–742

Pope CA, Dockery DW, Schwartz J (1995) Review of epidemiological evidence of health effects of particulate air pollution. Inhalation Toxicol 7(1):1–18

Radojević D, Antanasijević D, Perić-Grujić A, Ristić M, Pocajt V (2019) The significance of periodic parameters for ANN modelling of daily SO_2 and NO_x concentrations: a case study of Belgrade, Serbia. Atmos Pollut Res 10(2):621–628

Rehman MU, Kamal M, Tiwari SK (2025) Statistical analysis based on a two-staged SEM-SVM approach for environmental noise annoyance prediction and identification of influencing factors. Int J Human Capital Urban Manag 10(1)

Samet JM, Dominici F, Curriero FC, Coursac I, Zeger SL (2000) Fine particulate air pollution and mortality in 20 US cities, 1987–1994. N Engl J Med 343(24):1742–1749

Schneider P, Lahoz WA, van der A R (2015) Recent satellite-based trends of tropospheric nitrogen dioxide over large urban agglomerations worldwide. Atmos Chem Phys 15(3):1205-1220

Schwartz J (2005) How sensitive is the association between ozone and daily deaths to control for temperature? Am J Respir Crit Care Med 171(6):627–631

Seltzer KM, Shindell DT, Malley CS (2018) Measurement-based assessment of health burdens from long-term ozone exposure in the United States, Europe, and China. Environ Res Lett 13(10):104018

Shams SR, Jahani A, Kalantary S, Moeinaddini M, Khorasani N (2021) Artificial intelligence accuracy assessment in NO_2 concentration forecasting of metropolises air. Sci Rep 11(1):1805

Shang Z, Deng T, He J, Duan X (2019) A novel model for hourly $PM_{2.5}$ concentration prediction based on CART and EELM. Sci Total Environ 651:3043–3052

Silver B, Reddington CL, Arnold SR, Spracklen DV (2018) Substantial changes in air pollution across China during 2015–2017. Environ Res Lett 13(11):114012

Silvern RF, Jacob DJ, Mickley LJ, Sulprizio MP, Travis KR, Marais EA, Cohen RC, Laughner JL, Choi S, Joiner J et al (2019) Using satellite observations of tropospheric NO_2 columns to infer long-term trends in us NO_x emissions: the importance of accounting for the free tropospheric NO_2 backgrounds. Atmos Chem Phys 19:8863–8878

Tian X, An C, Nik-Bakht M, Chen Z (2022) Assessment of reductions in NO_2 emissions from thermal power plants in Canada based on the analysis of policy, inventory, and satellite data. J Clean Prod 341:130859

Tiwari SK, Kumaraswamidhas LA, Garg N (2022) Time-series prediction and forecasting of ambient noise levels using deep learning and machine learning techniques. Noise Control Eng J 70(5):456–471

Tiwari SK, Kumaraswamidhas LA, Patel R, Garg N, Vallisree S (2024) Traffic noise measurement, mapping, and modelling using soft computing techniques for mid-sized smart Indian city. Meas Sens 33:101203

Touloumi G, Katsouyanni K, Zmirou D, Schwartz J, Spix C, Ponce de Leon A, Tobias A, Quennel P, Rabczenko D, Bacharova L, Bisanti L, Vonk JM, Ponka A (1997) Short-term effects of ambient oxidant exposure on mortality: a combined analysis within the APHEA project. Am J Epidemiol 146(2):177–185

Van Der A RJ, Peters DHMU, Eskes H, Boersma KF, Van Roozendael M, De Smedt I, Kelder HM (2006) Detection of the trend and seasonal variation in tropospheric NO_2 over China. J Geophys Res Atmos 111(D12)

Wang C, Wang T, Wang P (2019) The spatial–temporal variation of tropospheric NO_2 over China during 2005 to 2018. Atmosphere 10(8):444

World Health Organization (2021) Review of evidence on health aspects of air pollution: REVIHAAP project: technical report. Technical Report. World Health Organization. Regional Office for Europe

Wu Y, Jing W, Liu J, Ma Q, Yuan J, Wang Y, Du M, Liu M (2020) Effects of temperature and humidity on the daily new cases and new deaths of COVID-19 in 166 countries. Sci Total Environ 729:139051

Xie Y, Wang W, Wang Q (2018) Spatial distribution and temporal trend of tropospheric NO_2 over the Wanjiang city belt of China. Adv Meteorol 2018(1):6597186

Xue T, Zheng Y, Tong D, Zheng B, Li X, Zhu T, Zhang Q (2019) Spatiotemporal continuous estimates of $PM_{2.5}$ concentrations in China, 2000–2016: a machine learning method with inputs from satellites, chemical transport model, and ground observations. Environ Int 123:345–357

Zhang K, Zhao C, Fan H, Yang Y, Sun Y (2020) Toward understanding the differences of $PM_{2.5}$ characteristics among five China urban cities. Asia-Pac J Atmos Sci 56:493–502

Zhang L, Lee CS, Zhang R, Chen L (2017) Spatial and temporal evaluation of long-term trend (2005–2014) of OMI retrieved NO_2 and SO_2 concentrations in Henan Province, China. Atmos Environ 154:151–166

Zhang Q, Zheng Y, Tong D, Shao M, Wang S, Zhang Y, Xu X, Wang J, He H, Liu W, Ding Y, Lei Y, Li J, Wang Z, Zhang X, Wang Y, Cheng J, Liu Y, Shi Q, Yan L, Geng G, Hong C, Li M, Liu F, Zheng B, Cao J, Ding A, Gao J, Fu Q, Huo J, Liu B, Liu Z, Yang F, He K, Hao J (2019) Drivers of improved $PM_{2.5}$ air quality in China from 2013 to 2017. Proc Natl Acad Sci 116(49):24463–24469

Zhang Y, Mo J, Weschler CJ (2013) Reducing health risks from indoor exposures in rapidly developing urban China. Environ Health Perspect 121(7):751–755

Zhou Y, Wu Y, Yang L, Fu L, He K, Wang S, Hao J, Chen J, Li C (2010) The impact of transportation control measures on emission reductions during the 2008 Olympic Games in Beijing, China. Atmos Environ 44(3):285–293

Zhu Y, Zhan Y, Wang B, Li Z, Qin Y, Zhang K (2019) Spatiotemporally mapping of the relationship between NO_2 pollution and urbanization for a megacity in Southwest China during 2005–2016. Chemosphere 220:155–162

Chapter 3
Understanding the Concept of Noise Pollution: Sources, Impacts, and Mitigation Strategies

Rohit Patel, Prasoon Kumar Singh, Shivam Saw, and Rakesh Kumar

Abstract This chapter thoroughly explores environmental noise, including its causes, impacts, and possible remedies. It identifies noise pollution as a significant environmental issue caused by human-related activities such as transportation, industrialization, and urbanization. Noise pollution has a wide range of effects, from hearing loss to cardiovascular disease, stress, and sleep difficulties. Furthermore, noise pollution disturbs ecosystems by conflicting with wildlife communication, breeding habits, and habitat choices. Numerous noise abatement strategies have been investigated to address these challenges. The chapter discusses ways to reduce noise at the source, such as managing vehicle noise, enhancing road surfaces, and encouraging quieter technologies. Transmission path reductions, such as noise barriers and zoning changes, are explored with receiver safeguard strategies, such as education campaigns and personal protective gear. The current chapter emphasizes the need for interdisciplinary collaboration among city planners, environmentalists, and policymakers to develop effective noise reduction techniques. Finally, the chapter emphasizes the importance of comprehensive measures addressing the intricate relationship between human actions and noise pollution.

Keywords Noise pollution · Sources of noise pollution · Effects of noise pollution · Noise mitigation measures · Noise control

---

R. Patel (✉) · P. K. Singh
Department of Environmental Science and Engineering, Indian Institute of Technology (Indian School of Mines), Dhanbad 826004, India
e-mail: rohitpatelp7@gmail.com

P. K. Singh
e-mail: pks0506@iitism.ac.in

S. Saw
Department of Chemical Engineering, National Institute of Technology, Durgapur 713209, India
e-mail: shivamsaw@gmail.com

R. Kumar
School of Engineering, Department of Mechanical Engineering, P P Savani University, Surat 394125, India
e-mail: rakesh.kumar@ppsu.ac.in

S. K. Tiwari et al. (eds.), *AI-Driven Environmental Pollution Management*, Climate Risks and Solutions, https://doi.org/10.1007/978-3-031-96243-1_3

3.1 Introduction

Noise pollution is the undesirable or unsafe levels of sound in the natural environment that interrupt daily life, primarily caused by human activity. These mainly consist of sounds from traffic, construction, manufacturing operations, and societal gatherings. Loudness, frequency, and length of exposure are all characteristics that can have varying effects on people and their surroundings (Cohen and Weinstein 1981). Persistent exposure to excessive noise levels can lead to many challenges, from slight discomfort to substantial health effects, confirming noise as a type of environmental pollution that deserves urgent attention (Murphy and King 2022).

The fast urbanization and industrialization rate has significantly raised noise pollution sources (Patel et al. 2024a; Tiwari et al. 2024). As cities grow to accommodate increasing populations and economic activity, the resultant infrastructure, including roadways, industries, and public transportation, often increases noise levels. The rise in automobile traffic, urban building, and workshop activities creates a constant discordance that pervades urban life, making it increasingly impossible for residents to find relief from the noise (Amir et al. 2023). This trend is witnessed globally, with areas of high population attempting to maintain a healthy equilibrium between development and a satisfactory standard of life for those who live there (Sonaviya and Tandel 2020).

Despite its broad existence, noise pollution is often overlooked among other related environmental problems, such as water and air pollution. This disregard can be due to the ethereal nature of sound, which is not as easily visible as other pollutants. Many people may regard excessive noise as a nuisance rather than a serious health risk. As a result, legislative measures and public awareness initiatives frequently prioritize improving air quality and water quality, keeping environmental noise unaddressed in public health talks. Noise pollution has extensive and complex repercussions for human health. High noise levels can cause various health and psychological problems, such as anxiety, stress, annoyance, and sleep disruption (Patel and Pandey 2017; Tiwari et al. 2023a). Long-term exposure can also lead to cardiovascular difficulties, cognitive deficits, and decreased quality of life (Hener 2022; Mishra et al. 2021a, b; Okokon et al. 2018). Besides human health, noise pollution can harm natural ecosystems. Wildlife frequently relies on sound to communicate, mate, and navigate. Increased noise levels can disturb these behaviors, resulting in changes in species distribution and population reductions (Kight and Swaddle 2011; Teff-Seker et al. 2022).

To mitigate the negative impacts of noise pollution in a region, it is essential to assess and forecast the noise levels of the region. This can be achieved by employing various noise prediction models (Tiwari et al. 2022). Therefore, evaluating the factors affecting noise levels and analyzing environmental noise sources and impacts using indices like L_{eq}, L_{10}, L_{50}, and L_{90}, are essential for developing an accurate noise prediction model. The development of noise prediction models involves various methodologies, including mathematical methods and techniques based on artificial intelligence and machine learning (Tiwari et al. 2023a, b).

This chapter aims to provide a thorough examination of noise pollution, addressing its sources, impacts on human well-being and ecological systems, and potential mitigation measures. It aims to highlight the importance of noise pollution as an environmental issue, frequently overlooked compared to air, water, and soil pollution. It emphasizes the need for collaborative work among city planners, environmentalists, and legislators to develop effective noise reduction methodologies. Moreover, the chapter also attempts to raise public awareness and implement regulatory initiatives to combat the negative impacts of noise pollution and enhance overall quality of life.

3.2 Basics of Sound and Sources of Noise Pollution

Sound is a form of energy emitted by a vibrating body, and upon reaching the ear, it causes the sensation of hearing through nerves. The sound produced by all vibrating bodies is not audible. The frequency limit of audibility is from 20 Hz to 20,000 Hz. The loudness of a sound is influenced by the amplitude of its wave, with greater amplitude corresponding to higher intensity. Another important aspect of a sound wave is its pitch, which is determined by its frequency. Sounds with frequencies below 20 Hz are termed infrasonic, while those beyond 20,000 Hz are called ultrasonic. Loudness is a key attribute of sound waves, representing an important characteristic of auditory perception. It is quantified in decibels (dB), with the human ear capable of detecting sounds within a range of 0 dB to 140 dB. However, exposure to sound levels between 120 and 140 dB can induce pain in the human ear.

The problem of noise pollution generally comprises of three interrelated elements including the source, receiver, and the path. The sound transmission path primarily involves the atmosphere through which sound travels. However, it can also include the structural materials of buildings containing the receiver, as depicted in Fig. 3.1.

The intensity of the sound declines considerably as the distance between the source and the observer increases. The noise level of sound is reduced as the distance from the source increases, according to the inverse square law. This principle indicates that sound intensity decreases proportionally to the square of the distance measured from its source. If the inverse square law is followed, doubling the distance from the sound source reduces the sound intensity by about 6 dB (Wiener and Keast 1959). The mathematical expression showing the relation between the sound intensity and the noise level is shown in Eq. 3.1.

$$\text{Sound Intensity} \propto \frac{1}{d^2} \tag{3.1}$$

The decrease in sound level 'L' in terms of dB can be estimated using Eq. 3.2 as follows.

$$L_n = L_o - 20 \log \frac{d_n}{d_o} \tag{3.2}$$

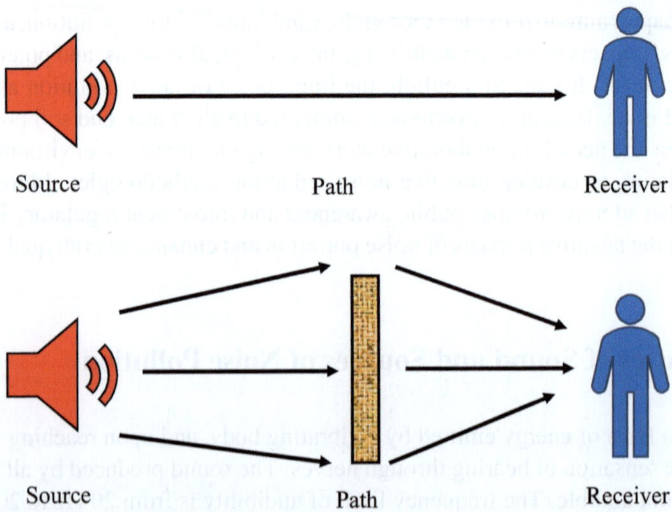

Fig. 3.1 Inter-relationship between the elements of noise

where:

L_n Represents the sound level at the new distance.
L_o Represents the sound level at the original distance.
d_n and d_o Represent the new and the original distance from the source, respectively.

It is widely recognized that the human ear can perceive varying noise levels, ranging from low to high degrees of noise pollution (Nazneen et al. 2020). Figure 3.2 illustrates the classification of noise levels according to distinct types of sound sources. This figure displays the loudness levels of various common noise sources, measured in dB, classified into low, moderate, loud, very loud, and extremely loud. At the lower end, sounds such as breathing, rustling leaves, and whispering fall into the "low" category, demonstrating that they are usually quiet and not disruptive. Moderate sound levels, ranging from 40 to 60 dB, incorporate noises like refrigerator hum, rainfall, and everyday conversation. While more evident, these sounds do not usually pose any threat to hearing. The "loud" category, encompassing 70 to 80 dB, includes noise sources like cars and trucks, which can become uncomfortable with prolonged exposure. Likewise, the "very loud" range, which spans 90 to 100 dB, is sound produced from hair dryers and helicopters, which can damage hearing over extended periods without protection. Finally, the "extremely loud" range, from 110 to 140 dB, includes intense noise sources like police sirens, jet engines, and rocket launches. Exposure to sounds at these levels is dangerous and can cause immediate hearing damage without protective measures. This categorization highlights the need to understand noise pollution's impact and the importance of protection in environments with high-decibel sounds.

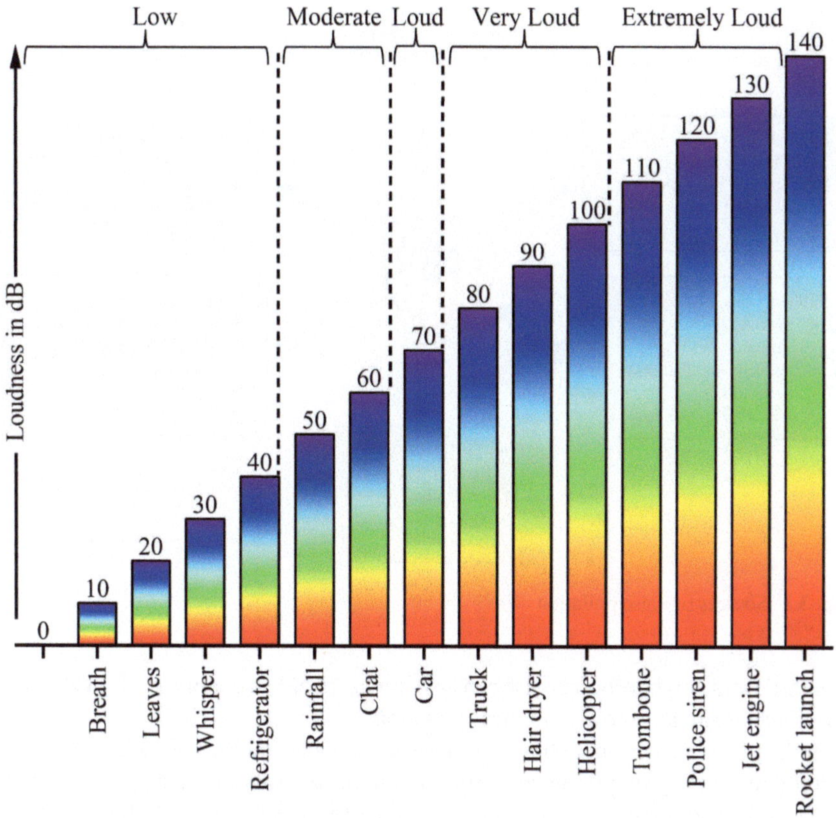

Fig. 3.2 Loudness of sound for various primary noise sources

Figure 3.3 depicts a comprehensive framework for classifying numerous sources of noise pollution, which is necessary for understanding the complexities of ambient noise and its harmful effects on the urban environment. Sources of Noise pollution can be divided into six categories: natural sources, industrial activities, residential areas, transportation, construction activities, and public events. Each of these groups uniquely adds to environmental noise levels, affecting both public health and urban planning.

Natural sources of noise, such as thunderstorms, strong wind, rain, volcanic eruptions, and wildlife, although not man-made, possess a substantial impact on environmental noise pollution. These natural occurrences, particularly in areas with extreme weather or rich wildlife, contribute to the background noise levels in rural and urban environments. In residential settings, regular activities such as using domestic appliances, lawn care tools, and the presence of pets lead to noise pollution, especially in highly populated regions where these noises can be prolonged and disruptive.

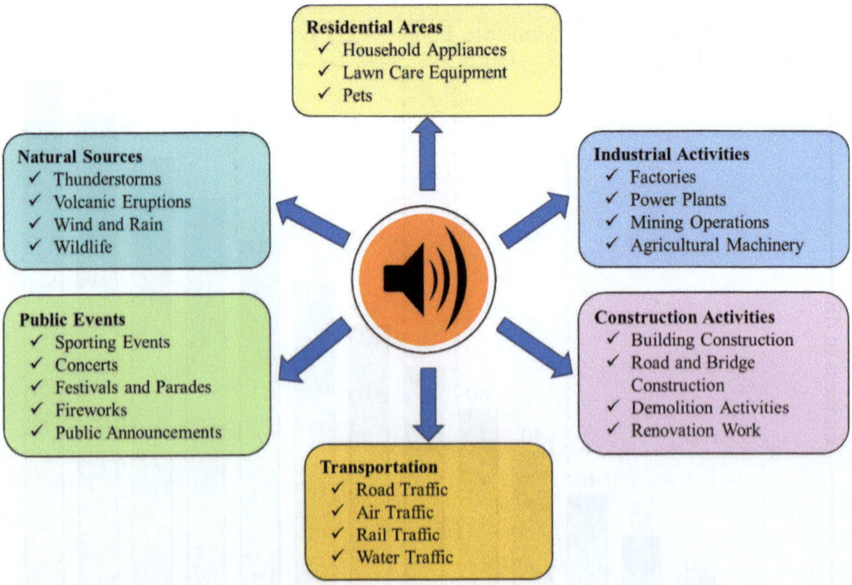

Fig. 3.3 Sources of noise pollution

Domestic and neighborhood noise can also impact people's quality of life. Therefore, noise management is critical in these situations.

Industrial activities, including factories, power plants, mining operations, and agricultural machinery, contribute to continuous, high-intensity noise, affecting workers and nearby communities. The noise generated by these activities often results in long-term exposure, making it a significant environmental concern. Construction activities, such as building and road construction, demolition, and renovation work, are typically characterized by intermittent but intense noise levels. These activities, while temporary, contribute significantly to noise pollution in urban spaces, especially near residential or commercial zones. Though periodic, public events such as sporting events, concerts, festivals, parades, fireworks, and public announcements often result in peak noise levels that can affect communities over short periods. These events, mainly when held in densely populated areas, can lead to significant temporary increases in environmental noise pollution.

Furthermore, transportation noise is a significant source of environmental noise pollution, with road, air, and rail traffic all producing different noise patterns. Traffic from roads is frequently the most prominent among them, contributing significantly to urban noise levels (Patel et al. 2022). The transportation sector contributes 70% to the overall generation of environmental noise pollution, as illustrated in Fig. 3.4. In the case of the transportation sector, a significant contribution is from road traffic noise, which contributes to nearly 70% of total traffic noise pollution (Mishra et al. 2021a, b; Patel et al. 2024b).

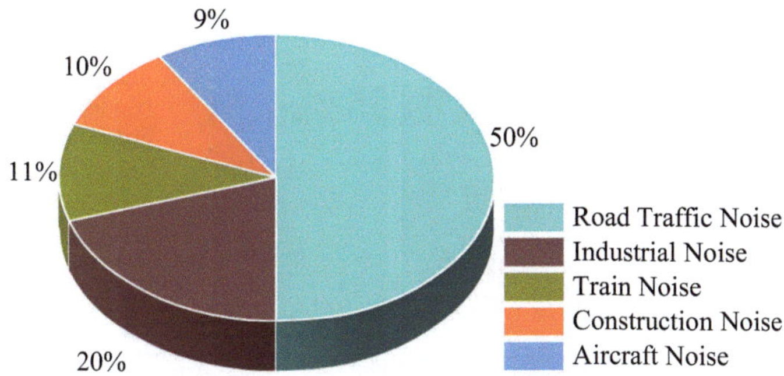

Fig. 3.4 Contribution of major noise sources in total noise generation in urban areas (Mishra et al. 2021a, b)

Finally, urban planners and environmental authorities must comprehend the numerous sources of noise pollution, both human generated and natural. This classification gives an organized approach to analyzing the causes and effects of noise pollution, allowing for the creation of focused mitigation solutions. By tackling each category through regulatory and technical interventions, noise management techniques can be considerably improved, improving the quality of life in rural and urban areas.

3.3 Adverse Effects of Noise Pollution

Noise pollution is a prevalent environmental concern with severe consequences for human health, the ecosystem, wildlife, and societies. Figure 3.5 highlights the broad and significant negative impacts of noise pollution. Extensive studies have demonstrated the negative consequences of excessive noise exposure on numerous aspects of human beings. The effects of noise pollution on human health can be broadly divided into two categories: auditory effects and non-auditory effects. Auditory effects refer to the direct impact of noise on the hearing system. Non-auditory effects are the broader health impacts of noise that do not directly affect hearing. Studies have frequently indicated a high association between noise exposure and hearing damage. Prolonged exposure to loud noise leads to irreparable hearing loss (Alves et al. 2020). In addition, environmental noise has been associated with an increased risk of hypertension and coronary artery disease. Research has revealed that noise exposure can stimulate the body's stress response, resulting in higher blood pressure and heart rate (Babisch 2019; Rabiei et al. 2021). Noise can also interrupt sleep cycles, causing exhaustion, irritation, and decreased mental performance. Also, persistent noise exposure has been linked to elevated levels of annoyance, stress, anxiety, and depression (Venet et al. 2023; Rehman et al. 2025).

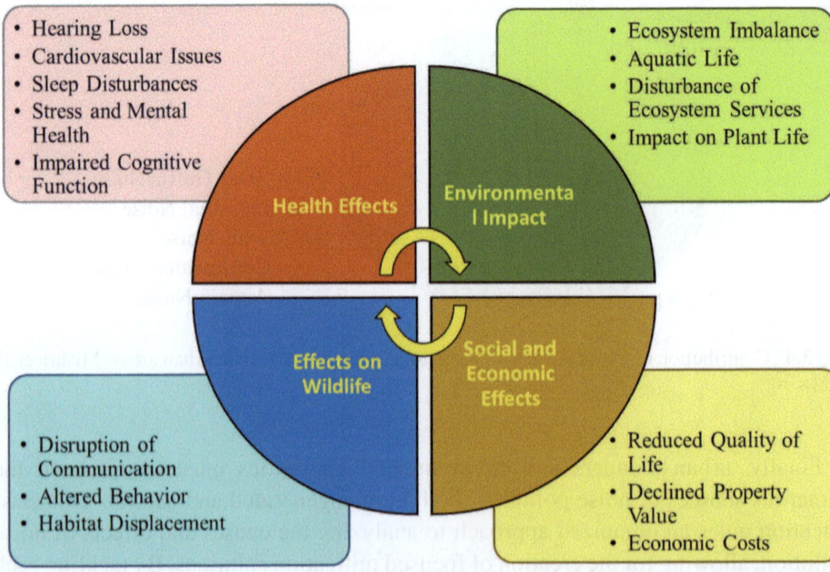

Fig. 3.5 Effects of noise pollution

Apart from human health, noise pollution can alter the natural equilibrium of ecosystems, impacting both terrestrial and aquatic life. Research has proven that noise may disrupt numerous species' communication, movement, and reproduction, causing population decreases and habitat loss (Rosa and Koper 2018). Noise can also reduce habitat quality, making it unsuitable for wildlife. According to previous studies, noise can disrupt wildlife communication, making it difficult for animals to find food, mates, and escape predators. Noise can also drive wildlife to leave their natural habitats searching for quieter locations, resulting in habitat dispersion and population decreases. Noise pollution can drastically degrade the quality of daily life for individuals and communities, resulting in lower well-being and contentment. Properties located in loud locations may lose value due to the detrimental impact on quality of life. Furthermore, noise pollution can have a substantial economic impact, including increased healthcare expenditures, lost productivity, and the requirement for noise reduction solutions.

3.4 Noise Pollution Management and Mitigation

Mitigating environmental noise pollution is a challenge that cannot be resolved immediately. Noise pollution poses a complicated challenge that necessitates an interdisciplinary response. Varied professionals must work together to reduce noise and its

Fig. 3.6 Mind map for noise pollution management approaches (Suhanek, 2023)

negative impacts significantly. Urban planners, architects, medical specialists, biologists, psychologists, and acoustic engineers must collaborate to utilize their strengths for a common goal: improving the overall quality of life. Figure 3.6 shows the mind map of various noise pollution management approaches.

Public awareness serves as an important foundation for noise abatement strategies. This stems from the necessity of engendering public recognition of noise as a significant environmental issue. Such recognition is paramount in garnering public acceptance and subsequent political support, both of which are essential for successfully implementing the majority of noise mitigation measures discussed hereafter. The public must be aware of noise mitigation and reduction via educational interventions. The primary groups for focused communication must be identified in the awareness-raising strategy. Table 3.1 presents several key target groups and subgroups for promoting noise awareness.

The mind map lists numerous strategies for reducing noise pollution, which is divided into three main categories:

Table 3.1 Groups and subgroups targeted for raising noise awareness (Kloth et al. 2008)

Target Group	Subgroup
Residents	Residents of city Workers of city Public transport users Cyclists and walkers Shop owners Drivers Parents of teenagers
Public transport operators	Public and private road, rail, and bus operators Public and private airlines
Planning and management sector	Progress control and forward planners Land use planners Environmental designers
Freight distribution sector	Truck drivers Logistics operators for the industry Workshop owners and associated business
Waste managing sector	Public and private waste management operators Drivers of waste collection fleet
Educational sector	Students Teachers Parents
Health sector	Hospital staff General practitioners Public and private health services Hospital patients
Media	Reporters Provincial and local newspapers Magazines precise to targeted peoples
NGOs	Community organizations Research organizations Environmental consultancy companies
Government/policymaker	Municipal councils National and regional authorities

- **Source Reduction** relates to strategies for reducing noise at the source, such as employing electric cars, which are quieter than conventional petrol or diesel-driven vehicles and designing buildings with sound insulation.
- **Path Reduction** is related to strategies for blocking the path of noise pollution, such as erecting barriers to block noise and smoothing out road surfaces.
- **Receiver Protection** corresponds to strategies for shielding people from the effects of noise pollution, such as educating the public about the risks of noise pollution and urging them to prevent prolonged exposure to loud noises. Furthermore, when noise pollution is unavoidable, such as in construction areas or workplaces, supplying and promoting personal protective equipment such as earplugs or earmuffs is essential.

3.4.1 Reduction at the Source

3.4.1.1 Regulation and Legislation

The best way to reduce noise pollution at the source while maintaining economic viability is to enact laws that set acceptable noise levels for manufacturing processes. For these restrictions to be effective, they must be strictly enforced, which calls for frequent audits, testing methods, and inspections to confirm that the regulations are obeyed.

3.4.1.2 Surface Property of the Road

Research indicates that porous and thin-layer asphalt are the most effective road surfaces for reducing noise emissions. However, future-generation surfaces may also reduce noise; thus, these alternatives should be explored further. Because thin-layer asphalt has been created with low noise emission, it may make an open surface texture more easily. This is accomplished by using small maximum aggregate sizes (between 6 and 8 mm), which smooths down the road surface (Bendtsen and Andersen 2005).

3.4.1.3 Low Noise Tires

The most recent tire-related European Union legislation seeks to improve road transport's economic viability, safety, and environmental sustainability by encouraging the use of safe, fuel-efficient, and low-noise tires. This law has been in effect since November 2012 and creates a standard for labeling tires with information on tire characteristics, such as external rolling noise. When buying tires, customers may use this labeling system to make intelligent choices that are safe for the environment and promote road safety.

3.4.1.4 Driving Conduct

Driver behavior is essential for reducing traffic noise. Drivers can help reduce the noise their vehicles produce by obeying speed limits, especially in residential areas where lower speeds are required for noise reduction. Furthermore, gentler driving techniques, such as gradual acceleration and stopping, help to create a quieter environment by reducing abrupt engine noise. Avoiding unnecessary horn use helps to reduce noise pollution on highways. Regular vehicle maintenance is also essential, as poorly maintained vehicles tend to make more noise, especially with malfunctioning exhaust systems or damaged tires.

Table 3.2 Changes in noise level due to speed reduction (Murphy and King 2011)

A decline in actual traffic speed (kmph)	Decline in noise level (in dBA) (for light vehicles)	Decline in noise level (in dBA) (for heavy vehicles)
130–120	1	–
120–110	1.1	–
110–100	1.2	–
100–90	1.3	1
90–80	1.5	1.1
80–70	1.7	1.2
70–60	1.9	1.4
60–50	2.3	1.7
50–40	2.8	2.1
40–30	3.6	2.7

3.4.1.5 Traffic Management

An effective traffic management strategy can reduce noise levels and create a healthier and more peaceful environment. There is an association between vehicular noise emissions and travel speed. While a general trend exists for propulsion noise to increase with engine revolutions per minute (RPMs), leading to elevated noise levels at higher gear selections and corresponding velocities, the connection does not show linear variation. For lower travel speeds generally below 30 kmph, noise generated through the engine becomes the dominant contributor, overshadowing the influence of speed on noise generation. Significant reductions in noise emissions are only observed when travel speed falls below 100 km/h. Interestingly, within the 100–130 km/h speed range, no appreciable reduction in noise levels is found, as shown in Table 3.2.

3.4.2 Reduction in Transmission Path

3.4.2.1 Land Use Planning

Land use planning can significantly help to minimize noise pollution. It prioritizes separating noisy places, such as industry and roadways, from residential areas and educational institutions. This can include encouraging walking and cycling through facilities such as bike lanes and establishing quiet zones comprising parks and noise-reducing infrastructure. Zoning restrictions can help to control noise by restricting the organization and establishing noise limits.

3.4.2.2 Interior Building Design

Efficient building design is essential for minimizing noise in indoor environments, with significant aspects including room configuration and the building's geometry and orientation. Regarding room layout, sections linked with less noise-sensitive events, such as kitchens and utility rooms, should be positioned closer to the noise source, such as a road. Rooms designed for more noise-sensitive activities, such as bedrooms and living areas, should be located at a higher distance from the noise source to reduce noise emissions and provide the best conditions for sleeping and resting.

3.4.2.3 Building/Facade Noise Insulation

The sound insulation of residential structures includes enhancing the isolation of windows and outer walls, those facing the building's face, and those directly exposed to external noise sources. This is usually the last choice for minimizing noise levels for people sensitive to noise. Although it can be highly efficient, it also requires massive investments (Tadeu and Mateus 2001).

3.4.2.4 Noise Barriers

Noise barriers have evolved as a popular method for reducing noise propagation. Since their initial adoption in California in 1968, their application has progressively increased in conjunction with continuous scientific advances in noise barrier performance and design. This demonstrates their increasing popularity as an effective technique for noise pollution mitigation (Ekici and Bougdah 2003).

Apart from the above-defined methods, a modern approach to mitigating noise pollution involves the application of the soundscape idea. This method comprehensively analyzes noise and its effects on human health. The concept of a soundscape refers to all audible elements that are detectable to humans and exist within a particular acoustic environment. The three main categories into which these auditory components are usually divided are bio-phony, geophony, and anthrophony. Bio-phony refers to the noise generated by living things, such as animals, bug chirps, and bird songs.

Similarly, Geophony encompasses natural sounds that are not biological, such as wind whispering leaves, streams, and seas moving, as well as geological events like earthquakes and thunder. However, Anthrophony comprises the sounds produced by human activity, such as industrial, transportation, and construction noise. Moreover, traffic noise is a crucial issue for legislators, urban planners, and public health officials to solve since it tends to have a more pervasive and continuous influence on urban populations than other kinds of noise pollution. Therefore, the issue of traffic noise pollution needs to be addressed and mitigated with careful planning and a comprehensive strategy.

3.4.3 Reduction at Receiver's End

Receiver protection emphasizes protecting people from the adverse effects of noise pollution when it is difficult to eradicate or minimize noise at the point of origin or along its pathway. It involves actions that directly safeguard people from exposure and reduce health hazards. One of the most efficient long-term techniques is to educate individuals about the risks of extended noise exposure and the significance of adopting precautions. Public campaigns, educational courses, and workplace instruction can help promote awareness about its possible health consequences, such as hearing impairment, anxiety, and cardiovascular problems. These courses of action encourage people to take specific steps to lessen their noise exposure.

Moreover, providing and encouraging the use of personal protective equipment is an essential component of receiver shielding in workspaces and other noisy environments. Noise-canceling headphones, earmuffs, and earplugs are helpful tools for lowering exposure to dangerous noise levels, particularly for employees in manufacturing, construction, and other sectors. Workplace safety rules frequently require using such devices, but developing a culture of regular usage ensures that workers stay safe even when noise levels fluctuate. Moreover, regular wellness checkups, primarily hearing tests, may assist in recognizing early indications of noise-induced hearing loss (NIHL). Regular noise monitoring in locations with continuously excessive noise levels can assist in identifying workers at risk. If noise-related hearing loss is diagnosed early, measures such as hearing aids, medication, or modifications in work routines can help prevent any more damage.

3.5 Conclusion

This chapter delves deeply into noise pollution, emphasizing its multifaceted causes and far-reaching consequences for human health and the environment. It identifies critical noise sources and the adverse human health consequences. The ecological impacts of noise, specifically the disruption of wildlife communication and habitat integrity, are highlighted, pointing to more significant environmental problems. Moreover, the discussion emphasizes the necessity for a multi-tiered strategy for noise pollution reduction. The underlying measures include source reduction initiatives, quieter technologies, and regulatory constraints. Nevertheless, controlling noise transmission via better urban planning and infrastructure is also essential. In situations in which noise cannot be eliminated, recipient protection through public awareness and the provision of personal protective equipment is critical.

Future research directions should be targeted to improve the noise prediction model, mainly with the help of artificial intelligence approaches, as well as the evaluation of methods for noise mitigation. Multidisciplinary collaboration is critical, necessitating constant cooperation among regulatory organizations, urban planners, environmental experts, and health specialists to construct complete noise control frameworks. The chapter concludes by emphasizing the importance of continuous support from the government and widespread public awareness for enhancing noise pollution regulation and reducing its adverse effects.

References

Alves JA, Paiva FN, Silva LT, Remoaldo P (2020) Low-frequency noise and its main effects on human health—a review of the literature between 2016 and 2019. Appl Sci
Amir S, Sadoway D, Dommaraju P (2023) Taming the noise: soundscape and livability in a technocratic city-state. East Asian Sci Technol Soc 17
Babisch W (2019) Cardiovascular effects of noise. In: Encyclopedia of environmental health, pp 543–552
Bendtsen H, Andersen B (2005) Experiences with thin noise reducing pavements. In: Forum acusticum budapest 2005: 4th European congress on acustics
Cohen S, Weinstein N (1981) Nonauditory effects of noise on behavior and health. J Soc Issues 37
Ekici I, Bougdah H (2003) A review of research on environmental noise barriers. Build Acoust
Hener T (2022) Noise pollution and violent crime. J Public Econ 215
Kight CR, Swaddle JP (2011) How and why environmental noise impacts animals: an integrative, mechanistic review. Ecol Lett
Kloth M, Vancluysen K, Clement F, Lars Ellebjerg P (2008) Practitioner handbook for local noise action plans: recommendations from the SILENCE project
Mishra A, Das S, Singh D, Maurya AK (2021a) Effect of COVID-19 lockdown on noise pollution levels in an Indian city: a case study of Kanpur. Environ Sci Pollut Res 28:46007–46019
Mishra RK, Nair K, Kumar K, Shukla A (2021) Dynamic noise mapping of road traffic in an urban city. Arab J Geosci 14
Murphy E, King EA (2011) Scenario analysis and noise action planning: modelling the impact of mitigation measures on population exposure. Appl Acoust
Murphy E, King EA (2022) Environmental noise pollution: noise mapping, public health, and policy. Noise Mapping, Public Health, and Policy, Environmental Noise Pollution
Nazneen S, Raza A, Khan S (2020) Assessment of noise pollution and associated subjective health complaints and psychological symptoms: analysis through structure equation model. Environ Sci Pollut Res 27
Okokon EO, Yli-Tuomi T, Turunen AW, Tiittanen P, Juutilainen J, Lanki T (2018) Traffic noise, noise annoyance and psychotropic medication use. Environ Int
Patel R, Kumar Singh P, Saw S (2022) Recent advancements in the challenges and strategies of globally used traffic noise prediction models. Environ Sci Pollut Res
Patel R, Pandey G (2017) A study of noise pollution at the campus of Madan Mohan Malaviya University of Technology, Gorakhpur, Uttar Pradesh (India). Int Res J Eng Technol 4:1258–1263
Patel R, Singh PK, Saw S (2024a) A modeling approach for suitability evaluation of traffic noise prediction under mixed traffic situation in mid-sized Indian cities. Innov Infrastruct Solut 9
Patel R, Singh PK, Saw S (2024b) Traffic noise modeling under mixed traffic condition in mid-sized indian city: a linear regression and neural network-based approach. Int J Math Eng Manag Sci 9:411–434

Rabiei H, Ramezanifar S, Hassanipour S, Gharari N (2021) Investigating the effects of occupational and environmental noise on cardiovascular diseases: a systematic review and meta-analysis. Environ Sci Pollut Res

Rehman MU, Kamal M, Tiwari SK (2025) Statistical analysis based on a two-staged SEM-SVM approach for environmental noise annoyance prediction and identification of influencing factors. Int J Hum Cap Urban Manag 10(1)

Rosa P, Koper N (2018) Integrating multiple disciplines to understand effects of anthropogenic noise on animal communication: ecosphere 9

Sonaviya DR, Tandel BN (2020) Integrated road traffic noise mapping in urban Indian context. Noise Mapp 7

Suhanek M (2023) Introductory chapter: management of noise pollution

Tadeu AJB, Mateus DMR (2001) Sound transmission through single, double and triple glazing. Experimental Evaluation Appl Acoust

Teff-Seker Y, Berger-Tal O, Lehnardt Y, Teschner N (2022) Noise pollution from wind turbines and its effects on wildlife: a cross-national analysis of current policies and planning regulations. Renew Sustain Energy Rev

Tiwari SK, Kumaraswamidhas LA, Garg N (2023a) Assessment of noise pollution and associated subjective health complaints in Jharia Coalfield. A structural equation model analysis. Noise Mapp, India, p 10

Tiwari SK, Kumaraswamidhas LA, Gautam C, Garg N (2022) An auto-encoder based LSTM model for prediction of ambient noise levels. Appl Acoust 195:108849

Tiwari SK, Kumaraswamidhas LA, Patel R, Garg N, Vallisree S (2024) Traffic noise measurement, mapping, and modeling using soft computing techniques for mid-sized smart Indian city. Meas Sensors 33:101203

Tiwari SK, Kumaraswamidhas LA, Prince, Kamal M, Rehman M (2023b) A hybrid deep leaning model for prediction and parametric sensitivity analysis of noise annoyance. Environ Sci Pollut Res 30

Venet T, Thomas A, Merlen L, Boucard S, Wathier L, Martin Remy A, Pouyatos B (2023) Parameters influencing auditory fatigue among professionals working in the amplified music sector: noise exposure and individual factors. Int J Audiol

Wiener FM, Keast DN (1959) Experimental study of the propagation of sound over ground. J Acoust Soc Am 31

Part II
Advanced Monitoring Techniques

Chapter 4
IoT Based Systems for Pollution Detection

Varun Tripathi, Alok Bhadauria, and Himanshu Vashishtha

Abstract Pollution is a critical issue for human beings to survive long-term in the current worldwide scenario. Scientists' efforts to control pollution in a systematic manner with the help of advanced technology in the current global scenario. In the current human living environment, majorly three forms of pollution include noise, water, and air pollution. These pollutions generate severe health conditions and work as a slow poison for the entire human race in the universe. The present chapter aims to eradicate the pollution problem from its roots, and attention has to be drawn so that the main source of pollution can be stopped, and a healthy life can be lived. The present chapter proposed a pollution detection approach for alarming human civilization using the Internet of Things principle. The developed approach has been built for monitoring the human society or area by considering the pollution level considering noise, water, and air contaminants. The discussion revealed that human beings could use an alternative way by alarming the system immediately by receiving pollution reports. The current chapter could help worldwide humanitarian life survive in a healthy and comfortable mode by avoiding pollution circumstances.

Keywords Noise pollution · Water pollution · Air pollution · Internet of things · Human survival

V. Tripathi (✉)
Uttaranchal Institute of Management, Uttaranchal University, Dehradun 248007, Uttarakhand, India
e-mail: cimme2009@gmail.com

A. Bhadauria
Department of Mechanical and Industrial Engineering, Manipal Institute of Technology Bengaluru, Manipal Academy of Higher Education, Manipal, Karnataka, India

H. Vashishtha
Department of Materials Science and Metallurgy, University of Cambridge, Cambridge, UK

S. K. Tiwari et al. (eds.), *AI-Driven Environmental Pollution Management*, Climate Risks and Solutions, https://doi.org/10.1007/978-3-031-96243-1_4

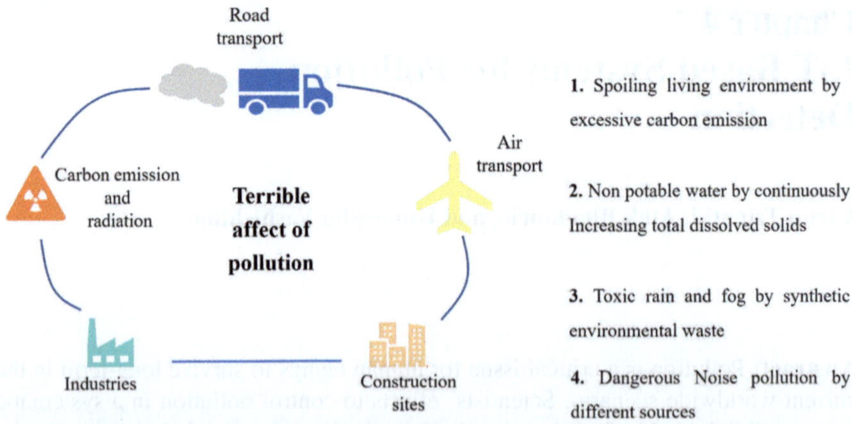

Fig. 4.1 Illustration of sources of pollution and results on the current earth scenario (Khan et al. 2018; Salman and Hasar 2023; Dey et al. 2004; Shao et al. 2006; Fuller et al. 2022)

4.1 Introduction

Pollution is creating a terrible condition for the future world lives. It is expanding daily and developing a crisis for the present living world. Pollution has several forms, including water, air, and noise pollution (Mishra 2008; Ahluwalia 2005; Plattenberg 2007). Worldwide reports have shown that pollution challenges human beings to survive healthy lives in the long term (Mohamed et al. 2021; Manisalidis et al. 2020). Pollution is present everywhere in the environment. Pollution is present in every human consuming thing, which is a massive threat to life. Pollution presence also deteriorates the Earth by poisoning the natural environment and other species (Almetwally et al. 2020; Von Schneidemesser et al. 2020; Piracha and Chaudhary 2022). The ever-increasing pollution is becoming a nightmare for humans, and they are shocked about how it will end. The effects of pollution are described in Fig. 4.1. The figure shows the terrible effects of pollution and helps us understand how pollution can take the Earth to a severe level and even destroy the idea of life on Earth.

4.2 Pollution Identification Sectors and IoT

The previous studies used different sectors to analyze the seriousness of the current pollution scenario. The public reports always showed that the well-wishers of humanitarian societies pointed to the sources with maximum chances of severe pollution (Hasan et al. 2019; Madhav et al. 2020; Colvile et al. 2001; Jariwala et al. 2017; Goines and Hagler 2007, Tiwari et al. 2023). Figure 4.2 describes the current pollution sources, sectors, and actions taken for their mitigation and elimination. A similar study has been reported by Saha et al. (2017), which discussed the identification of

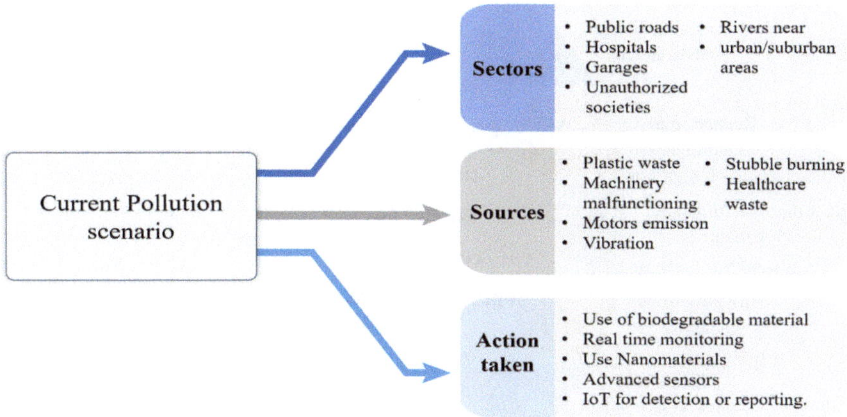

Fig. 4.2 Description of current pollution sectors, sources, and suggested actions

pollutants in nature using the Internet of Things. The study focused on preventing air and noise pollution using advanced technology. The study revealed that the pollution could controlled by efficiently monitoring and analyzing the contamination data using sensors.

The current reporting agencies showed a drastic growth in contaminant identification systems used for water, air, and surface pollutants using emerging technologies. IoT is a prominent technique for achieving a healthy living environment through flexible communication of pollutant element reports (Hasan et al. 2024; Almalki et al. 2023; Dhanwani et al. 2021). In the current scenario, intellectual persons support the integration of artificial intelligence and emerging technologies to provide sustainable living and operational setups. Ramdan et al. (2024) designed an air pollution monitoring and forecasting system for the chrome plating industry using real-time. The system used Artificial Intelligence and IoT sensors to detect air pollutants and provide those concentration levels. The study revealed that the designed approach could efficiently control air quality in industrial settings. The researchers continuously try to attain a pollution-free environment, but the reports show a need to focus on all pollution sources generated on Earth. The present chapter emphasizes the promotion of a zero-contaminated environment using emerging technologies. The study analyzed significant pollution sources, including noise, water, and air, to achieve a healthy human living environment. Figure 4.3 describes the contribution of IoT to pollution detection in the current scenario.

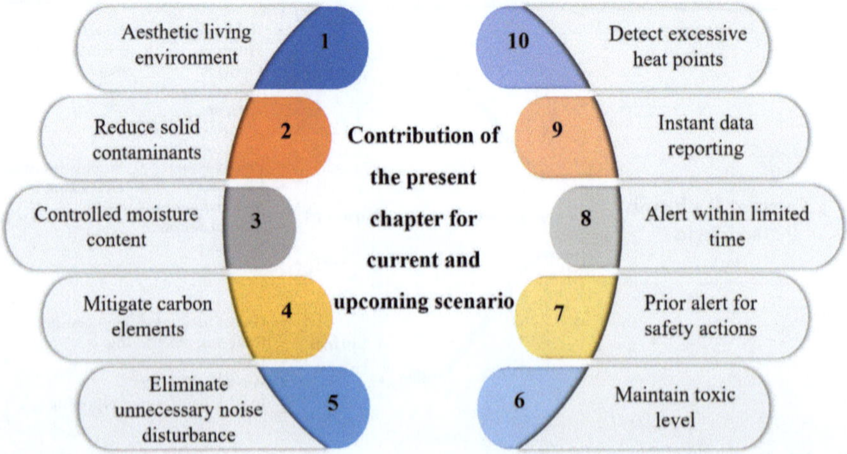

Fig. 4.3 Description of the contributions of the IoT (Ullo and Sinha 2020; Salman and Hasar 2023; Haji and Shallow 2021; Al Mamun and Yuce 2019; Guerrón et al. 2023; Barot and Kapadia 2020; Gangwar et al. 2023)

4.3 Development of Pollution Approach Using the Internet of Things

Global scenario needs a revolutionary action plan for tackling the pollution issues because it generates numerous severe health issues and challenges the human survival for long time. Pollution works as slow poison for human societies those located in cities and nearby townships. So, there is need to eliminate the pollution issues by implementing suitable and sustainable systems. However, researchers developed several remedies for defeating the pollution sources but report showed that the solutions were not sufficient to control the pollution. The present chapter reviews the pollution issues of current worldwide scenario and proposes an approach as shown in Fig. 4.4 for mitigating the pollutions effects on the human societies using the Internet of things principle. The proposed approach includes five factors: moisture, noise, toxic water, temperature, and carbon emission. First, moisture content must be in a standardized proportion for healthy body functional processes, and intelligent measuring devices can control it according to weather conditions. Second, water purity is essential because everyone knows 'water is life.' The clever merging of necessary minerals and eliminating toxic elements by the innovative filter system controls purity. Third is temperature; this factor can deteriorate any human thing in a few moments. The present approach reports heat variation in any place to analyze unexpected conditions. The fourth factor is carbon emission, which results from malfunctioned equipment, machinery, and motor cars, and it can control by communicating the conditions by alerting a hygienic environment. Finally, the fifth factor is the most important, noise, which severely affects human life. Noise can cripple the

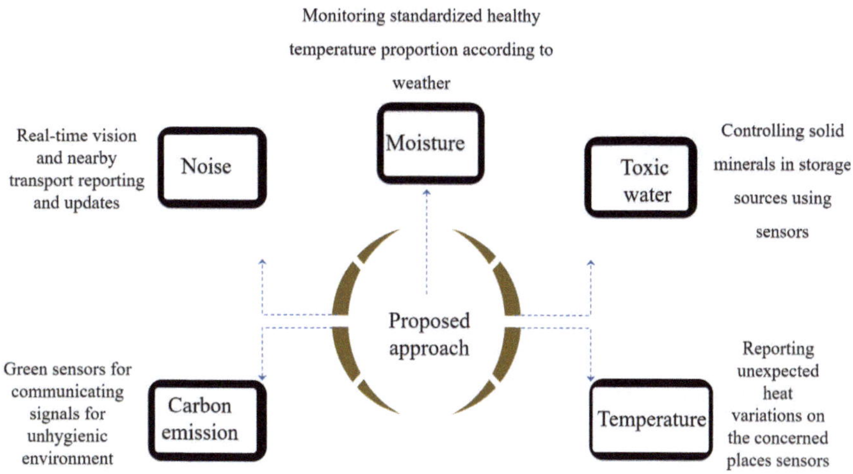

Fig. 4.4 Proposed approach

human and reduce the brain power to do work. Noise can be controlled by real-time reporting of convenience nearby societies using smart devices.

4.4 Summary

The proposed approach considered major pollution sources and analyzed the societies' reports before forwarding them to concerned management departments for further information. Humanitarian societies can feel comfortable physiologically and mentally after building the same approach in their society's welfare offices. The approach helps to inform society by providing suitable alerts within available constraints. Constraints mean barriers, including different health diseases, persons, age groups, and living organisms. The proposed approach works like an injection where all medicine fails or is not adopted by the human body, as shown in Fig. 4.5.

Humanitarian societies must wake up to approach long-lived survival by eliminating pollution sources. If pollution elimination remedies are not deployed right now, everything on earth will be destroyed soon. The present review wants to get intellectual attention to wake up to the upcoming crisis of natural resources. The developed approach can establish a cleaner and sustainable living environment. The Internet of Things plays a vital role in the current scenario for connecting and communicating with large human networks. It helps alert the concerned worldwide human within the micro version of times. The proposed approach helps detect pollution using advanced devices and provides a suitable action plan for tackling them within available barriers. It helps every age group according to their medical background by analyzing the various factors reports. Figure 4.6 describes the contribution of the present chapter considering significant pollution identified on Earth.

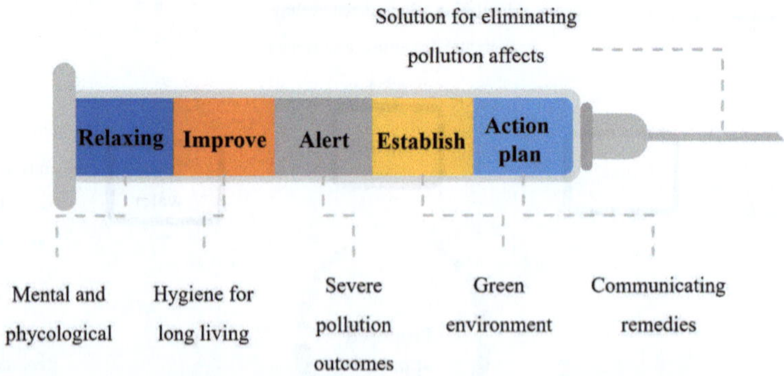

Fig. 4.5 Outcomes of the proposed approach

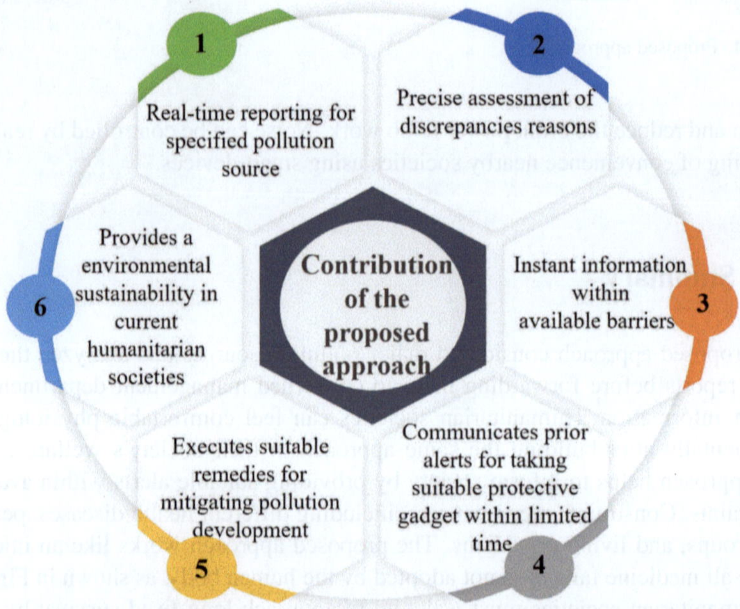

Fig. 4.6 Contribution of the proposed approach for upcoming humanitarian societies on earth

4.4.1 Role of the Internet of Things for Long-Term Living Environment

Regular worldwide health reports show that if human societies are not awakened immediately for taking suitable protection and security plan for environment deterioration then environment will soon perish. Intellectual personnel are vigilant but not looking for permanent key to close path for pollution generation and identification.

The current chapter focuses on resolving the pollution issues and their sources in humanitarian societies by informing the concerned authorities within available time limit. The study reviews the sources and parameters that must be considered in pollution detection. The Internet of Things helps in identifying real-time pollution level by suitable analysis precisely. The chapter proposes the Internet of Things-driven pollution detection approach for informing human civilization in a timely manner. The proposed approach establishes a centralized pollution monitoring system considering the different pollution level including noise, water, and air contaminants. The current study provides a long-living platform for eliminating environment deterioration and cleaner environment by avoiding pollution circumstances. The platform provides a cleaner and sustainable living environment within existing earth's condition. The chapter can direct the concerned pollution controlling teams for taking clever action plan for eliminating root source of pollutions.

4.4.2 Controlling Humanitarian Societies with Standardized Guidelines

Natural environment guideline is like lines written in stones. If there is any change in it then it works as a destruction. Worldwide humanitarian societies are worried because pollution enhancement in living environment. The pollution level drastically deteriorates aspiration of life by poisonous effect on living environment. There are some protocols of the earth's environment those must follow by living species. If they ignore them then it results in health deterioration and leads to the end. The current chapter focuses on pollution detection through the Internet of Things based pollution detection approach. The approach directs the concerned authorities for taking suitable action plan according to the identified pollution source. The action plan helps in eliminating pollution level including noise, water, and air. Worldwide health organization sets a limit of considered pollution including noise, water, and air for healthy human life. The proposed approach directs the pollution monitoring authorities for taking suitable prevention plan for blocking the identified sources of pollution. The humanitarian societies can live longer by implementing the proposed Internet of Things based pollution detection system. Humanitarian societies can establish self-awaken information system for long-living dreams within earth's environment protocol. The current chapter provides a cleaner and sustainable key for controlling pollution level before it takes horrible form.

4.5 Conclusion

The present chapter awakens humanitarian societies to a most important subject and is long-lived human survival. It is concluded that there is a need to mitigate pollution by living world societies. It can be made possible by an innovative action plan to tackle the severe pollution issues faced in the current worldwide scenario. Intellectual humans understand the problems, but it's time to awaken all humans and concerned societies to tackle the severe conditions. The present chapter proposes an approach for mitigating pollution in the living environment. The proposed approach can establish a sustainable human-living environment by mitigating the pollution sources within available constraints. The chapter considers significant factors, including moisture, noise, toxic water, temperature, and carbon emission. The medical scenario already reported these factors as challenging for long-lived human survival. The proposed approach can provide a cleaner and green environment with an innovative alert system. This approach introduces a novel pollution detection system for eliminating the severe issues responsible for unhygienic environment. Humanitarian societies can establish a protective living place through a pre-alert system using the Internet of Things principle. The proposed approach provides a fearless physiological environment for lifelong survival and enhances physical contribution to intellectual activities.

References

Ahluwalia VK (2005) In: Environmental pollution, and health. The Energy and Resources Institute (TERI)

Almalki FA et al (2023) Green IoT for eco-friendly and sustainable smart cities: future directions and opportunities. Mobile Netw Appl 28:178–202

Almetwally AA, Bin-Jumah M, Allam AA (2020) Ambient air pollution and its influence on human health and welfare: an overview. Environ Sci Pollut Res 27:24815–24830

Barot V, Kapadia V (2020) Air quality monitoring systems using IoT: a review. In: 2020 international conference on computational performance evaluation (ComPE). IEEE, pp 226–231

Colvile RN et al (2001) The transport sector as a source of air pollution. Atmos Environ 35(9):1537–1565

Dey AR, Kabir N, Effroymson D (2004) Noise pollution in Dhaka: current situation and suggestions for action. Bangladesh, Dhaka

Dhanwani R et al (2021) Smart earth technologies: a pressing need for abating pollution for a better tomorrow. Environ Sci Pollut Res 28(27):35406–35428

Fuller R et al (2022) Pollution and health: a progress update. The Lancet Planetary Health 6(6):e535–e547

Gangwar A et al (2023) The state-of-the-art in air pollution monitoring and forecasting systems using IoT, big data, and machine learning. Wireless Pers Commun 130(3):1699–1729

Goines L, Hagler L (2007) Noise pollution: a modem plague. South Med J 100(3):287–294

Guerrón RA et al. (2023) IoT sensor nodes for air pollution monitoring: a review. ACTA IMEKO 12(4):1–10. Available at: https://doi.org/10.21014/actaimeko.v12i4.1676

Haji SH, Sallow AB (2021) IoT for smart environment monitoring based on Python: a review. Asian J Res Comput Sci 9(1):57–70

Hasan Z et al (2024) Smart ventilation systems for real-time pollution control: a review of AI-driven technologies in air quality management. Front Appl Eng Technol 1(01):22–40

Hasan MK, Shahriar A, Jim KU (2019) Water pollution in Bangladesh and its impact on public health. Heliyon [Preprint] 8

Jariwala HJ et al (2017) Noise pollution and human health: a review. Indoor Built Environ 1(1):1–4

Khan J, Ketzel M, Kakosimos K, Sørensen M, Jensen SS (2018) Road traffic air and noise pollution exposure assessment—a review of tools and techniques. Sci Total Environ 634:661–676

Laha SR, Pattanayak BK, Pattnaik S (2022) Advancement of environmental monitoring system using IoT and sensor: a comprehensive analysis. AIMS Environ Sci 9(6):771–800

Madhav S et al. (2020) Water pollutants: sources and impact on the environment and human health. In: Sensors in water pollutants monitoring: role of material. pp 43–62

Mamun A, Yuce MA (2019) Sensors and systems for wearable environmental monitoring toward IoT-enabled applications: a review. IEEE Sens J 19(18):7771–7788

Manisalidis I et al. (2020) Environmental and health impacts of air pollution: a review. Front Public Health 8

Mishra PC (2008) In: Fundamentals of air and water pollution. APH Publishing

Mohamed AMO, Paleologos EK, Howari FM (2021) Noise pollution and its impact on human health and the environment. In: Pollution assessment for sustainable practices in applied sciences and engineering. Butterworth-Heinemann, pp 975–1026

Munera D, Aguirre J, Gomez NG (2021) IoT-based air quality monitoring systems for smart cities: a systematic mapping study. Int J Electri Comput Eng 11(4)

Pantelic J et al. (2022) Transformational IoT sensing for air pollution and thermal exposures. Front Built Environ 8

Piracha A, Chaudhary MT (2022) Urban air pollution, Urban Heat Island and human health: a review of the literature. Sustainability 14(15):9234. Available at: https://doi.org/10.3390/su14159234

Plattenberg RH (2007) In: Environmental pollution: new research. Nova Publishers

Ramadan MN et al. (2024) Real-time IoT-powered AI system for monitoring and forecasting of air pollution in industrial environment. Ecotoxicol Environ Safety 283

Saha HN et al. (2017) Pollution control using internet of things (IoT). In: 8th annual industrial automation and electromechanical engineering conference (IEMECON). IEEE, pp 65–68

Salman MY, Hasar H (2023) Review on environmental aspects in smart city concept: water, waste, air pollution and transportation smart applications using IoT techniques. Sustain Cities Soc 94

Schneidemesser V et al. (2019) How will air quality effects on human health, crops and ecosystems change in the future?. Philosoph Trans Royal Society A 378

Shao M et al. (2006) City clusters in China: air and surface water pollution. Front Ecol Environ 353–361

Tiwari SK, Kumaraswamidhas LA, Garg N (2023) Assessment of noise pollution and associated subjective health complaints in Jharia Coalfield, India: a structural equation model analysis. Noise Mapp 10(1):20220172

Ullo SL, Sinha GR (2020) Advances in smart environment monitoring systems using IoT and sensors. Sensors (Basel, Switzerland) 20(11):3113. Available at: https://doi.org/10.3390/s20113113

Von Schneidemesser E, Driscoll C, Rieder HE, Schiferl LD (2020) How will air quality effects on human health, crops and ecosystems change in the future?. Philos Trans Roy Soc A 378(2183):20190330

Chapter 5
Advanced Ammonia Detection Technologies Using Metal–Organic Frameworks (MOFs) as Environmental Monitoring

Mousumi Maji, Shivam Saw, Rajib Ghosh Chaudhuri, and Mrinal Kanti Mandal

Abstract Ammonia (NH_3) is a significant environmental pollutant, primarily produced in agricultural activities, industrial processes, and waste management. This leads to negative effects on air and water quality. The growing concerns regarding ammonia emissions require the development of efficient detection methods to monitor their concentration in different environments. Metal–organic frameworks (MOFs) is considered one of the promising materials for ammonia sensing due to their unique structural properties like tunable porosity, exceptional surface area, robust in nature and the capability for chemical modification. The AI-assisted MOF based advanced fabrication system successfully addressed the challenge in ammonia detection under humid conditions and archived accuracy of 90–98% in complex gas mixture of environment. The recent research presents a novel approach by combining superior properties of MOF materials and machine learning algorithm to achieve excellent sensing performance. This chapter highlights the effect of ammonia on human health and the environment and then discusses sensing methods and the latest innovations in sensor design. Insights into future directions for research and development are also discussed, emphasizing the potential of MOFs combined with Artificial Intelligence (AI) to advance environmental sustainability and enhance pollution remediation efforts.

M. Maji · S. Saw · R. G. Chaudhuri · M. K. Mandal (✉)
Department of Chemical Engineering, National Institute of Technology, Durgapur, India
e-mail: mkmandal.che@nitdgp.ac.in

M. Maji
e-mail: mm.22ch1106@phd.nitdgp.ac.in

S. Saw
e-mail: shivamsaw@gmail.com

R. G. Chaudhuri
e-mail: rgchaudhuri.che@nitdgp.ac.in

S. K. Tiwari et al. (eds.), *AI-Driven Environmental Pollution Management*, Climate Risks and Solutions, https://doi.org/10.1007/978-3-031-96243-1_5

Keywords Ammonia · Environment · Metal–organic frameworks (MOFs) · Sensing · Pollutant

5.1 Introduction

The necessity for effective environmental monitoring has become increasingly pressing in recent years due to the release of different types of toxic gases (CO_X, NH_3, NO_X, H_2S, SO_X, etc.) from industries, agricultural activities, and waste management. Ammonia is one of them which poses significant environmental and health risks. High ammonia concentrations in the environment contribute to eutrophication in water bodies, triggering excessive algal blooms that deplete oxygen levels and disrupt natural ecosystems, ultimately harming aquatic life (Wyer et al. 2022; Zhang et al. 2023a, b). Moreover, ammonia can contribute to soil acidification, impacting plant health and biodiversity. High levels of ammonia exposure show serious health risks in the human body (Fig. 5.1), particularly in the respiratory system, which leads to difficulty breathing and coughing. Long-term inhalation of ammonia can cause chronic bronchitis or asthma, reactive airway dysfunction syndrome (RADS), etc. According to the National Institute for Occupational Safety and Health (NIOSH), ammonia exposure to 300 ppm is immediately dangerous to life and health (Ammonia Facts 2024). As a result, monitoring and detecting ammonia are crucial for safeguarding the environment and ensuring human health. Metal oxides are currently the most used materials in commercial ammonia sensors, but existing detection devices face challenges like limited selectivity and the requirement for high operating temperature. Thus, there is an urgent need to develop advanced materials for detection of NH_3 with high selectivity at ambient temperatures.

Metal–organic frameworks (MOFs) as advanced material, distinguished by their unique properties like high porosity, high chemical stability and customizable functionalities, are regarded as a promising material for ammonia gas sensors. Additionally, MOFs have broad applications in environmental pollution management including air purification, noise level detection, heavy metal removal, precious metal recovery, etc. (López et al., 2021; Yuan et al., 2022). Most importantly the modified porosity and pre-design structure of MOFs can facilitate rapid sensing response under low operating temperature. Ammonia gas (NH_3) response by MOFs materials is the host–guest interaction between ammonia molecules and the active site of MOF materials have the potential to provide high selectivity for specific gas over gas mixture in the environment (Fang et al. 2018). Therefore, MOFs as a gas sensing material have attracting growing interest in recent years (Reddy et al. 2021; Yang et al. 2021). Despite notable progress in gas sensing with MOFs, their development specifically for NH3 detection remains in its early stages. Chen et al. developed a Ti/Zr- MOF which has attracted significant attention because of the high oxidation state of Zirconium (Zr) that forms a robust framework with an organic linker (Chen et al. 2024).

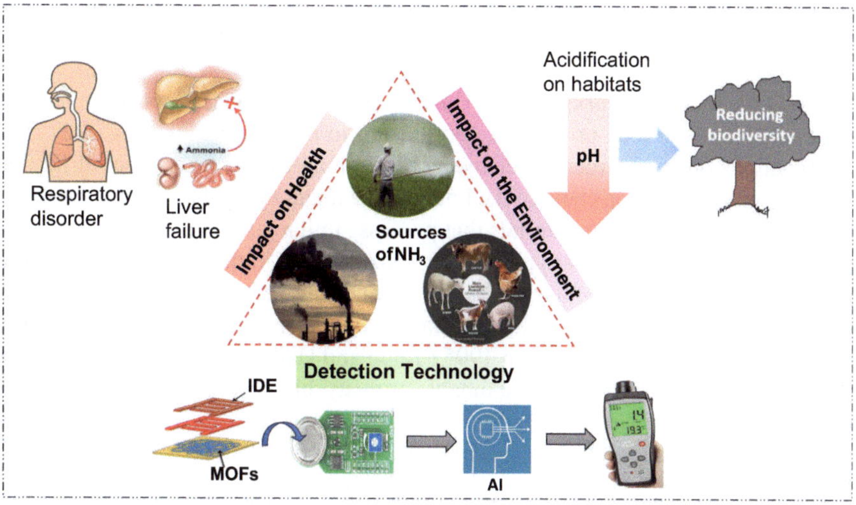

Fig. 5.1 The overall schematic of ammonia pollution and its detection technology for environmental pollution management

Recent advancements in MOF-based ammonia detection technologies have focused on enhancing the sensitivity, selectivity, and overall performance of sensors. Innovations such as the incorporation of rare-earth elements and the functionalization of MOFs with different organic linkers have demonstrated improved detection limits and stability, even under varying environmental conditions. Furthermore, the integration of MOFs sensors with electronic and data processing systems paves the way for smart sensing solutions in environmental monitoring. The integration of Artificial Intelligence (AI) technology marks a major advancement in ammonia detection, especially with the development of electronic nose (E-nose) systems, which have demonstrated the ability to enhance the accuracy and efficiency of ammonia sensors (Viejo et al., 2020; Moshayedi et al., 2023; Tiwari et al., 2023). These systems typically consist of an array of MOFs sensors along with AI algorithms such as Principal Component Analysis (PCA), Artificial Neural Networks (ANN) and long short-term memory (LSTM) for pattern recognition and data interpretation (Tiwari et al. 2022). Using large datasets of various gas sensor responses, AI algorithms can be trained to effectively detect specific gas and measure concentrations level in real time over complex gas mixture in environment. This advanced approach not only improves detection limits but also enables continuous and automated environmental monitoring, ensuring timely intervention and mitigation of ammonia pollution. This innovative synergy holds great promise for safeguarding both ecosystems and human health against the adverse effects of ammonia emissions.

5.2 Concept of Metal–Organic Frameworks (MOFs)

Metal–Organic Frameworks (MOFs) are a class of advanced porous materials composed of metal ions or clusters coordinated to organic ligands (Fig. 5.2). The combinations of these structural elements result in an enormous number of MOFs. By elongating the carbon chain of the linker while maintaining the original topology, it is feasible to synthesize materials that possess similar structural and symmetrical characteristics but exhibit variations in pore size. At the initial stage, MOF-5 (Côté et al. 2005), is composed of tetrahedral ZnO_4 units situated at the vertices of a cubic lattice, interconnected by terephthalate dianion linkers, and the term 'metal–organic framework' was first introduced in 1995 (Yaghi and Hailian 1995). This framework exhibits remarkable properties like tunable porosity, high crystallinity, high thermal and chemical stability, etc. that makes it a suitable material for various applications like gas sensing, drug delivery, catalyst, gas separation, and gas storage (Li et al. 2024). MOFs can be synthesized using various methods, ranging from traditional techniques to more specialized approaches, such as hydrothermal, solvothermal, microwave, ultrasonic, mechanochemical, and electrochemical processing. These frameworks are better than zeolite in multiple aspects, particularly because of their exceptionally high surface areas. Zeolites are constructed from tetrahedral units, with variations in their topology arising from a limited set of secondary structural elements. On the other hand, the secondary building units of MOFs consist of single metal ions and intricate clusters, or extended inorganic substructures that may be one, two, or three-dimensional frameworks. This versatility in MOFs contributes to their complex and diverse structures.

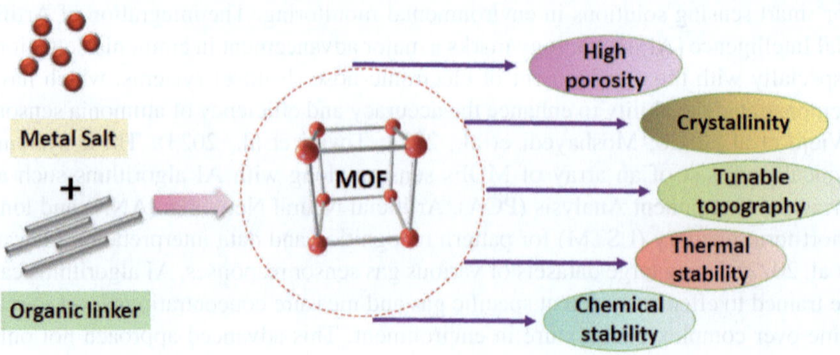

Fig. 5.2 The schematic of MOF structure and its properties

5.3 Ammonia Gas Sensing Methods

Ammonia sensors employ various technologies to detect and quantify the concentration of ammonia (NH_3) gas in different environments. The primary mechanism of the detection includes metal oxide chemiresistive sensing, electrochemical sensing, and optical sensing approaches. The various detection methods for ammonia gas sensing based upon sensing materials and recent advancements in this field are shown below.

5.3.1 Chemiresistive Method

Metal oxide chemiresistive gas sensors operate by measuring the electrical resistance of a sensing material when interaction with target gas molecules. The sensing mechanism initiates with the adsorption of gas molecules onto the surface of the chemiresistive material which is illustrated in Fig. 5.3a. The adsorption can occur in two forms: physisorption, where gas molecules are weakly bound through van der Waals forces, and chemisorption, where stronger chemical bonds form between the gas molecules and the sensing material. Upon adsorption, there is often a charge transfer between the gas molecules and the sensing material. This charge transfer can either donate or withdraw electrons from the material, depending on whether the adsorbed gas is reducing or oxidizing. For n-type semiconductors, oxidizing gases (e.g., NO_2) tend to withdraw electrons, thereby increasing resistance. Conversely, reducing gases (e.g., NH_3) donate electrons, reducing the resistance. The resultant charge transfer alters the conduction characteristics of the sensing material. In chemiresistive sensors, this change is primarily observed as a variation in electrical conductivity or resistance. The extent of this change is directly linked to the concentration of the target gas. For instance, oxidizing gases increase electron depletion in n-type materials, leading to higher resistance, whereas reducing gases increase electron accumulation, reducing the resistance (Lin et al. 2021).

Finally, the sensor's circuit measures the variation in electrical resistance, which is then converted into a detectable signal. This signal allows for the quantification of the target gas concentration. The entire system generally includes a transducer that translates the resistance variation into a readable output, typically in the form of a change in current that illustrated in Fig. 5.3b.

5.3.2 Electrochemical Method

The electrochemical gas sensing mechanism operates through a series of chemical reactions that generate an electrical signal directly proportional to the concentration of the target gas (Azad et al. 1992). Typically, an electrochemical sensor consists

(a)

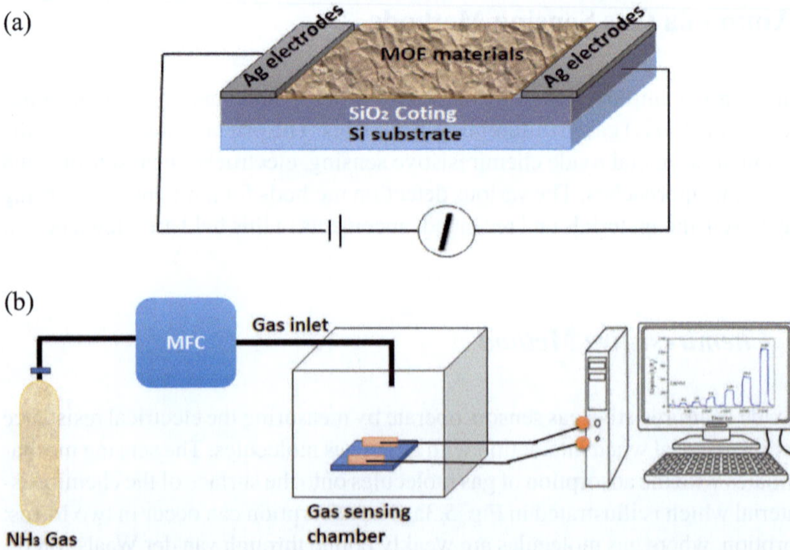

Fig. 5.3 **a** A schematic design of a chemiresistive gas detector **b** overall schematic representation of the gas sensing process

of at least three key electrodes: a working (or sensing) electrode, a counter electrode, and in many cases a reference electrode as shown in Fig. 5.4. A gas-permeable membrane allows target gas molecules to reach the electrodes while blocking contaminants. When the gas interacts with the working electrode, a chemical reaction occurs, producing an electrical response which is directly proportional to the gas concentration. Electrochemical sensors can be categorized based on the electrolyte type, such as solid-state or liquid-state. Commonly employed techniques include amperometry, which measures the current produced by reactions, and potentiometry, which detects voltage changes between electrodes. The NH_3 gas sensor used the potentiometric method, sensors work by measuring the potential difference between the sensing electrode and a counter electrode. When ammonia (NH_3) gas passes through the gas-permeable membrane and enters the electrolyte solution, an electrochemical reaction takes place at the sensing electrode. This reaction leads to the formation of nitrogen (N_2), hydrogen ions (H^+), and six electrons (e^-) as a result of ammonia oxidation, as represented in Eqs. 5.1 and 5.2

$$NH_3 \rightarrow N_2 + 6H^+ + 6e^- \tag{5.1}$$

$$O_2 + 4H^+ + 4e^- \rightarrow 2H_2O \tag{5.2}$$

The performance of this type of sensor heavily relies on the electrolyte, and most issues arise from its malfunction. Liquid electrolyte-based sensors face reduced lifespans as the electrolyte is gradually consumed during each detection. Whereas

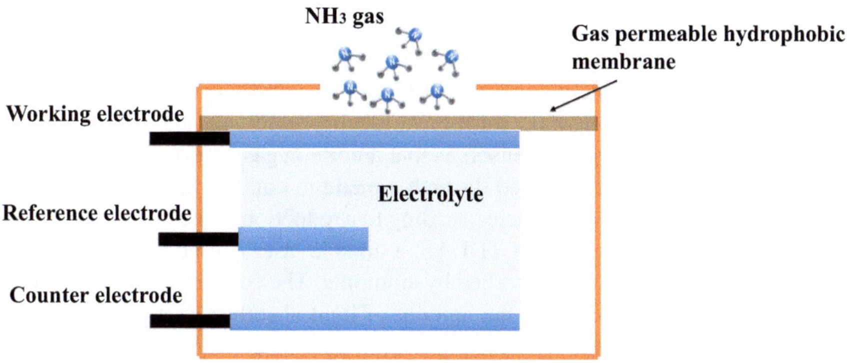

Fig. 5.4 A basic schematic design for electrochemical ammonia sensor

for the solid electrolyte-based sensors, the major issues are it's prone to electrolyte poisoning nature and high operating temperatures normally above 200 °C.

5.3.3 Optical Method

This technique is highly appealing for commercialization because it is simple to operate, offers excellent selectivity, fast response times, high sensitivity, and has a relatively long lifespan as shown in Table 5.1. Additionally, it is versatile for gas detection, functioning effectively across a wide temperature range (from room temperature up to 1500 °C) in even unstable and corrosive environments.

Table 5.1 Summary of sensing methods used in ammonia detection

Sensing methods	Principle	Advantages	Disadvantages
Metal oxide chemiresistive	Changes in resistance after gas adsorption on the surface of materials	Simple design, Low cost, High sensitivity, good response and recovery	Operating temperatures higher, Limited selectivity, Drift over time
Electrochemical	Chemical reaction between ammonia gas and sensor electrode, generating a current proportional to gas concentration	High sensitivity, wide measurement range, Real-time data monitoring, Low power consumption	Short lifespan, Sensitive to environmental factors, Requires regular calibration
Optical	Absorption of specific wavelengths of light by gas molecules is measured	Non-invasive, High sensitivity and selectivity, Fast response	High cost, Requires complex optical setups, Sensitive to interference

Optical ammonia gas sensors utilize the principle of light absorption to detect the presence and concentration of ammonia (NH_3) gas. These sensors typically rely on techniques such as tunable diode laser absorption spectroscopy (TDLAS), infrared (IR) absorption, and fiber optic sensors to identify ammonia gas in the air. The key mechanism behind optical sensors is that ammonia gas absorbs light at specific wavelengths. When light is passed through a medium containing NH_3 gas, the gas molecules absorb certain frequencies, leading to a reduction in the intensity of transmitted light. While in the case of TDLAS, a tunable laser is used to emit light at a wavelength that is selectively absorbed by ammonia. The concentration of ammonia can be determined by analyzing the amount of light absorbed as it passes through the gas. Generally, the NH_3 is detected in wavelength range from 1450 to 1560 nm. The measurement of NH_3 concentration by TDLAS methods applying the Lambert-Beer law show in Eq. 5.3.

$$I = I_0.\exp[-\alpha(v) \cdot l] \tag{5.3}$$

where

I represent the transmitted laser intensity at frequency v.
I_O represent the initial laser intensity (before passing through the absorbing medium).
α(v) represent the absorption coefficient at frequency v.
l represent the path length through the absorbing medium.

This equation describes how the light intensity decreases as it passes through the ammonia gas. The rate of absorption depends on the concentration of ammonia, the path length of the laser through the gas, and the absorption characteristics of ammonia at the specific wavelength. (Warland et al. 2001).

The absorption coefficient α(v) can be linked to the concentration of ammonia molecules using the following Eq. 5.4.

$$\alpha(v) = S \cdot N \cdot \Phi(v) \tag{5.4}$$

where

S is the line strength of the absorption, which is temperature-dependent.
N is the fractional concentration of ammonia in the gas.

Φ(v) is the line-shape function, which characterizes how the absorption varies with wavelength due to the molecular properties of ammonia. Thus, the absorption coefficient is directly proportional to the concentration of ammonia molecules in the gas. By knowing the absorption at specific wavelengths, the concentration of ammonia can be determined.

5.4 AI-Integration in MOF-Based Sensing Systems

The integration of Artificial Intelligence (AI) in Metal–Organic Framework (MOF)-based detection systems marks a notable progression in environmental monitoring, particularly for detecting ammonia (NH_3) and other hazardous gases. This integration enhances the efficacy of MOF sensors by utilizing AI's analytical and predictive capabilities, especially through machine learning (ML) techniques. Machine learning plays a vital role in improving the accuracy, sensitivity, and selectivity of MOF-based sensors (Zhang et al. 2023a, b). Machine learning algorithms, including Principal Component Analysis (PCA) and Linear Discriminant Analysis (LDA) are employed to process the features extracted from the sensor response profiles of MOF-based systems. These algorithms help in dimensionality reduction and class optimization, ensuring that the sensors can distinguish between different gases with high precision. For instance, in a study involving a graphene nano sensor functionalized with copper phthalocyanine, PCA and LDA were utilized to achieve perfect classification among NH_3, PH_3, and reference gases, with NH_3 displaying an isolated cluster due to its strong H-bonding interaction with the functionalized graphene (Huang et al. 2022). By training AI models on large datasets, the sensors can learn to identify ammonia concentrations accurately even at ultra-low levels (e.g., 100 ppb), as demonstrated in this study where the AI-enhanced MOF sensors achieved nearly 100% accuracy, sensitivity, and specificity for NH_3 detection at 500 ppb. Moreover, AI can also differentiate between ammonia and other interfering gases, ensuring reliable environmental monitoring (Wang et al. 2023).

Case studies highlight the successful implementation of AI in various MOF-based gas detection systems (Fig. 5.5). A prominent example is the machine-learning-assisted n-GaN-Au/PANI gas sensor array, which achieved a recognition accuracy of 94% for NH_3 detection. The n-GaN-Au/PANI sensor demonstrates exceptional performance with limit of detection is 10 ppb, outstanding resistance to moisture and a high response value of 195% for 200 ppm NH_3 at room temperature. (Han et al. 2024). This study highlights the transformative role of AI in improving gas sensor performance. In a recent advancement, an electrochemical nose system was designed using zeolitic imidazole framework (ZIF-8) material encapsulated with ferrocene (Fc@ZIF-8) for monitoring ammonia (NH_3) levels in breath. This innovative system shows promise for breathomics applications, capable of detecting low concentrations of ammonia gas (as low as 400 ppb) with remarkable sensitivity and specificity (Banga et al. 2021).

The integration of AI with MOF-based gas detection systems revolutionizes environmental monitoring by providing enhanced data analysis, improving sensitivity and selectivity, enabling automation and real-time monitoring, and facilitating predictive maintenance. These advancements promise more accurate and reliable monitoring of ammonia and other hazardous gases, contributing to better environmental and public health outcomes.

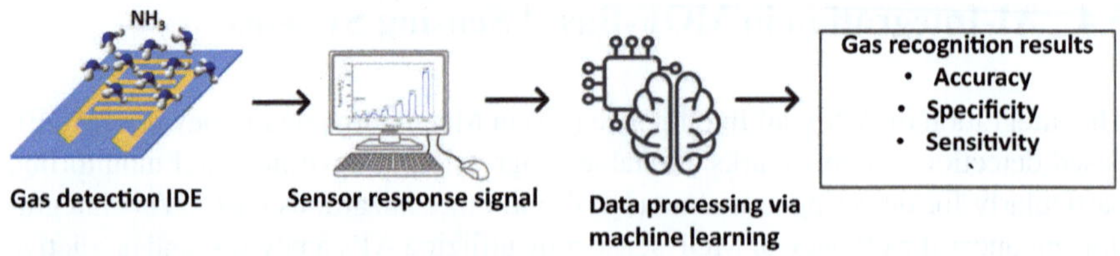

Fig. 5.5 Application of AI integration for Gas detection

5.4.1 Role of AI in Optimizing the Ammonia Detection

Recently AI technology is used in various sections for better performance, gas sensing is one of them where AI is used for optimizing the sensing mechanism. Till now many researches has been done on ammonia gas detection but very few where used AI technology for optimizing the detection methods (Huang et al. 2022; Weisbecker et al. 2024). In this case, AI analyses sensor data and dynamically calibrates sensors using machine learning (ML) models. This guarantees that sensors maintain their accuracy even when exposed to changing environmental factors, including variations in humidity or temperature. It can also analyze the historical data and predict future trends of ammonia concentration. This predative analysis is specifically useful in industrial and environmental monitoring, where early detection of ammonia leaks or build-ups is critical for safety and regulatory compliance. Another role is to enhanced sensitivity and selectivity of material towards ammonia gas by simulating various scenarios then optimized sensor material and configuration. The important role of AI technology is real time monitoring integrated with Internet of Things (IoT) devices which enable real time monitoring the sensing data by remote monitoring and control. This is particularly helpful for large scale industrial setup (Asha et al. 2022).

5.5 Development of Ammonia (NH_3) Sensor Using Metal–Organic Frameworks (MOFs) Materials

Although significant advancements in utilizing MOFs for gas sensing, their application in NH_3 detection remains in its early stages. There are several challenges to overcome to optimize MOFs for this purpose. One major issue is that many MOFs tend to experience structural deterioration upon NH_3 adsorption. Additionally, they often exhibit a relatively low binding affinity for ammonia molecules (Ma et al. 2022). This structural instability and weak adsorption hinder the practical application of MOFs as NH_3 sensors, making it challenging to develop robust and efficient MOF-based sensors. Another challenge is the incomplete understanding of the mechanisms underlying NH_3 sensing in MOFs. This limitation arises largely from the absence of advanced methods to thoroughly study adsorption kinetics and perform detailed molecular simulations. Understanding the interactions at the atomic level between NH_3 molecules and MOF structures remains a critical area of ongoing research. Currently, the MOF materials reported for NH_3 sensing often require doping with additional materials to improve their performance. Examples include Cu-BTC@GO and PANI/UiO-66, which necessitate time consuming fabrication procedures under high operating temperature (Lin et al. 2021). These additional steps complicate the sensor design and increase the cost and complexity of the manufacturing process. The use of pure MOF materials for NH_3 sensing remains uncommon due to the high costs and limited availability of appropriate ligands. Most of the reported examples, such as Cu-HHTP and Cu_3HITP_2, suffer from expensive and hard-to-find ligands, leading to increased operating costs and limiting widespread adoption. As a result, there is a significant need for designing cost-efficient and high-quality MOF materials for NH_3 sensing, especially under ambient temperature. Recent process has focused on developing robust and reliable MOFs that can operate efficiently at room temperature. (Wang et al. 2023).

These materials also exhibit high stability, rapid response times, and high selectivity, making them promising candidates for future NH_3 sensing technologies. Furthermore, ongoing research aims to enhance the intrinsic properties of MOFs through innovative synthesis methods and functionalization techniques. Enhancing the thermal and chemical stability of MOFs, along with fine-tuning their pore structures, can greatly improve their performance in sensing applications. Additionally, the integration of MOFs with other functional materials is being explored to overcome existing limitations and improve the overall performance of NH_3 sensor. By addressing these challenges and leveraging recent technological advancements, MOFs hold great potential for the development of advanced NH_3 detection technologies for environmental monitoring.

5.6 Conclusion

The advanced ammonia detection technologies utilizing Metal–Organic Frameworks (MOFs) combined with Artificial Intelligence (AI) offer significant potential for environmental monitoring. MOFs, with their highly ordered crystalline frameworks and exceptional surface area, diverse chemical functionalities serve as highly effective materials for ammonia adsorption. The integration of AI enhances the accuracy, selectivity, and real-time monitoring capabilities of these sensors by enabling advanced data analysis and predictive analytics. Continued research in MOF materials and AI techniques is essential to overcome challenges like cross-sensitivity and long-term stability, ensuring that these systems contribute effectively to managing ammonia emissions and promoting a healthier environment. Future advancements in machine learning will further improve data differentiation and predictive accuracy, optimizing real-time monitoring for environmental pollution control.

References

Ammonia Facts (2024) stonemountaintechnologies.com. https://stonemountaintechnologies.com/ammonia-facts

Asha P, Natrayan LBTJRRGS, Geetha BT, Beulah JR, Sumathy R, Varalakshmi G, Neelakandan S (2022) IoT enabled environmental toxicology for air pollution monitoring using AI techniques. Environ Res 205:112574

Azad AM, Akbar SA, Mhaisalkar SG, Birkefeld LD, Goto KS (1992) Solid-state gas sensors: a review. J Electrochem Soc 139(12):3690

Banga I, Paul A, Muthukumar S, Prasad S (2021) ZENose (ZIF-based electrochemical nose) platform for noninvasive ammonia detection. ACS Appl Mater Interfaces 13(14):16155–16165

Chen Z, Shi Z, Zhang P, Gao L, Liu B, Qiao L (2024) Trace ammonia detection realized by mixed Ti-Zr metal-organic-frameworks and its application in exhaled breath sensing at room temperature. Sens Actuators Rep 100216

Cote AP, Benin AI, Ockwig NW, O'Keeffe M, Matzger AJ, Yaghi OM (2005) Porous, crystalline, covalent organic frameworks. Science 310(5751):1166–1170

Fang X, Zong B, Mao S (2018) Metal–organic framework-based sensors for environmental contaminant sensing. Nano-Micro Lett 10:1–19

Han D, Wang Y, Wang Y, Duan Q, Li D, Ge Y, He X, Zhao L, Wang W, Sang S (2024) Machine-learning-assisted n-GaN-Au/PANI gas sensor array for intelligent and ultra-accurate ammonia recognition. Chem Eng J 495:153705

Huang S, Croy A, Panes-Ruiz LA, Khavrus V, Bezugly V, Ibarlucea B, Cuniberti G (2022) Machine learning-enabled smart gas sensing platform for identification of industrial gases. Adv Intell Syst 4(4):2200016

Jagan Mohan Reddy A, Katari NK, Nagaraju P, Hussain Reddy K, Surendra Babu MS (2021) Fabrication and deployment of nanodisc ZnO@ ZIF-8, ZnO@ NA and ZnO@ INA core–shell MOFs: enhanced NH 3 and HCHO gas sensing. J Mater Sci Mater Electron 32:7827–7840

Li D, Yadav A, Zhou H, Roy K, Thanasekaran P, Lee C (2024) Advances and applications of metal-organic frameworks (MOFs) in emerging technologies: a comprehensive review. Global Chall 8(2):2300244

Lin Y, Li WH, Wen Y, Wang GE, Ye XL, Xu G (2021) Layer-by-layer growth of preferred-oriented MOF thin film on nanowire array for high-performance chemiresistive sensing. Angew Chem Int Ed 60(49):25758–25761

Lopez YC, Viltres H, Gupta NK, Acevedo-Pena P, Leyva C, Ghaffari Y, Gupta A, Kim S, Bae J, Kim KS (2021) Transition metal-based metal–organic frameworks for environmental applications: a review. Environ Chem Lett 19:1295–1334

Ma Y, Lu W, Han X, Chen Y, Da Silva I, Lee D, Sheveleva AM, Wang Z, Li J, Li W, Fan M (2022) Direct observation of ammonia storage in UiO-66 incorporating Cu (II) binding sites. J Am Chem Soc 144(19):8624–8632

Moshayedi AJ, Sohail Khan A, Hu J, Nawaz A, Zhu J (2023) E-nose-driven advancements in ammonia gas detection: a comprehensive review from traditional to cutting-edge systems in indoor to outdoor agriculture. Sustainability 15(15):11601

Tiwari SK, Kumaraswamidhas LA, Gautam C, Garg N (2022) An auto-encoder based LSTM model for prediction of ambient noise levels. Appl Acoust 195:108849

Tiwari SK, Kumaraswamidhas LA, Prince K, M. and Rehman, M.U. (2023) A hybrid deep leaning model for prediction and parametric sensitivity analysis of noise annoyance. Environ Sci Pollut Res 30(17):49666–49684

Viejo CG, Fuentes S, Godbole A, Widdicombe B, Unnithan RR (2020) Development of a low-cost e-nose to assess aroma profiles: an artificial intelligence application to assess beer quality. Sens Actuators B Chem 308:127688

Wang S, Fu Y, Wang T, Liu W, Wang J, Zhao P, Ma H, Chen Y, Cheng P, Zhang Z (2023) Fabrication of robust and cost-efficient Hoffmann-type MOF sensors for room temperature ammonia detection. Nat Commun 14(1):7261

Warland JS, Dias GM, Thurtell GW (2001) A tunable diode laser system for ammonia flux measurements over multiple plots. Environ Pollut 114(2):215–221

Weisbecker H, Shanahan J, Liu Y, Zhang L, Xie W, McDow S, Lambert N, Huang A, Sopp S, You W, Bai W (2024) AI assisted multimodal breath sensing system with semiconductive polymers for accurate monitoring of ammonia Biomarkers. Adv Mater Technol 9(9):2301884

Wyer KE, Kelleghan DB, Blanes-Vidal V, Schauberger G, Curran TP (2022) Ammonia emissions from agriculture and their contribution to fine particulate matter: a review of implications for human health. J Environ Manage 323:116285

Yaghi OM, Li H (1995) Hydrothermal synthesis of a metal-organic framework containing large rectangular channels. J Am Chem Soc 117(41):10401–10402

Yang T, Gao L, Wang W, Kang J, Zhao G, Li D, Chen W, Zhang H (2021) Berlin green framework-based gas sensor for room-temperature and high-selectivity detection of ammonia. Nano-Micro Letters 13:1–13

Yuan H, Li N, Fan W, Cai H, Zhao D (2022) Metal-organic framework based gas sensors. Adv Sci 9(6):2104374

Zhang TX, Li MR, Liu C, Wang SP, Yan ZG (2023a) A review of the toxic effects of ammonia on invertebrates in aquatic environments. Environ Pollut 336:122374

Zhang Z, Wang L, Lee C (2023b) Recent advances in artificial intelligence sensors. Adv Sens Res 2(8):2200072

Chapter 6
Integrating of AI with Pollution Monitoring for Traffic Status Prediction

Deepak Kumar Rakesh and Parikesh Dhal

Abstract Urbanization has significantly exacerbated traffic congestion and air pollution, posing severe environmental and public health challenges. This study introduces an integrated framework that leverages machine learning models, including Seasonal Autoregressive Integrated Moving Average (SARIMA), Support Vector Machines (SVM), and Long Short-Term Memory (LSTM), to predict pollutant levels and monitor traffic congestion. By employing real-time sensor data and predictive analytics, the system identifies critical congestion levels based on emissions of carbon monoxide (CO) and nitrogen oxides (NOx). The analysis demonstrates the superior performance of SARIMA in forecasting long- term trends, while LSTM effectively captures dynamic fluctuations. This interdisciplinary approach provides scalable solutions for urban pollution management, emphasizing the potential of artificial intelligence to enable sustainable smart city ecosystems.

Keywords Traffic congestion · Air pollution · Machine learning models · Sensor-based analytics · Sustainable urban management

6.1 Introduction

Traffic congestion and air pollution are critical challenges exacerbated by urbanization and rapid industrial growth. Traffic congestion is not only frustrating and wasteful but also has significant economic and environmental implications. Various solutions have been proposed, such as expanding roadways, promoting public transportation, and reducing private vehicle usage. However, these strategies often depend on regional demographics, lifestyle, and the efficacy of governance policies. Sensors at roundabouts offer innovative methods for assessing real-time traffic status by

D. K. Rakesh
University of Economics and Health Sciences, Warsaw, Poland
e-mail: dk.rakesh@vizja.pl

P. Dhal (✉)
GITA Autonomous College, BBSR, Gadajagasora, India
e-mail: parikeshdhal@gmail.com

S. K. Tiwari et al. (eds.), *AI-Driven Environmental Pollution Management*, Climate Risks and Solutions, https://doi.org/10.1007/978-3-031-96243-1_6

monitoring vehicular emissions such as CO, and NOx. These pollutants, emitted during combustion processes, are harmful and directly correlate with congestion levels. Analyzing historical traffic patterns helps in decision-making for congestion mitigation.

Simultaneously, air pollution from vehicles and industries is a growing concern with severe impacts on public health and ecological balance. Major pollutants include CO, NO_2, PM2.5, PM10, SO_2, O_3, and volatile organic compounds.

Studies reveal that prolonged exposure to fine particulate matter increases risks of cardiovascular and respiratory diseases, while NO_2 is linked to reproductive health issues. In India, industrial activities contribute 51% of total pollution, with alarming consequences for human health and the economy. WHO estimates attribute seven million annual deaths to air pollution globally, with China and India bearing a significant burden. The economic impact is equally daunting; in China, particulate matter pollution cost$29.18 billion in 2004, while urban air pollution in India continues to impair GDP growth.

Integrating real-time traffic data with air quality analysis offers a pathway to address these intertwined issues. Machine learning models like SVM and LSTM enhance the prediction of pollution by analyzing pollutant trends, providing actionable insights (Tiwari et al. 2022a, b, 2023a, 2022a). Meanwhile, sensor-based traffic systems classify congestion levels and help policy makers implement targeted interventions. A comprehensive framework combining these technologies highlights the potential for sustainable urban management, mitigating the dual threats of traffic congestion and air pollution.

This study integrates machine learning and sensor-based analytics to address the above issues effectively. Machine learning models such as SARIMA, Support Vector Machines (SVM), and Long Short-Term Memory (LSTM) were employed to predict the Air Quality Index (AQI), with SVM demonstrating superior accuracy in capturing pollutant trends like CO and NOx over six years. Simultaneously, real-time sensor data collected at circular roundabouts assessed vehicular emission sand traffic flow, enabling the classification of congestion levels based on pollutant thresholds and facilitating data-driven traffic management. Artificial Intelligence (AI) further enriches this integration by enabling real-time environmental monitoring, pollution source detection, and bridging technological gaps, particularly in addressing rural–urban disparities (Tiwari et al. 2023a, b). By leveraging predictive analytics and real-time monitoring, this interdisciplinary approach highlights the potential for scalable solutions in urban pollution and congestion management, fostering sustainable smart city ecosystems.

The structure of the paper is as follows: Sect. 6.2shows the literature survey of the paper. Section 6.3 discusses the proposed method followed by results and discussion in Sect. 6.4. Finally, we conclude this article in Sect. 5.

6.2 Literature Survey

Urbanization and industrial growth have significantly worsened traffic congestion and air pollution, leading to critical environmental and public health concerns. Recent research has highlighted the effectiveness of advanced technologies, particularly machine learning and sensor-based systems, in addressing these challenges. For instance, De Vito et al. (2008) explored the use of electronic noses for real-time air quality monitoring, while Duan et al. (2023) proposed hybrid models combining ARIMA, CNN, and LSTM for improved air quality forecasting. Similarly, Zaini et al. (2022) demonstrated the potential of deep learning models for predicting PM2.5 levels, and Méndez et al. (2023) provided a comprehensive review of machine learning adaptability in air quality predictions.

Vehicle emissions, particularly nitrogen oxides (NOx) and carbon monoxide (CO), remain major contributors to urban air pollution. Ge et al. (2023) utilized machine learning to monitor NOx emissions from heavy-duty vehicles based on on-board diagnostics (OBD) data, while Falai and Misul (2023) developed a data-driven model for estimating NOx emissions in diesel engines. Giovannardi et al. (2024) explored AI-powered virtual sensing techniques to monitor pollutant emissions at vehicle tailpipes, introducing innovative approaches to emission control in high-performance vehicles.

Machine learning has also been extensively applied in forecasting air quality indices (AQI). Natarajan et al. (2024) proposed an optimized model for AQI prediction across different regions, and Deepan and Saravanan (2024) combined SARIMA and LSTM models to capture both seasonal and dynamic pollution trends. Comparative studies by Sofiah et al. (2024) revealed that SARIMA excels in long-term trend analysis, whereas LSTM is more effective in capturing short-term fluctuations.

6.2.1 Feature Selection and AI-Driven Air Pollution Monitoring

Feature selection is a critical step in enhancing air quality prediction and traffic pollution forecasting models. Rakesh and Jana (2022) [introduced a mutual information-based feature selection approach that improves predictive performance by selecting relevant features. Additionally, Rakesh et al. (2023) proposed ranking-based stability measures to enhance the consistency of feature selection algorithms for dynamic air pollution datasets.

Feature selection techniques have also been successfully applied in fraud detection and anomaly detection. Rakesh and Jana (2023) utilized an improved differential evolution algorithm to detect fraudulent transactions, demonstrating methodologies that could be adapted for identifying irregular pollution patterns and traffic emission anomalies. Furthermore, Shukla and Rakesh (2020) introduced a dynamic ensemble-based feature selection model for fraud detection, which could be leveraged for real-time urban pollution prediction, enhancing model adaptability and efficiency.

6.2.2 Integration of AI, Feature Selection, and Traffic Analytics

The integration of AI, feature selection, and real-time traffic data analytics has significantly advanced urban traffic management strategies. Wang et al. (2024) employed big data methods to map vehicle emissions using real-time traffic data, enabling policymakers to design targeted interventions. Sensor-based monitoring systems, as demonstrated by De Vito et al. (2008) and Wang et al. (2024), provide real-time insights into congestion levels and pollutant concentrations, aiding in effective urban planning and traffic management.

These studies collectively underscore the transformative potential of machine learning and sensor-based technologies in addressing traffic congestion and air pollution in urban environments. They enhance prediction accuracy, facilitate real-time monitoring, and support scalable AI-driven solutions for sustainable urban development. Future research should focus on cost-effective, scalable solutions that address regional disparities and integrate IoT and advanced sensor networks to improve urban pollution and congestion management systems.

6.3 Proposed Method

The proposed methodology is divided into two sections. First, we explain the framework designed for identifying traffic congestion levels at circular roundabouts, followed by the integration of machine learning into the proposed framework.

6.3.1 Proposed Framework

To better understand the proposed framework, we present a virtual realization of the system, simplifying the illustration by making certain data assumptions. Figure 6.2 depicts a single circular roundabout lane with four entry points: Points A, B, C, and D, representing the West, North, East, and South entrances, respectively. Suppose vehicles $V1$, $V2$, and$V3$ are traveling from points A to B, C to A, and D to C, respectively. If a vehicle completes one full revolution of the roundabout at an average speed of X km/hr or x m/sec in t seconds, the number of rotations it can complete in one minute at xm/sec is calculated as $R = \frac{60}{t}$. Consequently, the roundabout can accommodate R vehicles completing one rotation in t seconds at xm/sec. The distance traveled by one vehicle in t seconds is d meters, and the total distance covered by R vehicles in one minute is R^2meters.

Assuming the emission rates of CO and NOx are c gm/meter and o gm/meter, respectively, the total emissions from R^2 vehicles traveling dR meters aredR^2cgm for CO anddR^2ogm for NOx. This implies in one minute, R^2 vehicles revolving around the roundabout emi tdR^2 c gm of CO and dR^2 o gm for NOx. If the roundabout has a safe spacing of S meters between vehicles, the N vehicle can complete one rotation in t seconds. The number of rotations these N vehicles can complete in one minute is:

$$N \times R = NR \tag{6.1}$$

Thus, $N\ R$ vehicles can pass through the roundabout in one minute, covering a total distance of $N\ Rd$ meters. The corresponding CO and NOx emissions are $N\ Rdc$ gm and $N\ Rdc$ gm, respectively. These emission values serve as the base line for defining three congestion levels, determined by varying the spacing S. The critical points for congestion levels are defined as follows:

- **Critical Point 1 (S1 = 2S meters)**: Represents the first congestion threshold.
- **Critical Point 2 (S2 = S meters)**: Represents the second congestion threshold.
- **Critical Point 3 (S3 = S/2 m)**: Represents the third congestion threshold.

The congestion levels are then categorized based on CO and NOx emissions relative to these critical points:

- **Level1 (Clear Traffic)**: $CO \leq CP1(CO)$ and $NOx \leq CP1(NOx)$.
- **Level 2 (Increasing Traffic)**: $CP1(CO) < CO \leq CP2(CO)$ and $CP1(NOx) < NOx \leq CP2(NOx)$.
- **Level 3 (Almost Congested)**: $CP2(CO) < CO \leq CP3(CO)$ and $CP2(NOx) < NOx \leq CP3(NOx)$.
- **Level4(Congested)**:$CO > CP3(CO)$ and $NOxCP3(NOx)$.

These levels are determined by the relationship between vehicle spacing and emissions. As the number of vehicles increases, the average spacing between them decreases, leading to higher emissions. This framework allows for the classification of traffic congestion based on real-time emission data, providing a basis for traffic management decisions.

Single Roundabout Lane: Consider a roundabout with a center island diameter of 30 m, an inscribed circle diameter of 40 m, and a lane width of 10 m. The average vehicle length is assumed to be 4 m. The number of rotations a vehicle can complete in one minute at a given speed is calculated as $\frac{60}{23} = 2.60$. This means that two vehicles can complete a full rotation, and one vehicle can complete half a rotation in one minute at the specified speed, assuming a one-rotation gap between vehicles. As the number of vehicles increases, the spacing between them decreases, leading to

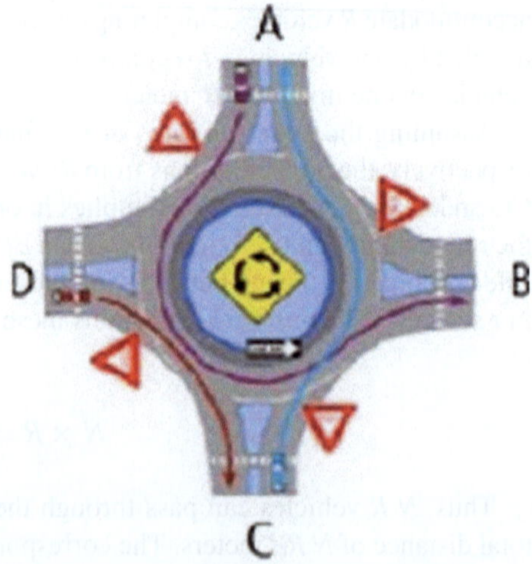

Fig. 6.1 Circular roundabout

higher emissions. The critical points for emissions are calculated based on different spacing values:

- **Critical Point 1 (S = 2 m)**: 21 vehicles can be accommodated, completing 55 rotations in one minute, emitting 14.51 gm of CO and 0.210 gm of NOx.
- **Critical Point 2 (S = 1 m)**: 25 vehicles can be accommodated, completing 65 rotations in one minute, emitting 17.15 gm of CO and 0.249 gm of NOx.
- **Critical Point 3 (S = 0 m)**:31 vehicles can be accommodated, completing 81 rotations in one minute, emitting 21.38 gm of CO and 0.310 gm of NOx.

These calculations provide the basis for defining congestion levels on a single-lane roundabout. Similar calculations can be extended to double-lane roundabouts, where the number of vehicles and emissions are proportionally higher. This framework enables the classification of traffic congestion based on real-time emission data, facilitating data-driven traffic management decisions (Fig. 6.1).

Double Round about Lane: To extend the framework to a double-lane roundabout, we consider the same principles as the single-lane scenario but account for the increased capacity and emissions due to the additional lane. Figure 6.3 illustrates a double-lane roundabout with the same dimensions as the single-lane version: a center island diameter of 30 m, an inscribed circle diameter of 40 m, and a lane width of 10 m. The average vehicle length remains 4 m, and the safe spacing between vehicles is denoted as *S*.

In a double-lane roundabout, the number of vehicles that can be accommodated is effectively doubled compared to a single lane. The number of rotations a vehicle can complete in one minute at a given speed is calculated as $\frac{60}{23} = 2.60$ This means that two vehicles can complete a full rotation, and one vehicle can complete half a rotation in one minute at the specified speed, assuming a one- rotation gap between vehicles. For a double-lane roundabout, the number of vehicles and rotations are

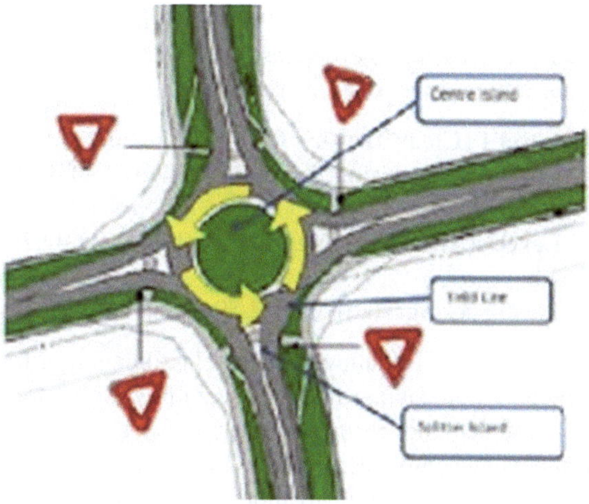

Fig. 6.2 Single circular roundabout lane

Fig. 6.3 Double circular round about lane

scaled accordingly. The critical points for emissions are calculated based on different spacing values, similar to the single-lane scenario but adjusted for the double-lane capacity:

- **Critical Point 1 (S = 2 m)**: 42 vehicles can be accommodated (21 vehicles per lane), completing 109 rotations in one minute, emitting 28.77 gm of CO and 0.417 gm of NOx.
- **Critical Point 2 (S = 1 m)**: 50 vehicles can be accommodated (25 vehicles per lane), completing 130 rotations in one minute, emitting 34.31 gm of CO and 0.498 gm of NOx.
- **Critical Point 3 (S = 0 m)**: 62 vehicles can be accommodated (31 vehicles per lane), completing 161 rotations in one minute, emitting 42.50 gm of CO and 0.617 gm of NOx.

These calculations provide the basis for defining congestion levels on a double-lane roundabout. The congestion levels are categorized based on CO and NOx emissions relative to these critical points:

- **Level 1 (Clear Traffic)**: CO 28.77 gm and NOx 0.417gm.
- **Level 2 (Increasing Traffic)**: 28.77gm CO 34.31gm and 0.417gm NOx 0.498 gm.
- **Level3 (Almost Congested)**: 34.31gm CO 42.50 gm and 0.498gm NOx 0.617 gm.
- **Level4 (Congested)**: CO 42.50gmand NOx 0.617gm.

This framework allows for the classification of traffic congestion based on real-time emission data, providing a scalable solution for traffic management in double-lane roundabouts. By leveraging these critical points, policymakers and urban planners can implement targeted interventions to mitigate congestion and reduce emissions, contributing to more sustainable urban environments (Tables 6.1 and 6.2).

The decision-making being used for machine learning depends on four levels that have been defined based on the different critical points of single and double circular roundabout lane as

Level 1: Traffic is Clear

$CO \leq (CP1CO)$

$NOx \leq (CP1NOx)$

Level 2: Traffic is Increasing gradually but safe to travel

$(CP1CO) > CO \leq (CP2CO)$

Table1 Single circular roundabout lane

Levels	CO emission	NOx emission	Result concluded
Level 1	≤ 14.51	≤ 0.210	Traffic is clear
Level 2	14.51 < And ≤ 17.51	0.210 < And ≤ 0.249	Traffic is increasing but safe to travel
Level 3	17.51 < And ≤ 21.38	0.249 < And ≤ 0.310	Traffic is almost congested
Level 4	21.38 <	0.310 <	Traffic is congested

Table 6.2 Double circular roundabout lane

Levels	CO emission	NOx emission	Result concluded
Level 1	≤ 28.77	≤ 0.417	Traffic is clear
Level 2	28.77 < And ≤ 34.31	0.417 < And ≤ 0.498	Traffic is increasing but safe to travel
Level 3	34.31 < And ≤ 42.50	0.498 < And ≤ 0.617	Traffic is almost congested
Level 4	42.50 <	0.617 <	Traffic is congested

(CP1NOx) > NOx ≤ (CP2NOx)

Level 3: Traffic is almost Congested

(CP2CO) > CO ≤ (CP3CO)

(CP2NOx) > NOx ≤ (CP3NOx)

Level 4: Traffic is Congested

(CP3CO) > CO (CP3NOx) > NOx.

6.3.2 Integration of Machine Learning

When considering modeling air quality, it is challenging to forecast using an environment quality model with the required precision. In this work, we embed thepredictionsfromthemachinelearningmodelsintotheproposedframework to identify traffic status. Figure 6.4 shows the architectural diagram for model selection. The diagram shows a dataset that passes through various steps, such as preprocessing, train/test split, application of ML models, evaluation of these models, and selection of the best-performing ML model. Once the best performance is selected on a given dataset, the combination strategy is applied to identify the traffic status as shown in Fig. 6.5. Further, we explain the combination strategy in Illustration1. According to the illustration, if the condition in Level 1 is met, it will generate a value of 1. Subsequently, Level 2 will generate a value of 2, and soon. Finally, we extract the maximum value generated by the levels. This final value indicates the level of traffic; for instance, a value of 1 signifies clear traffic, a value of 2 indicates increasing traffic but still safe for travel, and so on.

Next, we brief about different ML models used in this work.

SARIMA The ARIMA (Auto Regressive Integrated Moving Average) model is a popular method for time series forecasting. It predicts future values using past observations, specifically through it slag sand lagged prediction errors. ARIMA consists of two components: Auto Regression (AR) and Moving Average (MA). Its parameters, *p*, *d*, and *q*, represent the auto regressive order, degree of differencing, and moving average order, respectively. The model's equation is expressed as:

$$Y_t = \phi_1 Y_{t-1} + \ldots + \phi_p Y_{t-p} + \theta_1 \epsilon_{t-1} + \ldots + \theta_q \epsilon_{t-q} + \epsilon_t \quad (6.2)$$

where Y_t is the observed value, ϕ and θ are coefficients, and ϵ_t is the error term.

SARIMA (Seasonal ARIMA) extends ARIMA to handle time series with seasonal patterns. It adds parameters *P*, *D*, and *Q* for seasonal autoregression, differencing, and moving average, along with *s* for the seasonal period. SARIMA is expressed as$(p, d, q)(P,D,Q)_s$.

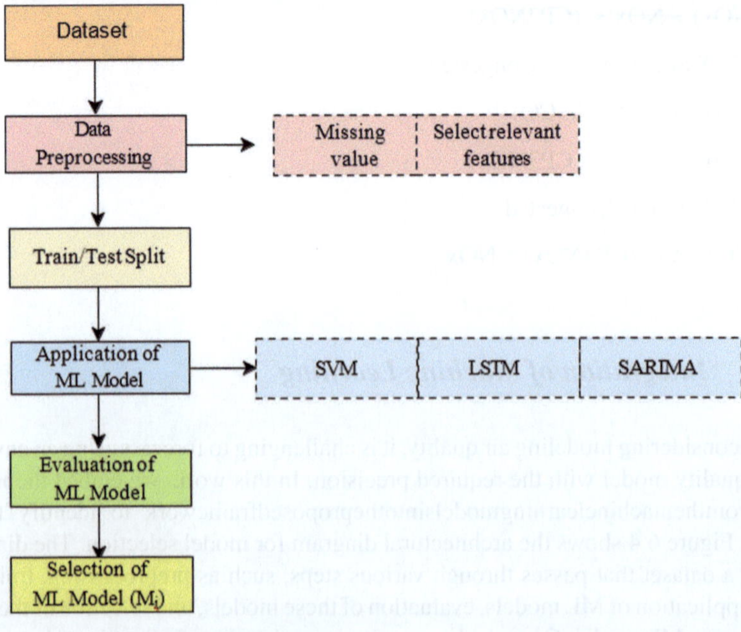

Fig. 6.4 An architecture for model selection

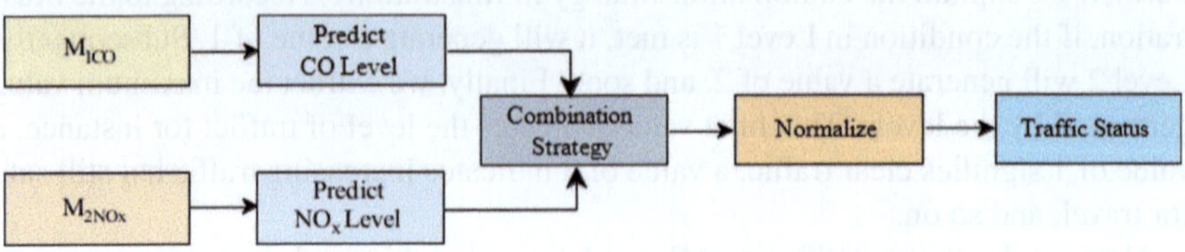

Fig. 6.5 An architecture for combination strategy for traffic status prediction

Its components include: -**Ordinary AR/MA: **Polynomials $\phi(B)$and $\theta(B)$ define non-seasonal components.—**Seasonal AR/MA: ** Polynomials $\Phi(B^s)$ and $\Theta(B^s)$ define seasonal patterns.

By combining these, SARIMA effectively models complex time series with trends and seasonality.

SVM Support Vector Machines (SVM) are widely used machine learning algorithms, particularly for supervised learning tasks in both regression and classification. In regression tasks, the SVM approach is commonly referred to as Support Vector Regression (SVR). The model identifies a hyper plane, or straight line, that best fits the data points. SVM is also effective in handling non-linearly separable data by mapping it to higher-dimensional spaces using kernel functions such as Radial Basis Function (RBF), Gaussian, and Polynomial kernels. This adaptability makes SVM well-suited for solving complex problems where data is not linearly separable.

SVM has demonstrated exceptional performance in various applications, including air pollution prediction, showcasing its robustness in handling challenging scenarios.

LSTM The forecasting model is built using a Long Short-Term Memory (LSTM) Recurrent Neural Network (RNN) architecture. In this model, the LSTM RNN layers consist of LSTM units as the fundamental components. The LSTM unit's cell state plays a crucial role, as it stores the information carried by the unit. The unit operates through gates, which control the addition or removal of information from the cell state. It also features a hidden state that tracks the prior data of the observed sequence. The three gates—forget gate, input gate, and output gate—are essential parts of the LSTM unit. The forget gate determines which part of the cell state should be discarded based on the current input and hidden state. The input gate calculates how much new information should be incorporated into the cell state by considering the current input, hidden state, and input gate weights. Lastly, the output gate controls the influence of the cell state on the current output, using the output gate weights, hidden state, and current input.

Illustration 1

Combination strategy			
Level 1	*If True ? 1:0* (a_1)	*if True ? 1: 0* (b_1) $max(a_1, b_1) = ab_{11}$	$max\ (ab_{11}, ab_{22}, ab_{33}, ab_{44})$
	$X_{CO} <= M_{1CO}$	$X_{NOx} <= M_{1NOx}$	
Level 2	*If True ? 2: 0* (a_2)	*if True ? 2: 0* (b_2) $max(a_2,b_2) = ab_{22}$	
	$M_{1CP1CO} <= X_{CO}$	$M_{1CP1NOx} <= X_{NOx}$	
	$<= M_{1CP2CO}$	$<= M1_{CP2NOx}$	
Level 3	*If True ? 3:0* (a_3)	*if True ? 3: 0* (b_3) $max(a_3,b_3) = ab_{33}$	
	$M_{1CP2CO} <= X_{CO}$	$M_{1CP2NOx} <= X_{NOx}$	
	$<= M_{1CP3CO}$	$<= M_{1CP3NOx}$	
Level 4	*If True ? 4: 0* (a_4)	*if True ? 4: 0* (b_4) $max(a_4,b_4) = ab_{44}$	
	$M_{1CP3CO} <= X_{CO}$	$M_{1CP3NOx} <= X_{NOx}$	

6.3.3 Results and Discussion

The proposed methodology considers the data released by the U.S. Environment Protection Agency in July 2000 (De Vito et al 2008). Estimates the average passenger car emission of components including hydrocarbons, carbon monoxide, and NOx by light-duty vehicle, light-duty truck, and medium-duty passenger vehicles. This work considers the two main components, carbon monoxide and nitrogen oxide (Figs. 6.6, 6.7 and 6.8).

Fig. 6.6 SARIMA model

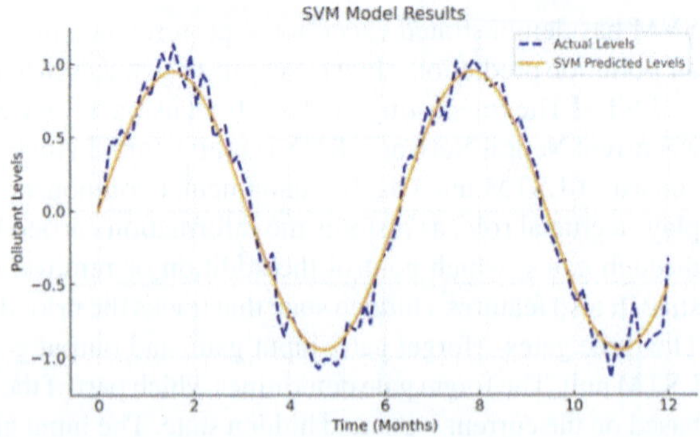

Fig. 6.7 SVM

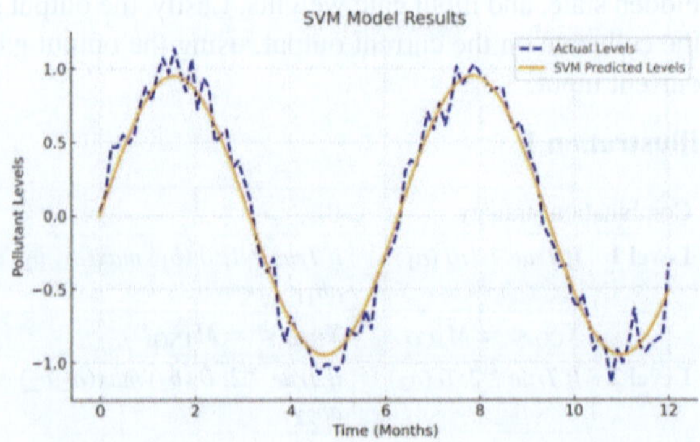

Fig. 6.8 LSTM

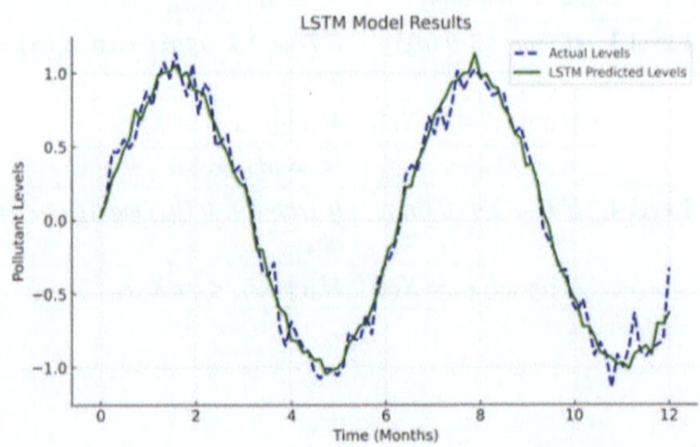

The performance of the SARIMA, SVM, and LSTM models was assessed using Mean Absolute Error (MAE) and Mean Squared Error (MSE) as evaluation metrics. Lower values of these metrics indicate better performance. The results of the analysis showed that SARIMA achieved the lowest MAE of 0.0826 and MSE of 0.0104, demonstrating the highest accuracy among the models. LSTM followed closely with a slightly higher MAE of 0.0829 and MSE of 0.0110, while SVM exhibited the highest error values with a MAE of 0.0876 and MSE of 0.0117.

SARIMA (Seasonal ARIMA) proved to be the most accurate model, excelling in datasets with clear periodic patterns such as pollutant emissions varying by season or time of day. Its reliance on historical data and predefined seasonality makes it highly effective for seasonal trends, though it may struggle with unexpected anomalies or irregular patterns. LSTM (Long Short-Term Memory) performed comparably to SARIMA, with strengths in capturing temporal dependencies and short-term fluctuations, making it ideal for real-time pollutant monitoring. Its ability to handle non-linear and complex time-series relationships further underscores its adaptability to dynamic data. However, LSTM is computationally intensive and requires a larger dataset for optimal training. SVM (Support Vector Machines), while being a simpler and more computationally efficient model, showed the highest error rates, indicating its limitations in handling temporal dependencies and sudden fluctuations. Despite these limitations, SVM is effective for capturing broader trends and non-linear relationships, making it suitable for medium-term forecasting.

In terms of accuracy, SARIMA outperformed the other models, making it the best choice for datasets with strong seasonal characteristics. Its reliance on predefined seasonality, however, may limit its adaptability to real-time anomalies. LSTM, on the other hand, demonstrated comparable accuracy and is better suited for real-time applications and data with complex, non-linear patterns, providing scalability for dynamic urban environments. SVM, while less accurate, offers a computationally efficient option for medium-term trends, particularly in applications with limited resources. SARIMA is best suited for long-term strategic planning where seasonal trends dominate. LSTM is ideal for real-time pollution monitoring and short-term forecasting with dynamic data, while SVM is appropriate for scenarios with limited computational resources or when medium- term trends are sufficient.

6.3.4 Relation to Traffic Emissions

SARIMA, SVM, and LSTM models were evaluated using Mean Absolute Error (MAE) and Mean Squared Error (MSE) metrics to determine their suitability for predicting traffic emissions of CO and NOx.

SARIMA demonstrated the highest accuracy, with the lowest MAE (0.0826) and MSE (0.0104), excelling in datasets with clear seasonal trends. LSTM closely followed, handling real-time and dynamic scenarios effectively but requiring more computational resources. SVM, while less precise, provided computational efficiency and general trend classification, though it struggled with detailed transitions and anomalies. Emission thresholds for CO and NOx, such as those from Level1 (Traffic is Clear) to Level4 (Traffic Congested) in single and double lanes, align with the strengths of each model. SARIMA is well-suited for modeling gradual emission changes across levels, such as CO 14.51 g and NOx $\leq$ 0.210 g for Single Lane. LSTM excels in capturing sharp fluctuations in higher congestion levels (e.g., CO

> 21.38 g or NOx > 0.310 g for Single Lane), leveraging its ability to handle non-linear relationships. SVM efficiently classifies broader trends but lacks the precision needed for detailed predictions.

SARIMA is ideal for long-term trend forecasting in predictable scenarios, LSTM is best for real-time and complex dynamic conditions, and SVM offers a simpler, resource-efficient option for medium-term classification. The choice of model depends on the specific requirements of traffic monitoring and emission prediction scenarios.

6.4 Conclusion

This study demonstrates the efficacy of integrating machine learning models with real-time sensor analytics to address urban traffic congestion and air pollution. The SARIMA model excels in handling seasonal trends, making it suitable for long-term strategic planning, whereas LSTM proves effective in capturing real- time dynamics and short-term fluctuations. SVM, while less precise, offers a computationally efficient option for medium-term analysis. By quantifying emissions of CO and NOx under varying traffic scenarios, the proposed framework provides actionable insights for proactive traffic management and pollution control. This approach underscores the potential of AI-driven technologies to create sustainable and intelligent urban environments. Future work aims to enhance the system's adaptability for diverse urban settings and further explore its integration with advanced IoT ecosystems.

References

De Vito S, Massera E, Piga M, Martinotto L, Di Francia G (2008) On field calibration of an electronic nose for benzene estimation in an urban pollution monitoring scenario. Sens Actuators B Chem 129(2):750–757

Deepan S, Saravanan M (2024) Air quality index prediction using seasonal autoregressive integrated moving average transductive long short-term memory. ETRI J e12658

Duan J, Gong Y, Luo J, Zhao Z (2023) Air-quality prediction based on the ARIMA-CNN-LSTM combination model optimized by dung beetle optimizer. Sci Rep 13(1):12127

Falai A, Misul DA (2023) Data-driven model for real-time estimation of NOx in a heavy-duty diesel engine. Energies 16(5):2125

Ge Y, Hou P, Lyu T, Lai Y, Su S, Luo W, He M, Xiao, L. (2023). Machine learning-aided remote monitoring of NOx emissions from heavy-duty diesel vehicles based on OBD data streams. Atmosphere 14(4):651

Giovannardi E, Brusa A, Petrone B, Cavina N, Tonelli R, Kitsopanidis I (2024) AI-based virtual sensing of gaseous pollutant emissions at the tailpipe of a high-performance vehicle. SAE Int J Engines 17(03-17-04-0029)

Méndez M, Merayo MG, Núñez M (2023) Machine learning algorithms to forecast air quality: a survey. Artif Intell Rev 56(9):10031–10066

Natarajan SK, Shanmurthy P, Arockiam D, Balusamy B, Selvarajan S (2024) Optimized machine learning model for air quality index prediction in major cities in India. Sci Rep 14(1):6795

Rakesh DK, Jana PK (2022) A general framework for class label specific mutual information feature selection method. IEEE Trans Inf Theory 68(12):7996–8014

Rakesh DK, Jana PK (2023) An improved differential evolution algorithm for quantifying fraudulent transactions. Pattern Recogn 141:109623

Rakesh DK, Anwit R, Jana PK (2023) A new ranking-based stability measure for feature selection algorithms. Soft Comput 27(9):5377–5396

Shukla S, Rakesh D (2020) Dynamic ensemble based feature selection model for credit card fraud detection. In: 2020 IEEE 17th India council international conference (INDICON). IEEE, pp 1–6

Sofiah NA, Tania KD, Meiriza A, Wedhasmara A (2024) A comparative assessment SARIMA and LSTM models for the Gurugram air quality index's knowledge discovery. In: 2024 International conference on electrical engineering and computer science (ICECOS). IEEE, pp 26–31

Tiwari SK, Kumaraswamidhas LA, Garg N (2022a) Time-series prediction and forecasting of ambient noise levels using deep learning and machine learning techniques. Noise Control Eng J 70(5):456–471

Tiwari SK, Kumaraswamidhas LA, Garg N (2023) Strategies and implications of noise pollution monitoring, modelling, and mitigation in urban cities. In: Handbook of metrology and applications. Springer Nature Singapore, Singapore, pp 1571–1593

Tiwari SK, Kumaraswamidhas LA, Gautam C, Garg N (2022b) An auto-encoder based LSTM model for prediction of ambient noise levels. Appl Acoust 195:108849

Tiwari SK, Kumaraswamidhas LA, Prince, Kamal M, Rehman MU (2023) A hybrid deep leaning model for prediction and parametric sensitivity analysis of noise annoyance. Environ Sci Pollut Res 30(17):49666–49684

Wang Y, Wang H, Zhang B, Liu P, Wang X, Si S, Xue L, Zhang Q, Wang Q (2024) High-resolution mapping of on-road vehicle emissions with real-time traffic datasets based on big data. EGUsphere 2024:1–26

Zaini NA, Ean LW, Ahmed AN, Abdul Malek M, Chow MF (2022) PM2. 5 forecasting for an urban area based on deep learning and decomposition method. Sci Rep 12(1):17565

Part III
Artificial Intelligence in Pollution Management

Chapter 7
Artificial Intelligence Application in Air and Water Pollution Management: Sources, Monitoring, and Control Measures

Shivam Saw, Mrinal Kanti Mandal, Rohit Patel, and Namrata Pandey

Abstract Air and water pollution pose significant threats to environmental sustainability and human health, driven by various anthropogenic activities. This study offers an in-depth analysis of the key origins, monitoring methods, and control strategies for addressing air and water pollution. Air pollution is largely attributed to industrial emissions, vehicular exhaust, and agricultural activities, while water pollution arises from untreated industrial effluents, agricultural activities, and improper waste disposal. Advanced monitoring systems, such as real-time sensors and Artificial Intelligence (AI)-driven models, offer enhanced capabilities for detecting pollutants, predicting trends, and tracing pollution sources. Effective control measures, including emission reduction technologies, sustainable agricultural practices, and stringent regulatory frameworks, are critical for mitigating the adverse effects of pollution. This paper emphasizes the importance of integrating modern technologies with regulatory policies to achieve sustainable pollution management. Such approaches not only address current pollution levels but also provide proactive solutions for long-term environmental protection.

Keywords Air pollution · Water pollution · Monitoring · Control measure · AI & ML

S. Saw (✉) · M. K. Mandal
Department of Chemical Engineering, National Institute of Technology, Durgapur, India
e-mail: shivamsaw@gmail.com

M. K. Mandal
e-mail: mkmandal.che@nitdgp.ac.in

R. Patel
Department of Environmental Science and Engineering, Indian Institute of Technology (Indian School of Mines), Dhanbad, India
e-mail: rohitpatelp7@gmail.com

N. Pandey
Faculty of Civil Engineering, Shri Ramswaroop Memorial University, Barabanki, India
e-mail: pandeynamrata11@gmail.com

S. K. Tiwari et al. (eds.), *AI-Driven Environmental Pollution Management*, Climate Risks and Solutions, https://doi.org/10.1007/978-3-031-96243-1_7

7.1 Introduction

Environmental pollution, encompassing air, water, and noise pollution, is a serious challenge that modern society faces. The rapid pace of industrialization, urbanization, and population growth has exacerbated the release of pollutants into the environment, resulting in severe consequences for human health, ecosystems, and climate stability (Saw et al. 2023a). Of the three major types of pollution, noise pollution is often overlooked by researchers, as it is not typically regarded as having a significant impact on the environment (Patel et al. 2022; Tiwari et al. 2024). However, noise pollution is gaining recognition as a growing concern because of its negative impact (Patel et al., 2024). Air pollution, marked by the presence of harmful pollutants like particulate matter, nitrogen oxides, sulfur dioxide, and volatile organic compounds, poses serious threats to air quality and is associated with respiratory issues, cardiovascular ailments, and early mortality. The top ten countries that contribute the most to common pollutants include Bangladesh, which holds the highest PM2.5 concentration, reaching approximately 79.9 $\mu g/m^3$ (Fig. 7.1). This significant level of air pollution is largely driven by industrial activities, brick kilns, and vehicular emissions. The dense population and rapid urbanization exacerbate these issues, leading to poor air quality and associated health risks (McMichael et al., 2000). Pakistan follows closely with a PM2.5 concentration of around 73.3 $\mu g/m^3$. Similar to Bangladesh, Pakistan's air quality is impacted by industrial, vehicular pollution, and the burning of biomass and crop residues. The situation is particularly severe in urban areas like Lahore and Karachi. India is another country grappling with high levels of air pollution, with a PM2.5 concentration exceeding 60 $\mu g/m^3$. India is another country grappling with increased levels of air pollution, with a PM2.5 concentration of 54.5 $\mu g/m^3$. Cities like Delhi often experience hazardous air quality, especially during the winter months when crop residue burning is prevalent. The following countries are also included in the like Tazakastian, Burkina Faso, United Arab Emirates, Egypt, Nepal, and the Republic of Congo. Their respective PM 2.5 concentrations are 49 $\mu g/m^3$, 46.5 $\mu g/m^3$, 43 $\mu g/m^3$, 42.5 $\mu g/m^3$, 42.2 $\mu g/m^3$, and 40.8 $\mu g/m^3$ (World Air Quality Report, 2023).

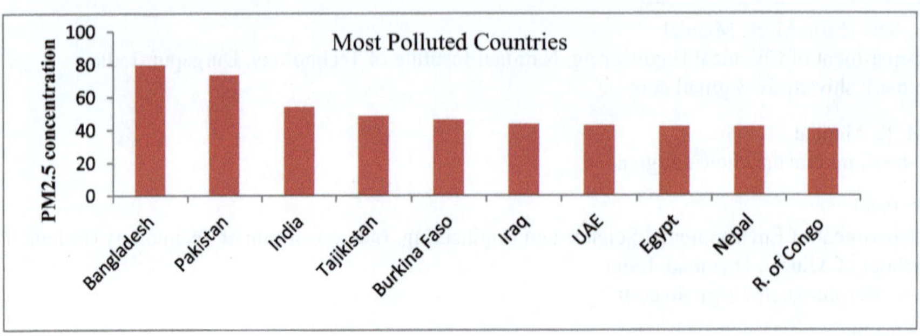

Fig. 7.1 Top polluted countries according to PM2.5 concentration

Conversely, water pollution occurs when water bodies are polluted with toxic chemicals, heavy metals, pathogens, and excessive nutrients. This leads to the deterioration of aquatic ecosystems, a decline in biodiversity, and risks to human health due to polluted drinking water sources (Khatri and Tyagi 2015; Saw et al. 2023a). It is crucial to understand the sources of pollution, as well as effective monitoring and control measures, to mitigate its impact. Identifying pollution sources, whether point sources like effluents from industrial activities or diffuse sources like agricultural runoff is the first step in devising appropriate regulatory and technological interventions. Monitoring, both through traditional methods and cutting-edge techniques like remote sensing and sensor networks, provides critical data that informs the design and implementation of control measures. Control measures range from regulatory frameworks and emission standards to innovative technological solutions like scrubbers, wastewater treatment plants, and sustainable practices such as renewable energy and water recycling. In current years, the addition of Artificial Intelligence (AI) into environmental management has emerged as a transformative approach. AI offers powerful tools for predicting pollution levels, analyzing complex environmental data, and optimizing pollution control strategies. Machine learning algorithms, for example, can analyze large datasets collected from multiple sources to predict air and water quality trends with high accuracy. AI-driven systems can also enhance real-time monitoring, allowing for quicker and more effective responses to pollution events. The use of AI in environmental management not only improves the precision of pollution monitoring and control but also facilitates the development of proactive strategies to prevent pollution before it occurs (Popescu et al. 2024). As environmental challenges grow more complex, the role of AI in pollution management will become increasingly important, offering new possibilities for safeguarding human health and preserving the environment.

This chapter highlights the diverse sources of air and water pollution, along with their monitoring methods and control strategies. It also delves into the increasing importance of real-time monitoring enabled by Artificial Intelligence (AI) and machine learning technologies, which have gained significant recognition across various industries for their revolutionary capabilities.

7.2 Sources

7.2.1 Sources of Air Pollution

Natural events can have a significant impact on air pollution, and in many cases, these processes are beyond human control. One of the primary natural sources of air pollution is volcanic eruptions, which release large quantities of gases, such as carbon dioxide (CO_2), sulfur dioxide (SO_2), and particulate matter into the atmosphere (Gaffney and Marley, 2003). These emissions impacted on air quality both regionally and globally and contribute significantly to the formation of smog, acid rain,

Fig. 7.2 Major pollutants from air pollution sources

Types	Sources	Major Pollutant
Vehicles		**CO**
Factories		**NH_3**
Power Plant		**SOx**
Forest Fire		**NOx**
Mining		**PM 2.5**

and respiratory diseases. Another major natural source of air pollution is wildfires, which emit substantial amounts of smoke, particulate matter, and volatile organic compounds (VOCs). Wildfire emissions can spread over great distances, impacting air quality in regions far from their source and influencing global atmospheric conditions (Reid et al. 2016). Human actions are the primary drivers of air pollution, particularly in urban and industrial regions. These pollutants and Vehicular emissions are another major anthropogenic source of air pollution. Burning fossil fuels in internal combustion engines generates pollutants such as unburnt carbon, hydrocarbons (HC), oxides of nitrogen, and particulate matter (PM) which negatively impact air quality and contribute to global warming (Kampa and Castanas 2008). Additionally, agricultural activities, such as the release of ammonia (NH_3) from fertilizers and methane (CH_4) from livestock, contribute to air pollution and have significant environmental and health consequences (Montzka et al. 2011). Figure 7.2 illustrates the various types of pollutant sources along with their major emitted pollutants.

7.2.2 Sources of Water Pollution

Industries contribute substantially to water pollution by releasing heavy metals, toxic chemicals, and organic waste directly into water bodies. These pollutants can negatively impact water quality and aquatic life. For example, metal or chemical processing factories often release wastewater containing harmful substances like lead, mercury, and other hazardous compounds. The intensive use of fertilizers, pesticides, and herbicides in agriculture significantly contributes to water pollution (Saw et al. 2023b; Mahato et al. 2023). Rainwater washes these chemicals into rivers, lakes, and groundwater, causing contamination. Nutrient pollution, mainly from nitrogen and phosphorus, results in eutrophication, triggering harmful algal blooms and reducing oxygen levels in water bodies (Xu et al. 2020). Urban areas generate substantial volumes of wastewater from households, including sewage and greywater, which contain pathogens, organic matter, and various contaminants. Without adequate treatment, this wastewater poses significant risks to water bodies. Additionally, urban stormwater runoff, such as streets, parking lots, and rooftops, introduces pollutants including oils, heavy metals, and debris into aquatic ecosystems. The lack of natural infiltration due to these impervious surfaces exacerbates pollution by channeling runoff directly into drainage systems that discharge into rivers and lakes (David and Mota 2022). Mining operations, especially in developing regions, often result in the release of toxic substances, including arsenic, cyanide, and heavy metals, into nearby water bodies (Prasad et al. 2022). The process of acid mine drainage, where sulfide minerals are exposed to air and water, can lead to long-term contamination of waterways. The disposal of waste materials, including plastics, chemicals, and radioactive substances, into oceans is another significant source of water pollution. Marine dumping leads to the degradation of marine ecosystems and affects the health of aquatic species.

7.3 Monitoring

7.3.1 Monitoring of Air Quality

The initial step in deploying air quality monitors involves selecting appropriate locations that take into consideration environmental factors, population density, and potential pollution sources (Munir et al. 2019). Once locations have been determined, suitable air quality monitoring instruments called Respirable Dust Sampler (RDS) must be chosen, which must meet the standards set by regulatory bodies to ensure they can adequately measure pollutants, that includes PM, NO_2, and SO_2. Before deployment, all monitoring instruments must undergo calibration to ensure that measurements are precise and reliable by aligning the instruments with recognized standards. Following the setup and calibration, continuous data collection commences. Consistency in data collection is crucial for identifying trends and making informed decisions. Finally, the data is integrated into the AQI, which categorizes air quality into different levels (Good to Very Unhealthy) based on established thresholds (WHO 2023; Wang et al. 2023). Figure 7.3 outlines the process of calculating the Air Quality Index (AQI), a critical indicator for assessing ambient air pollution levels.

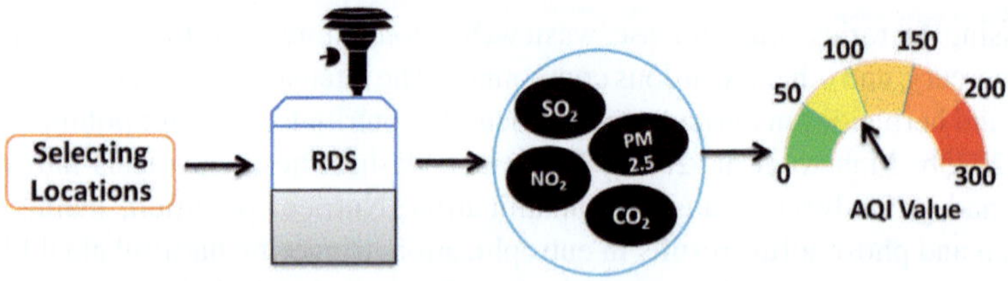

Fig. 7.3 Process of calculation of AQI value

Finally, the collected data is analyzed to evaluate adherence to air quality standards. The results are then reported to regulatory authorities and the public. Analysis helps in understanding air quality trends, pollutant sources, and the effectiveness of air quality management strategies.

7.3.2 *Monitoring of Water Quality*

Water sample monitoring is a critical process in environmental science, water quality management, and public health (Kirschke et al. 2020). The procedure typically begins with careful sample collection, where appropriate sampling locations are selected to represent the water body accurately. Clean, sterile containers are used, and samples are labeled with essential information such as date, time, location, and sample ID. Immediate on-site measurements of field parameters, such as temperature, pH, salinity, dissolved oxygen (D.O.), electrical conductivity, and turbidity, are conducted using calibrated portable multiparameter instruments. Sample preservation is crucial to maintain the integrity of the water samples. This may involve adding chemical preservatives for specific tests and keeping samples cool, usually at around 4 °C. A chain of custody form is completed to ensure proper tracking of sample handling and transfer. Samples are then transported to the laboratory as quickly as possible while maintaining appropriate temperature conditions. In the laboratory, various physicochemical analyses are performed, including physical tests (e.g., color, odor), chemical tests (e.g., Organic and Inorganic compounds, Heavy metals), and microbiological tests (e.g., total coliforms, E. coli). Quality control measures, such as running blank samples and duplicates, are implemented to ensure the accuracy of results. Data analysis involves comparing results to established water quality standards or guidelines and identifying any trends or anomalies.

The final steps include preparing a comprehensive report detailing methods, results, and interpretations, and initiating appropriate actions if any parameters exceed acceptable limits. Establishing a regular monitoring schedule based on the water body's characteristics and intended use is essential for ongoing water quality management. It's noted that specific procedures may vary depending on the type of water being analyzed and the particular parameters of interest. Therefore, it is crucial to follow standardized methods published by recognized authorities in the field.

7.4 Control Measures

7.4.1 Control Measures for Major Air Pollutants

Air pollution control measures are vital for addressing pollutants including NO_x, CO, SO_2, PM). To control PM, baghouse filters and electrostatic precipitators (ESPs) are widely used in industrial processes to capture fine particulates, while cyclones and wet scrubbers are effective for larger particles (Tang et al. 2024) shown in Fig. 7.4. Vehicle emission standards also play a significant role in curbing PM emissions from transportation. SO_2 emissions are primarily managed through flue gas desulphurization (FGD), where SO_2 is removed from exhaust gases using chemical absorbents like limestone (Yelebe et al. 2013). The shift to low-sulfur fuels and stringent regulations on industrial emissions further reduces SO_2 levels. Selective catalytic reduction (SCR) systems are often used to manage NOx emissions by transforming them into nitrogen and water through a reaction with ammonia. (Kim et al., 2020). Additionally, low NOx burners in power plants and exhaust gas recirculation (EGR) systems in vehicles help manage NOx emissions by controlling combustion temperatures. CO emissions are mitigated by catalytic converters, which convert CO into carbon dioxide and water in vehicle exhaust systems. Substituting to another fuels, such as compressed natural gas (CNG), also contributes to lower CO emissions. VOCs are managed through vapor recovery systems at fuel storage facilities, the use of low-VOC products in industrial settings, and thermal oxidizers, which destroy VOCs before they are released (Khan and Ghoshal 2000; Reid et al. 2016). Table 7.1 presents the various treatment technologies, their target pollutants, and the principles of their operation.

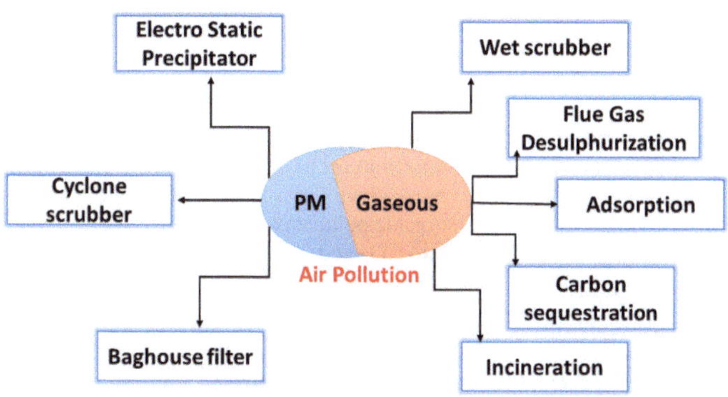

Fig. 7.4 Various control measures of air pollutant

Table 7.1 Treatment technology of common air pollutants

Treatment technology	Target pollutants	Principle of operation	Applications	References
Electrostatic precipitator	Particulates, dust, smoke	Charges particles and collects them on oppositely charged plates using electrostatic forces	Industrial processes, power plants, incinerators	(Liu et al. 2023; Marchand et al. 2016)
Baghouse filters	Particulates (dust, soot, smoke)	Uses fabric filter bags to capture particulate matter from airstreams as they pass through	Cement plants, steel mills, and other manufacturing units	(Maduna and Tomašić 2017)
Wet scrubbers	SOx, NOx, particulate matter, volatile compounds	Utilizes a liquid to eliminate pollutants from a gas stream through absorption or chemical reactions	Chemical plants, steel mills, power plants	(Liu et al., 2023)
Activated carbon adsorption	Volatile Organic Compounds (VOCs), odors	Uses activated carbon to adsorb pollutants from gas streams by trapping them in a porous structure	VOC control in chemical, petrochemical, and manufacturing industries	(Fayyad and Iváncsy 2024)
Catalytic converters	NOx, CO, hydrocarbons	Converts harmful gases into less harmful substances (e.g., CO_2, N2, H2O) using a catalyst	Automobile exhaust systems	(Caputo and Pelagagge 2001)
Thermal oxidizers	VOCs, hazardous air pollutants (HAPs)	Uses high temperatures to break down organic pollutants into CO_2 and water vapor	Chemical production, waste treatment facilities	(Testoni, 2011)

7.4.2 *Control Measures of Water Pollutants*

Effective control measures for major water pollutants, including heavy metals, nutrients, pathogens, and organic pollutants, are crucial for safeguarding water quality and public health. Screening, coagulation, flocculation, aeration chambers, and secondary treatment are critical components of wastewater treatment processes. Screening removes large debris and solids, such as plastics and rags, to protect downstream equipment. Coagulation involves adding chemicals (e.g., alum or ferric chloride) to neutralize charges on suspended particles, promoting their aggregation, while flocculation gently mixes the water to form larger flocs that can be easily removed. The aeration chamber introduces oxygen to support aerobic microorganisms that degrade organic matter, a key step in biological treatment. Finally, secondary treatment, often involving activated sludge or biofilm processes, further reduces organic content and suspended solids, ensuring effluent quality meets regulatory standards before discharge or reuse. Together, these processes ensure effective wastewater purification and environmental protection.

To mitigate heavy metal contamination, techniques such as adsorption using activated carbon, ion exchange, and chemical precipitation are commonly employed. These methods effectively remove metals like lead, cadmium, and mercury from wastewater (Razzak et al. 2022). Nutrient pollution, primarily from agricultural runoff, can be controlled through practices such as constructed wetlands and vegetated swales that naturally filter excess nitrogen and phosphorus (Yang and Lusk 2018). For pathogen control, chlorination, UV disinfection, and advanced oxidation processes (AOPs) are effective methods used in water treatment facilities to eliminate harmful microorganisms. In the case of organic pollutants, technologies such as bioremediation, where microorganisms degrade organic contaminants, and membrane filtration are increasingly being utilized. Furthermore, regulatory measures, including the implementation of the Water Quality Monitoring Program, play a critical role in monitoring and managing water quality to ensure compliance with standards. The most common technology used for wastewater treatment are presented in Table 7.2. Figure 7.5 illustrates the sustainable recycling of wastewater using a basic treatment process.

7.5 Application of AI and machine learning in air and water pollution management

AI and ML are becoming vital tools in air and water pollution management, offering increased accuracy and efficiency in monitoring, prediction, and control strategies. In air pollution management, AI models, such as neural networks and deep learning, are used to predict air quality by analyzing large datasets from meteorological conditions, traffic patterns, and industrial emissions. These models help forecast concentrations of pollutants like PM2.5, NOx, and ozone, allowing authorities to implement proactive measures to mitigate exposure (Rautela and Goyal 2024). AI also aids in identifying emission sources by analyzing complex datasets and distinguishing between pollution from vehicles, industries, and other sources. In water pollution management, AI techniques are employed to optimize water treatment processes by predicting the efficiency of different treatment methods. Machine learning algorithms can analyze water quality parameters such as pH, turbidity, and dissolved oxygen, identifying contamination events in real time. Additionally, AI-driven systems are

Table 7.2 Various treatment technologies for removal of water pollutants

Treatment technology	Description	Pollutants Removed	References
Activated sludge process	A biological treatment method where microorganisms digest organic pollutants in wastewater	Organic matter, BOD, nitrogen, phosphorus	(Dold and Ekama 1981)
Constructed wetlands	Natural treatment systems that use wetland vegetation to filter and degrade contaminants	Nutrients (nitrogen and phosphorus), heavy metals, pathogens	(Liu et al. 2015)
Membrane filtration	Makes semi-permeable membranes to filter contaminants from water by utilizing differences in size and charge	Suspended solids, bacteria, viruses, heavy metals	(Hube et al. 2020)
Reverse osmosis (RO)	A process that uses pressure to push water through a membrane, removing dissolved solids and contaminants	Salts, heavy metals, pesticides, organic contaminants	(Malaeb and Ayoub 2011)
Activated carbon adsorption	Uses activated carbon to adsorb organic compounds and some heavy metals from water	Organic compounds, chlorine byproducts, heavy metals	(Bhatnagar et al., 2013)
Ultraviolet (UV) disinfection	Employs UV light to kill pathogens in water by disrupting their DNA	Bacteria, viruses, protozoa	(Guo et al. 2022)
Bioremediation	Utilizes microorganisms to degrade organic pollutants	Hydrocarbons, heavy metals, nutrients	(Shukla et al. 2010)
Ozonation	Uses ozone gas to oxidize organic contaminants and disinfect water	Organic compounds, bacteria, viruses, taste, and odor	(Malik et al. 2020)

used to model pollutant transport in water bodies, enabling a better understanding of how contaminants move through aquatic ecosystems and assisting in source tracking (Lowe et al. 2022). These applications demonstrate the significant potential of AI and ML to improve pollution management by enhancing accuracy, reducing costs, and facilitating data-driven decision-making (Fig. 7.6).

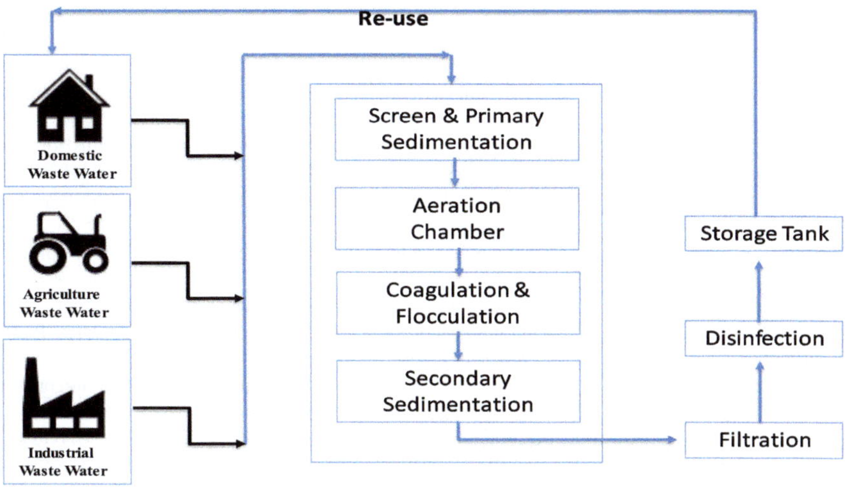

Fig. 7.5 Technology used for treatment of wastewater

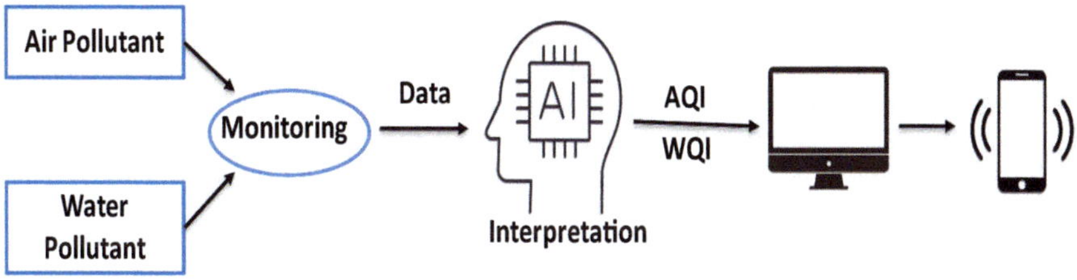

Fig. 7.6 AI and IoT-based applications for monitoring air and water quality

7.6 Conclusion

Air and water pollution present formidable challenges to sustainability and public health, underscoring the urgent need for effective monitoring and mitigation strategies. Industrial activities, agricultural practices, and rapid urbanization have significantly increased pollutant levels, demanding advanced solutions for detection and control. While traditional monitoring methods remain valuable, the integration of artificial intelligence (AI) has revolutionized pollution management. AI's capacity to analyze large datasets in real-time enhances predictive accuracy and enables proactive interventions, optimizing measures such as emission controls, filtration systems, and wastewater treatment. By leveraging AI-driven technologies, more efficient and timely actions can be implemented, substantially reducing environmental contamination and fostering healthier ecosystems for future generations.

References

Bhatnagar A, Hogland W, Marques M, Sillanpää M (2013) An overview of the modification methods of activated carbon for its water treatment applications. Chem Eng J 219:499–511

Caputo AC, Pelagagge PM (2001) Economic comparison of emission control systems for glass manufacturing furnaces with heat recovery. J Air Waste Manage Assoc 51(7):1009–1020

David LM, Mota TM (2022) Quality assessment of small urban catchments stormwater models: a new approach using Old metrics. Hydrology 9(5):87

Dold PL, Ekama GA (1981) A general model for the activated sludge process. In: Water pollution research and development. Pergamon, pp 47–77

Fayyad MB, Iváncsy T (2024) Numerical study of the influence of using semicircular corrugated collecting plate and inverted semicircular corrugated on the electrostatic precipitator. Period Polytech Electr Eng Comput Sci 68(2):124–134

Guo K, Wu Z, Chen C, Fang J (2022) UV/chlorine process: an efficient advanced oxidation process with multiple radicals and functions in water treatment. Acc Chem Res 55(3):286–297

Hube S, Eskafi M, Hrafnkelsdóttir KF, Bjarnadóttir B, Bjarnadóttir MÁ, Axelsdóttir S, Wu B (2020) Direct membrane filtration for wastewater treatment and resource recovery: a review. Sci Total Environ 710:136375

Kampa M, Castanas E (2008) Human health effects of air pollution. Environ Pollut 151(2):362–367

Khan FI, Ghoshal AK (2000) Removal of volatile organic compounds from polluted air. J Loss Prev Process Ind 13(6):527–545

Khatri N, Tyagi S (2015) Influences of natural and anthropogenic factors on surface and groundwater quality in rural and urban areas. Front Life Sci 8(1):23–39

Kim HS, Kasipandi S, Kim J, Kang SH, Kim JH, Ryu JH, Bae JW (2020) Current catalyst technology of selective catalytic reduction (SCR) for NOx removal in South Korea. Catalysts 10(1):52

Kirschke S, Avellán T, Bärlund I, Bogardi JJ, Carvalho L, Chapman D, Dickens CW, Irvine K, Lee S, Mehner T, Warner S (2020) Capacity challenges in water quality monitoring: understanding the role of human development. Environ Monit Assess 192:1–16

Liu H, Luo S, Yu L, Zhao H, Liang J, Guo Y, Cui Y, Shan S, Liu X, Li J (2023) Study on multi-pollutant test and performance index determination of wet electrostatic precipitator. Separations 10(10):536

Liu R, Zhao Y, Doherty L, Hu Y, Hao X (2015) A review of incorporation of constructed wetland with other treatment processes. Chem Eng J 279:220–230

Lowe M, Qin R, Mao X (2022) A review on machine learning, artificial intelligence, and smart technology in water treatment and monitoring. Water 14(9):1384

Maduna K, Tomašić V (2017) Air pollution engineering. Phys Sci Rev 2(12):20160122

Mahato JK, Saw S, Kant N, Patel R, Yadav B (2023) GIS-based spatial variability and water quality index (WQI) modeling of trace elements in groundwater of Jamshedpur, Eastern India. J Geol Soc India 99(3):415–420

Malaeb L, Ayoub GM (2011) Reverse osmosis technology for water treatment: State of the art review. Desalination 267(1):1–8

Malik SN, Ghosh PC, Vaidya AN, Mudliar SN (2020) Hybrid ozonation process for industrial wastewater treatment: Principles and applications: a review. J Water Process Eng 35:101193

Marchand M, Lemont F, Reess T, de Ferron AS, Souakri S (2016) Electrostatic precipitator: technological development and applications to thermal incineration processes in CEA. Int J Plasma Environ Sci Technol 10(2):157–160

McMichael AJ, Smith KR, Corvalan CF (2000) The sustainability transition: a new challenge. Bull World Health Organ 78:1067–1067

Montzka SA, Dlugokencky EJ, Butler JH (2011) Non-CO_2 greenhouse gases and climate change. Nature 476(7358):43–50

Munir S, Mayfield M, Coca D, Jubb SA (2019) Structuring an integrated air quality monitoring network in large urban areas–Discussing the purpose, criteria and deployment strategy. Atmos Environ X 2:100027

Patel R, Kumar Singh P, Saw S (2022) Recent advancements in the challenges and strategies of globally used traffic noise prediction models. Environ Sci Pollut Res 29(32):48168–48184

Patel R, Singh PK, Saw S (2024) Traffic noise modeling under mixed traffic condition in mid-sized Indian city: a linear regression and neural network-based approach. Int J Math Eng Manage Sci 9(3)

Popescu SM, Mansoor S, Wani OA, Kumar SS, Sharma V, Sharma A, Arya VM, Kirkham MB, Hou D, Bolan N, Chung YS (2024) Artificial intelligence and IoT driven technologies for environmental pollution monitoring and management. Front Environ Sci 12:1336088

Prasad D, Singh PK, Mahato JK, Saw S (2022) Hydrogeochemical characterization of groundwater in fire and non-fire zones of Jharia coal field, Eastern India, using water quality index (WQI), hierarchical cluster analysis (HCA), and human health risk. Arab J Geosci 15(9):927

Rautela KS, Goyal MK (2024) Transforming air pollution management in India with AI and machine learning technologies. Sci Rep 14(1):20412

Razzak SA, Faruque MO, Alsheikh Z, Alsheikhmohamad L, Alkuroud D, Alfayez A, Hossain SMZ, Hossain MM (2022) A comprehensive review on conventional and biological-driven heavy metals removal from industrial wastewater. Environ Adv 7. https://doi.org/10.1016/j.envadv.2022.100168

Reid CE, Brauer M, Johnston FH, Jerrett M, Balmes JR, Elliott CT (2016) Critical review of health impacts of wildfire smoke exposure. Environ Health Perspect 124(9):1334–1343

Saw S, Singh PK, Mahato JK, Patel R, Shikha D (2023) A modeling approach for the suitability evaluation and human health risk assessment of heavy metals dispersion in groundwater resources. Environ Dev Sustain 1–25

Saw S, Singh PK, Mahato JK, Patel R, Deoli V, Singh L (2023b) Application of GIS-based DRASTIC model and its validation with solute transport model in the assessment of groundwater vulnerability index: a case study from coal mining region of India. Environ Qual Manage 33(2):265–276

Shukla KP, Singh NK, Sharma S (2010) Bioremediation: developments, current practices and perspectives. Genet Eng Biotechnol J 3:1–20

Tang W, Wu CW, Lin SL, Wu JL, Huang SW, Song M (2024) Enhanced mitigation of inhalable particles and fine particle-bound PAHs from a novel hazardous waste-power plant candidate. Environ Pollut 343:123220

Testoni AL (2011) Energy efficient removal of volatile organic compounds (VOCs) and organic hazardous air pollutants (o-haps) from industrial waste streams by direct electron oxidation. Advanced Electron Beams Inc., Wilmington, MA

Tiwari SK, Kumaraswamidhas LA, Patel R., Garg N, Vallisree S (2024) Traffic noise measurement, mapping, and modeling using soft computing techniques for mid-sized smart Indian city. Meas Sens 33:101203

Wang Y, Huang L, Huang C, Hu J, Wang M (2023) High-resolution modeling for criteria air pollutants and the associated air quality index in a metropolitan city. Environ Int 172:107752

World Health Organization (2023) WHO ambient air quality database, 2022 update: status report. World Health Organization

Xu D, Lee LY, Lim FY, Lyu Z, Zhu H, Ong SL, Hu J (2020) Water treatment residual: a critical review of its applications on pollutant removal from stormwater runoff and future perspectives. J Environ Manage 259:109649

Yang YY, Lusk MG (2018) Nutrients in urban stormwater runoff: current state of the science and potential mitigation options. Curr Pollut Rep 4:112–127

Yelebe ZR, Samuel RJ, Yelebe BZ (2013) SO_2 removal from flue gas using gas-solid treatment process. J Eng Appl Sci 5(2)

Chapter 8
Role of Artificial Intelligence and Machine Learning in Pollution Management

Anubhava Srivastava, Shruti Bharadwaj, Raziqa Masood, Rakesh Dubey, and Susham Biswas

Abstract Air pollution management has become a very big issue these days due to the rapid increase in industrialization and urbanization. Combining Artificial Intelligence (AI) and Machine Learning (ML) with the pollution management system points to a different perspective of air quality and contributes to the efficient and accurate monitoring of air. This paper investigates the use of AI and ML in pollution management by tracking and con-trolling air pollution levels, mainly using Google Earth Engine (GEE) for the monitoring of air pollutants such as nitrogen dioxide (NO_2) and carbon dioxide (CO_2). Enriching the data obtained from GEE, machine learning algorithms use satellite data in almost real-time, conducting predictive modeling and pattern recognition for pollutant dispersion. The use of AI-based techniques, linting through supervised classification and regression models, presents the origin of pollution, the prediction of high-risk areas, and the formulation of successful mitigation strategies. This research unveils the capabilities of AI and ML in producing innovative data-driven measures that can be used to fight the air pollution problem, hence ensuring sustainable environmental management.

Keywords Google Earth Engine · Classification · Machine learning · GIS

A. Srivastava · S. Bharadwaj (✉)
Department of Computer Science & Engineering, Sharda University, Greater Noida, India
e-mail: pgi18001@rgipt.ac.in

A. Srivastava
e-mail: anubhavacse@gmail.com

S. Bharadwaj
Centre for SeNSE, Indian Institute of Technology, New Delhi, India

R. Masood
Department of Computer Science & Engineering, Integral University, Lucknow, India

R. Dubey
Department of Mechanical Engineering, SITRC Nasik, Nasik, Maharashtra, India

S. Biswas
Department of Computer Science Engineering, RGIPT Amethi, Uttar Pradesh, Amethi, India

S. K. Tiwari et al. (eds.), *AI-Driven Environmental Pollution Management*, Climate Risks and Solutions, https://doi.org/10.1007/978-3-031-96243-1_8

8.1 Introduction

Air pollution: is one of the great problems that this world is faced with today. As humanity is getting more industrialized, more urbanized, and less forested, it is evident that the atmosphere concentration of the harmful gases keeps growing and polluting it. These pollutants, such as nitrogen dioxide (NO_2), carbon dioxide (CO_2), methane (CH_4), sulfur dioxide (SO_2), and particulate matter (PM2.5 and PM10), environment and the human health toll on the atmosphere. To decrease the issues of air pollution, the system must be highly monitored, accurately analyzed, and the prevention strategies have to be taken in advance. The use of Artificial Intelligence (AI) and Machine Learning (ML) have been used on a rising scale as of late in pollution management, thus allowing for large datasets to be processed, for pattern recognition to be able to be performed, and for real-time monitoring to occur, respectively. Furthermore, the combination of platforms like Google Earth Engine (GEE) (Gorelick et al. 2017) and the application of technological innovations has led to the development of more efficient and scalable environmental data analysis, environmental pollutant tracking tools, and pollution models for alleviation through prediction (Wubalem 2021; Srivastava 2025).

8.1.1 The Role of Artificial Intelligence and Machine Learning in Pollution Management

Air pollution is the primary environmental challenge of humanity today. As industrialization, urbanization, and deforestation have increased, the concentration levels of harmful pollutants in the atmosphere have also started to grow. Such pollutants, among others, allow large volumes of NO_2, CO_2, CH_4, SO_2, and particulate matter (PM2.5 and PM10) which can cause severe injuries to human health as well as environmental damage. Successful air pollution management is enacted through the surveillance of air pollution, data analytics, and practiced abstention based on the information. Development of Artificial Intelligence (AI) and Machine Learning (ML) has been recognized as a useful tool in the management of pollution in pollution management technology. AI and ML are supported with the automation of the imposed large data processing as well as the observation of digitized patterns and real time intelligence. This is besides the fact that platforms like Google Earth Engine (GEE) (Gorelick et al. 2017) have allowed innovative, far-reaching, and scalable solutions to the environmental data analysis, monitoring pollutant levels, and predictive modeling for pollution management.

AI and ML technologies have brought a breakthrough to pollution management by giving the highly developed tools to process a variety of large data from remote sensing systems, ground-based sensors, and meteorological data (Bharadwaj et al. 2020; Bharadwaj et al. 2024). Such solutions are fundamental in identifying the pollution hotspots by examination of the pollutant plume dispersion patterns and the future pollutant level forecast, among others. AI and ML as automated objects of data gathering and doing it is the direct cause of more efficient decision-making and faster effectiveness against imminent pollution problems.

The use of Machine Learning models, specifically supervised learning algorithms, accounts for the effectiveness of air pollution data patterns analysis. Methods such as decision trees, random forests, gradient boosting machines, and deep learning models are being used for pollutant classification, concentration prediction, and air quality condition forecasting. The models analyze past pollution data, meteorological data (e.g., temperature, humidity, wind speed), and land-use patterns, to acquire predictive abilities (Bar et al. 2022).

8.1.2 Google Earth Engine for Pollution Monitoring

Google Earth Engine (GEE) is a platform in the cloud that was designed to work with very large satellite and environmental data by processing and analyzing it. It affords you access to a huge archive of datasets brought by satellites of different missions, like Landsat, MODIS, Sentinel, and others (Srivastava et al. 2022). The relevance of GEE in pollution management is realized by the fact that it permits researchers and policymakers to watch the development of pollutant levels in a specific place over a certain period of time and also the relationship between different places that are being surveyed with high resolution in both space and time (Srivastava et al. 2023a, b, c).

GEE is now made flexible with the extension of the artificial intelligence and machine learning power to it through the algorithms which are then able to, in turn do the analysis of the satellite pictures and extract important information about air quality. To illustrate, we can also source data from satellites like the TROPO-spheric Monitoring Instrument (TROPOMI) carried by Sentinel-5P to show the status of NO_2, CO_2, methane, and ozone which in the case of the prominent gases they are mostly emitted through the generator of electricity, manufacturing, traffic, and so on, as well as by the wildfires, and volcanic activities such as the rest natural processes (Dash et al. 2023; Srivastava et al. 2023a, b, c). GEE, researchers and policymakers can follow the movement of air pollutants around the world and see how they affect countries across the borders that help them to propose a proper intervention plan. Furthermore, GEE suitably transforms from a very local to a global level of interrelation thanks to such things as clouds, atmospheric chemistry, land-use, and population data. In connection to this, even meteorological data may also be included to help the better formation of pollution models.

8.1.3 Monitoring Nitrogen Dioxide (nO_2) and Carbon Dioxide (CO_2)

Nitrogen dioxide (NO_2) is by far the most common air pollutant released in vehicle, power plants, and industrial facilities through the burning of fossil fuels (Dubey et al. 2021a, b; Dubey et al. 2025). Elevation of high NO_2 levels in the atmosphere results in lung diseases such as asthma and bronchitis and is the cause of ground-level ozone (O_3) which is yet another harmful pollutant.

Carbon dioxide (CO_2), though in the concentrations commonly found in the atmosphere may be only indirectly harmful to humans, is a major driver of the climate change phenomenon. CO_2 emissions from fossil fuels, de-forestation, and other human activities lead to the thermal insulation of the Earth's atmosphere, which in turn becomes warmer through the greenhouse effect. The importance of the CO_2 tracking for the air pollution analysis also comes from the fact that it is essential for addressing the broader environmental issues like climate change.

Satellite information from instruments like TROPOMI can be obtained through General Electric Environment and used to track NO_2 and CO_2 levels in nearly real-time (Srivastava et al. 2023a, b, c; Srivastava et al. 2023a, b, c; Srivastava and Biswas 2023). The use of machine learning algorithms on these data allows us to recognize shifts in pollutants, the origins of pollutants, and future concentrations. More than that, AI models can be programmed to discern between jams caused by traffic, the output of factories, and air conditions that usually goes with NO_2 emanations.

Furthermore, CO_2 control is the process of understanding the carbon cycle and the plan for emission reductions. The AI can combine the satellite data track with ground-based measurements to assess the success of carbon capture technologies, reforestation programs and other climate change mitigation programs.

8.1.4 The Impact of Wildfires on Air Pollution

Due to climate change and land use practices Wildfires have become a more widespread and disruptive natural phenomenon. These fires release huge amounts of pollutants into the atmosphere, including CO_2, particulate matter (PM2.5 and PM10) (Hay et al. 2016), methane, and volatile organic compounds (VOCs). Smoke and smog from wildfires can travel hundreds or thousands of kilometers. which affects air quality over a large area. The environmental impact of wildfires on air pollution is significant. Burning forests and grasslands release stored carbon. This results in higher CO_2 levels in the atmosphere. and accelerate global warming. Particulate matter emitted by wildfires poses serious health risks. Including respiratory and heart disease. Wildfires can also damage the ecosystem. Reducing biodiversity and changing land use patterns (Viana et al. 2019). This leads to long-term environmental degradation. AI and ML play an important role in monitoring and mitigating the impact of wildfires on air pollution. By integrating satellite data with meteorological and land use data. Machine learning models can predict the likelihood of wildfire events based on factors such as temperature, humidity, wind patterns and vegetation type. The Google Earth tool is a powerful platform for tracking wildfires. Tools such as MODIS and VIIRS can use satellite data to detect active fires. Map the burned area and assess emissions (Yang et al. 2021). By applying AI algorithms to these datasets Researchers can predict fire behavior Assess the environmental impact of wildfires and develop management strategies.

8.2 Related Work

Artificial intelligence (AI) and machine learning (ML) are changing the landscape of air pollution management. Including nitrogen dioxide (NO_2), carbon dioxide (CO2), methane (CH4), traditional pollution tracking methods that rely heavily on ground-based sensors Often confined to geographic and temporal boundaries, AI and ML, when used in conjunction with satellite platforms such as Google Earth Engine (GEE), provide global coverage - overcoming these limitations by Providing data analysis services GEE's vast satellite imagery repository combined with cloud computing (Viana et al. 2019; Bar et al. 2022). It allows researchers to track pollution trends in real time. Studies such as "Measuring Machine Learning Performance" by Srivastava and others have shown how ML algorithms such as Random Forest (RF) and Gradient Tree Boosting (GTB) improve the accuracy of air quality forecasts (Masood et al. 2024). These models use large data sets collected from satellites like NASA's Sentinel-5P and OCO-2, which measure pollutants like NO_2, CO_2 and more with unprecedented accuracy. By analyzing this data, AI models can identify pollution hotspots. Link greenhouse gas emissions to specific industrial activities or urban areas. Future pollution follows past trends. and meteorological factors - levels and forecasts. Combining AI with satellite data has significant advantages over traditional methods (Srivastava 2025). A key strength lies in the ability of AI models to process large amounts of environmental data and detect patterns that cannot be easily detected by For example, GEE allows satellite data to be combined with AI-powered models to track pollution over large geographic areas (Li et al. 2020), and the spatial details of satellite data can be captured (Liu et al. 2020). Combined with the temporal resolution provided by AI models, a very accurate and dynamic understanding of air quality is possible, for example in urban environments where greenhouse gas emissions are high, using these AI models based on traffic patterns, weather conditions, and emissions data. past gas Can predict peak pollution times These insights help policymakers implement targeted interventions, such as traffic bans during peak hours or enforcing stricter emissions controls in industrial areas. To reduce pollution more efficiently (Viana et al. 2019; Srivastava 2024; Srivastava and Biswas 2023). In addition, becoming a major source of air pollution, AI and ML play an important role in understanding the environmental impacts of natural phenomena such as forest fires. Forest fires release large amounts of particulate matter (PM2.5, PM10), carbon monoxide (CO).) and carbon dioxide gas (CO_2) released into the atmosphere. This not only contributes to the deterioration of local air quality. but also, global climate change. Research on Anubhav Srivastava's "Forest Change Survey" highlights the importance of AI in tracking emissions and spread from wildfires (Srivastava et al. 2022; Srivastava et al. 2023a, b, c). AI models can detect smoke and Distribution of pollution using GEE satellite data from sources such as MODIS and VIIRS. These models provide inval.3uable information about the spread of pollutants from wildfires. This makes it possible to assess the air quality in and around affected area immediately Being able to track these pollutants over time is important in issuing health advisories (Vilasan and Kapse 2022). Especially for at-risk populations who are exposed to hazardous air quality conditions. In addition to tracking current pollution levels from wildfires, AI models are used to predict the likelihood of future wildfires based on past data. Vegetation density and meteorology Prediction models have proven effective in identifying areas at high risk of wildfire. This allows officials to provide resources. combustion control Strategic Vegetation Removal and firefighting efforts Allocating and being able to implement preventative measures, such as a proactive approach to the intensity and frequency of wildfires (Vreugdenhil et al. 2022), will reduce, thereby reducing their contribution to air pollution. AI-powered models can also assess long-term environmental impacts of wildfires, such as the amount of carbon dioxide released into the atmosphere and the resulting impacts on the global carbon cycle. The combination of AI, ML and satellite data from platforms like GEE marks a new era in air pollution management. It provides unprecedented accuracy and scalability in detecting human and natural sources of pollution (Bocher et al. 2019; Dubey et al. 2020; Zafar et al. 2020; Dubey et al. 2021a, b). It allows real-time tracking of pollution. AI-powered models to predict future trends Promote policy implementation. In addition, climate change As wildfires increase in frequency and severity, the role of AI in predicting, monitoring, and mitigating their impact on air quality is increasingly important (Srivastava and Sharma 2024; Joshi et al. 2020; Wubalem 2021; Dubey et al. 2025). This comprehensive approach not only increases our ability to respond to immediate air quality concerns; Instead, it supports long-term efforts to reduce pollution and mitigate the effects of global climate change.

8.3 Methodology

The process of utilizing Artificial Intelligence (AI) and Machine Learning (ML) techniques in controlling air pollution, specifically by Google Earth Engine (GEE), goes through a few essential steps. To begin with, the pertinent satellite data from platforms like Sentinel-5P, NASA's Orbiting Carbon Observatory (OCO-2), MODIS, and VIIRS, which provide the respective resolution images and data of airborne substances, such as nitrogen dioxide (NO_2), carbon dioxide (CO_2), carbon monoxide (CO), and the particulates (PM2.5 and PM10) to the integration with a data scientists' regime. The cloud-based environment of GEE procures the borderless flow of the information from the satellite and automatically integrates this data into AI/ML models. Data captured, often over years, or even decades, is pre-processed in the signal's liquid crystals so as to remove the noise and correct the variations of the atmosphere (Rakesh Dubey et al. 2021a, b).

Through the use of machine learning, such as Random Forest (RF), Gradient Tree Boosting (GTB), and Long Short-Term Memory (LSTM), that the information presented is analyzed. Specific criteria determined their selection from a range of models including capacity to process large datasets quickly, high efficiency in constructing relationships of pollutants to environmental factors-for example, weather conditions, land use, and the activity of human beings (Srivastava 2024). After the models are trained using these data, the models generate a regression model where known hazardous levels of pollution are set as the target variable, and the input variables are the satellite data, the atmosphere, and the geographic feature. For monitoring wildfire pollution, MODIS and VIIRS satellites data is used to detect live fire incidents and monitor the movement of smoke and the life hazards. For the prediction of the spread of these pollutants, AI models are used, taking wind speed, temperature, humidity, and topographical features into account. Besides, the wildfire risk is modeled predicting with Random Forest or other classification approaches by integrating information from historical fire outbreaks, the vegetation kind, and climate conditions. Thereby the identification of high-risk zones as well as the development of prevention measures like controlled burns or the allocation of firefighting resources are possible.

Finally, the step of the presented methodology deals with the usage of the prediction models for forecasting the future environmental pollution levels that occur with the current and historical data inputs. These models are trained on GEE's satellite data and meteorological datasets, so that they can make predictions about the pollution trends that allow air quality management to improve. Such predictions are evaluated against the ground sensors that are used as the control group and the satellite imaging data are the actual tools.

8.4 Discussion

AI is coming a long way in providing tools and solutions that can be well used in the visualized data spatial, like Google Earth Engine (GEE) (Hashmi et al. 2024). These models can effortlessly handle many data and richly visualize information; thus, they can have a clear picture of the atmospheric pollution occurring in real-time. For that reason, they are outstanding assets to confront the environmental problems of the world. GEE being the first among equals when satellite data are combined with real-time processing the remote accessing becomes transparent. The process is scalable, so it meets a variety of needs; the tasks range in complexity and involve, for example, streetscapes or entire spatial arrangements.

The ability of ML models, e.g., Random Forest and Gradient Tree Boosting, also to predict the areas with high pollution as well as the forecast of future air quality is probably the most important aspect of it. The models visualize the quality of the air first, using historical pollution data and actual satellite observations, which are then updated in real-time. This knowledge serves as a guide for policymakers and urban planners to timely interventions, such as the controlling of factory pollution and making the traffic regulations to be less causing pollution to become high peaks. Besides, the models are not just some computers calculating inputs, they take into account the weather conditions, land use alterations, and human sector which makes them closer to happenings in the real world, thus, allow them to be better at predictions. Apart from other aspects, wildfire pollution detection monitoring is among the areas where both AI and ML technology are demonstrating the greatest impact. The emission of great pollutants that can be harmful to human beings and other organisms into the atmosphere is caused by wildfires. This leads to the deterioration of local air quality as well as the production of greenhouse gases at a global level that are very harmful. AI models can monitor the dispersion of wildfire emissions in real-time by employing the satellite data from MODIS, and VIIRS, among other sources, which are the applications of AI. On the other hand, satellites are able to take images of the earth that can be analyzed by computers. It is one of the ways through which we benefit a lot from the data satellites take with the help of artificial intelligence (AI). At present, some of the city parks already have solar panels installed in their machines to power the place. This is where the term solar or photovoltaic comes in. They are places that use solar or photovoltaic as a source of power. This aids in measuring the variability of the grasses according to time along with their drying after receiving rain. This also holds for the influence of the herbivores' activities excretions or grazing activities on the grasses' re-growth rate, with different types of grass having different levels of resistance or tolerance to drought. In fact, none of this would be possible without the help of the greens and other parts of the tree. The more effort people make, the more damage they face that can turn out to be irreparable.

8.4.1 Climate Analysis

Studying climate dynamics has become essential to understanding environmental change and its impact on human health and ecosystems. Google Earth Engine (GEE) enables comprehensive analysis of climate data. By combining satellite observations with powerful geospatial tools, using AI and machine learning (ML) methods, this data can now be interpreted with higher accuracy. This is especially true when mapping air pollution. This approach provides important insights into how climate factors such as rainfall and temperature affect air pollution. Therefore, helps in environmental monitoring and decision making.

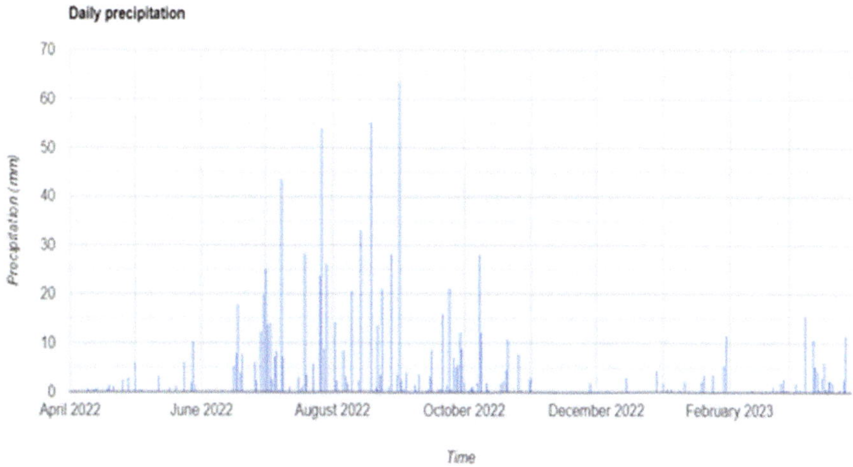

Fig. 8.1 Monthly rain level over any selected study area in mm

8.4.2 *Rain Analysis*

Rainfall acts as an important regulating factor in releasing particles and other pollutants in the form of PM2.5, NO_2, and other kinds of pollutants into the air. By using the TRMM and GPM satellites, time series and spatial rainfall data can be extracted by researchers, which will help AI/ML models to re-veal the relationship between rainfall variables and subsequent changes in the concentrations of air-borne pollutants. High rainfall contributes to the depleted concentrations of airborne pollutants. For example, by using machine learning algorithms such as random forests, the impact of rainfall on pollutant concentration can be forecast, thereby revealing how natural precipitation events cleanse the atmosphere as shown in Fig. 8.1.

8.4.3 *Temperature Analysis*

Temperature fluctuations directly affect air quality, influencing both the formation and dispersion of pollutants. Higher temperatures, for example, can accelerate chemical reactions in the atmosphere, leading to higher concentrations of ground-level ozone (O_3), while colder temperatures may increase the persistence of pollutants such as particulate matter as shown in Fig. 8.2. GEE enables to carry out temperature data analysis for large regions, and AI/ML techniques can be used to design the interaction model of changing temperatures with pollutant behavior. Techniques like gradient boosting or support vector machines would enable scientists to predict how heat waves or the colder season affects pollution, or worded another way, let them describe the climate-pollution interaction with much greater precision.

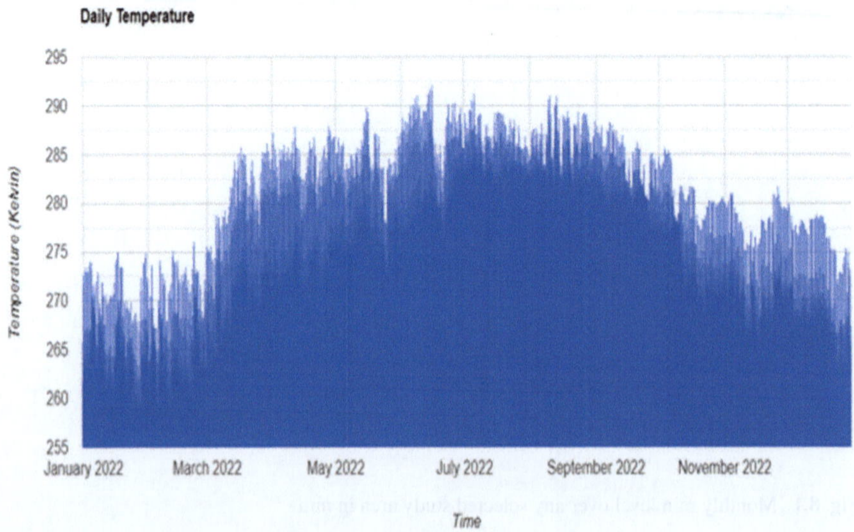

Fig. 8.2 Daily temperature over any selected study area in Kelvin

8.4.4 Greenhouse Gas Monitoring

Monitoring greenhouse gases is essential in determining the impacts caused globally, as well as to human health because of climate change. Such gases involve carbon dioxide (CO_2), and nitrogen dioxide (NO_2). The rising concentration of these gases is associated with the increasing temperatures, airborne pollution and degradation in quality. Google Earth Engine (GEE) provides an integrated platform for GHG emission monitoring by analyzing data from satellites and atmospheric models. Thus, the merging of AI and ML together forms an effective approach to map, track, and predict greenhouse gas concentrations, thereby providing valuable insights into mitigation strategies and pollution control measures.

8.4.5 Carbon Dioxide (CO_2) Monitoring

Carbon dioxide is among the primary greenhouse gases responsible for global warming. Monitoring its concentration is essential in identifying sources of release, with the additional sources primarily being industrial activities, deforestation, and combustion of fossil fuel. This can be accomplished through satellite mission data such as those from OCO-2, which can be analyzed using GEE. Re-searchers can map CO_2 hotspots by utilizing AI/ML algorithms such as random forests or neural networks in predicting future trends of emissions. Correlations between levels of

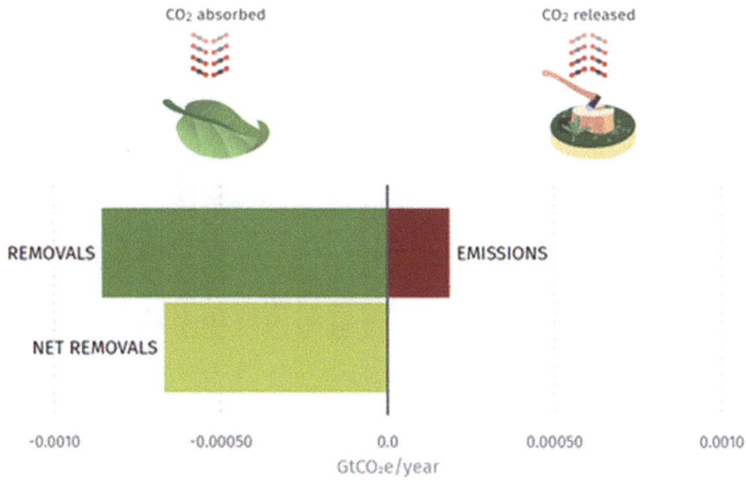

Fig. 8.3 CO_2 absorbed and Emitted computed over any geographical region

CO_2 and human activities can be determined through these models, allowing for a more precise definition of carbon emissions and their environmental impact.

8.4.6 *Nitrogen Dioxide (NO_2) Monitoring*

Because nitrogen dioxide is harmful as an air pollutant and is a precursor to ground-level ozone and smog, the availability of satellite product data, like those from the Sentinel-5P mission, may offer very clear pictures of NO_2 concentration over many parts of the globe as shown in Fig. 8.3. These data can be processed via AI/ML techniques to identify patterns or trends in NO_2 emissions and connect those to sources such as vehicular traffic, industrial processes, or agricultural activities. Such machine learning-based decision tree models, including gradient boosting and support vector machines, help predict the im-pacts of NO_2 pollution levels on air quality and, importantly, map the contribution of NO_2 emissions towards local and global air pollution challenges as shown in Figs. 8.4 and 8.5.

8.4.7 *Wildfires Prediction*

Air pollution due to wildfires: Wildfires are the most important contributors to air pollutants, emitting huge and detrimental gases such as CO_2, carbon monoxide (CO), nitrogen oxides (NOx), and fine particulate matter (PM2.5), which indicate a considerable lowering of air quality. Using Google Earth Engine, scientists can now monitor

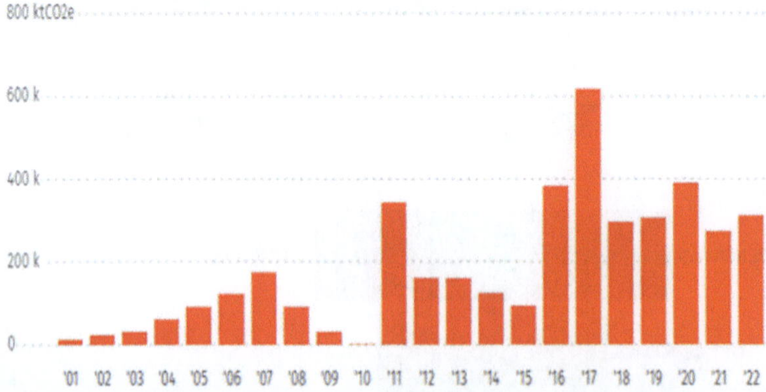

Fig. 8.4 Total emitted green house gases computed over any geographical region

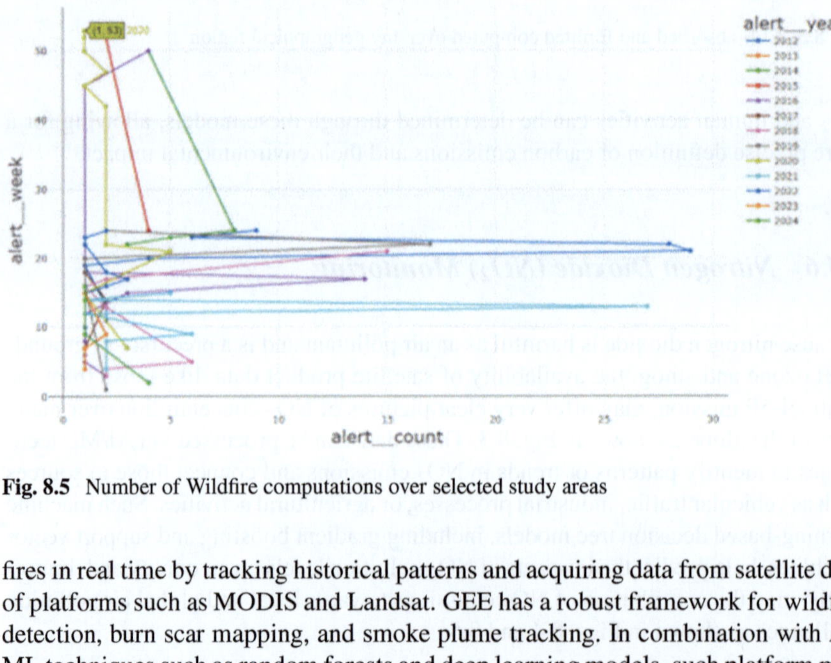

Fig. 8.5 Number of Wildfire computations over selected study areas

fires in real time by tracking historical patterns and acquiring data from satellite data of platforms such as MODIS and Landsat. GEE has a robust framework for wildfire detection, burn scar mapping, and smoke plume tracking. In combination with AI/ML techniques such as random forests and deep learning models, such platform may enable accurate modeling of the behavior, intensity, and emissions of fires. These AI/ML models can be used to draw a spatial distribution of pollutants that are emitted during wildfires; this will help in understanding the contribution of fires toward local and regional air quality degradation. The combination of GEE with AI/ML provides invaluable information on the fire-related transient as well as long-term impacts on atmospheric pollution, which becomes an essential tool for more efficient management and mitigation strategies. The computational output of Wildfires Prediction over Google Earth Engine Code Editor is shown in Fig. 8.6.

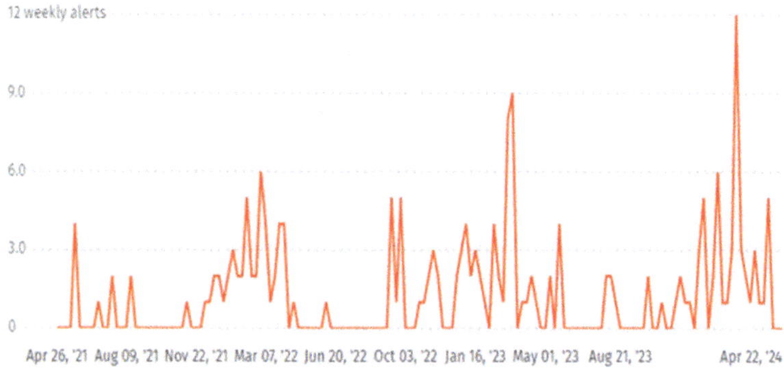

Fig. 8.6 Weekly alerts computed by Google Earth Engine

8.5 Conclusion

The AI and ML technologies when they are used in platforms such as Google Earth Engine make a significant impact in the problem of air pollution including its management and monitoring path. They do that by providing a very detailed description of sources, trends, and future potential for public health and environmental dis-asters, thus empowering the authorities with strong knowledge base to make the right decisions. As the case is with evolving and maturing technologies, the issue of data quality and computational resources becomes the main question for global air quality management. Furthermore, the development of tools to monitor and predict the influence of wildfires on air quality turns particularly important given the threats that climate change brings about. That's not to forget vast amounts of data and specialized knowledge that AI-driven technology possesses, without which our long-term environmental practice will suffer.

References

Bar S, Parida BR, Pandey AC, Kumar N (2022) Pixel-based long-term (2001–2020) estimations of forest fire emissions over the Himalaya. Remote Sens 14(21)

Bharadwaj S, Deepika K, Kumar A, Jaiswal S, Miglani S, Singh D, Fartyal P, Kumar R, Singh S, Singh MP, Gaidhane AM, Kumar B, Jha V (2024) Exploring the artificial intelligence and its impact in pharmaceutical sciences: insights toward the horizons where technology meets tradition. Chem Biol Drug des 104(4):e14639

Bharadwaj S, Dubey R, Biswas S (2020) Determination of the best location for setting up a transmission tower in the city. In: Proceedings of the 2020 International conference on smart innovations in design, environment, management, planning and computing, ICSIDEMPC 2020, pp 63–68

Bocher E, Guillaume G, Picaut J, Petit G, Fortin N (2019) Noisemodelling: an open source GIS based tool to produce environmental noise maps. ISPRS Int J Geo Inf 8(3)

Dash P, Sanders SL, Parajuli P, Ouyang Y (2023) Improving the accuracy of land use and land cover classification of Landsat data in an agricultural watershed. Remote Sens 15(16):4020

Dubey R, Bharadwaj S, Deepika K, Singh A, Siddiqui A, Ali H, Farooqui A (2025) Geospatial analysis and machine learning for vehicular mobility patterns on Indian two-way roads: leveraging geotagged microphone data and modified CNN classifier. Commun Comput Inf Sci 2194:69–81

Dubey R, Bharadwaj S, Zafar MI, Bhushan Sharma V, Biswas S (2020) Collaborative noise mapping using smartphone. Int Arch Photogramm Remote Sens Spat Inf Sci ISPRS Arch 43(B4):253–260

Dubey R, Bharadwaj S, Zafar MI, Biswas S (2021a) Collaborative air quality mapping of different metropolitan cities of India. Int Arch Photogramm Remote Sens Spat Inf Sci ISPRS Arch 43(B4–2021):87–94

Dubey R, Bharadwaj S, Zafar MI, Mahajan V, Srivastava A, Biswas S (2021) GIS mapping of short-term noisy event of Diwali night in Lucknow city. ISPRS Int J Geo Inf 11(1):25

Gorelick N, Hancher M, Dixon M, Ilyushchenko S, Thau D, Moore R (2017) Google Earth Engine: planetary-scale geospatial analysis for everyone. Remote Sens Environ 202:18–27

Hashmi T, Chauhan G, Kumar N, Srivastava A, Sharma H (2024) Exploring artificial intelligence & machine learning in precision agriculture. In: 2024 international conference on emerging technologies and innovation for sustainability (EmergIN), Greater Noida, India, 2024, pp 320–324. https://doi.org/10.1109/EmergIN63207.2024.10961728

Hay JE, Easterling D, Ebi KL, Kitoh A, Parry M (2016) Introduction to the special issue: observed and projected changes in weather and climate extremes. Weather Clim Extrem 11:1–3

Joshi G, Pal B, Zafar I, Bharadwaj S, Biswas S (2020) Developing intelligent fire alarm system and need of UAV. Lect Notes Civ Eng 51:403–414

Li W, Zhao D, Shen Y, Zhang K (2020) Modeling Australian TEC maps using long-term observations of australian regional GPS network by artificial neural network-aided spherical cap harmonic analysis approach. Remote Sens 12(23):3851

Liu Y, Liu L, Yan Y (2020) Network topology change detection based on statistical process control. In: Proceedings of the 2020 4th High performance computing and cluster technologies conference & 2020 3rd International conference on big data and artificial intelligence. ACM, New York, NY, USA, pp 145–151

Masood R (2024) AI-driven real-time monitoring of earthquakes using google earth engine. In: 2024 International conference on artificial intelligence and quantum computation-based sensor application (ICAIQSA), pp 1–5. https://doi.org/10.1109/ICAIQSA64000.2024.10882382

Srivastava A, Bharadwaj S, Dubey R, Sharma VB, Biswas S (2022) Mapping vegetation and measuring the performance of machine learning algorithm in LULC classification in the large area using sentinel-2 and landsat-8 datasets of Dehradun as a test case. Int Arch Photogramm Remote Sens Spatial Inf Sci XLIII-B3-2022: 529–535. https://doi.org/10.5194/isprs-archives-XLIII-B3-2022-529-2022

Srivastava A, Biswas S (2023) Analyzing land cover changes over Landsat-7 data using Google Earth Engine. In: Proceedings of the 3rd International conference on artificial intelligence and smart energy, ICAIS 2023, Feb, pp 1228–1233. https://doi.org/10.1109/ICAIS56108.2023.10073795

Srivastava A, Dubey R, Biswas S (2023a) Comparison of sentinel and Landsat data sets over Lucknow region using gradient tree boost supervised classifier. In: Noor A, Saroha K, Pricop E, Sen A, Trivedi G (eds) Proceedings of third emerging trends and technologies on intelligent systems. ETTIS 2023. Lecture notes in networks and systems, vol 730. Springer, Singapore. https://doi.org/10.1007/978-981-99-3963-3_18

Srivastava A, Umrao S, Biswas S (2023b) Exploring forest transformation by analyzing spatial-temporal attributes of vegetation using vegetation indices. Int J Adv Comput Sci Appl 14(5):1110–1117. https://doi.org/10.14569/IJACSA.2023.01405114

Srivastava A, Umrao S, Biswas S, dubey R, Zafar MI (2023c) FCCC: Forest cover change calculator user interface for identifying fire incidents in forest region using satellite data. Int J Adv Comput Sci Appl 14(7):948–959. https://doi.org/10.14569/IJACSA.2023.01407103

Srivastava A (2024) Temporal analysis of multi-spectral instrument level and surface reflectance data sets for seasonal variation in land cover dynamics by using Google Earth Engine. Geod Cartogr 50(4):162–178. https://doi.org/10.3846/gac.2024.20106

Srivastava A, Sharma H (2024) AI-driven environmental monitoring using Google Earth Engine. In: IoT sensors, ML, AI and XAI: empowering a smarter world. Springer Nature Switzerland, Cambridge, pp 375–385

Srivastava A (2025) Science of the total environment quantifying forest degradation rates and their impact on environmental condition in Dehradun, India. Sci Total Environ 992:179987. https://doi.org/10.1016/j.scitotenv.2025.179987

Viana CM, Girão I, Rocha J (2019) Long-term satellite image time-series for land use/land cover change detection using refined open source data in a rural region. Remote Sens 11(9):1104

Vilasan RT, Kapse VS (2022) Evaluation of the prediction capability of AHP and F-AHP methods in flood susceptibility mapping of Ernakulam district (India). Nat Hazards 112(2):1767–1793

Vreugdenhil M, Greimeister-Pfeil I, Preimesberger W, Camici S, Dorigo W, Enenkel M, van der Schalie R, Steele-Dunne S, Wagner W (2022) Microwave remote sensing for agricultural drought monitoring: Recent developments and challenges. Front Water 4

Wubalem A (2021) Landslide susceptibility mapping using statistical methods in Uatzau catchment area, Northwestern Ethiopia. Geoenvironmental Disasters 8(1):1

Yang F, Guo J, Zhang C, Li Y, Li J (2021) A regional zenith tropospheric delay (ZTD) model based on GPT3 and ANN. Remote Sens 13(5):838

Zafar MI, Dubey R, Bharadwaj S, Kumar A, Paswan KK, Srivastava A, Tiwary SK, Biswas S (2023) GIS based road traffic noise mapping and assessment of health hazards for a developing urban intersection. Acoustics 5(1):87–119. https://doi.org/10.3390/acoustics5010006

Chapter 9
Harnessing Machine Learning and Deep Learning Techniques for Environmental Pollution Prediction: A Comprehensive Review

Masood U. Rehman

Abstract Environmental pollution poses significant challenges to public health and well-being. Accurate prediction and monitoring of pollution levels are essential to mitigate its adverse effects. Recent advancements in machine learning (ML) and deep learning (DL) techniques have revolutionized predictive modelling by enabling robust analysis of complex and high-dimensional environmental data. The present study provides a comprehensive examination of ML and DL applications in predicting various pollution types including air, water, noise, and soil pollution. Key ML and DL methodologies and their performance metrics are discussed to highlight their effectiveness and limitations. ML techniques such as regression models, decision trees, and ensemble methods offer simplicity and efficiency while DL architectures, including convolutional neural networks (CNNs) and recurrent neural networks (RNNs), excel in processing spatial and temporal data. Integration with Internet of Things (IoT) devices for real-time monitoring and the emergence of explainable AI (XAI) further enhance the applicability of these models.

Keywords Environmental pollution · Machine learning · Deep learning · Prediction models · Air quality · Water quality · Noise pollution · Smart sensors

9.1 Introduction

Environmental pollution is an escalating global challenge with far-reaching adverse consequences for public health and well-being. It manifests in various forms such as air, water, noise, and soil pollution. These environmental pollutions are intricately linked to rapid urbanization, industrialization, and unsustainable human practices across the globe. Air pollutions are characterized by pollutants such as particulate matter (PM2.5 and PM10), nitrogen oxides (NOx), and ozone (O3) (Jyethi 2016).

M. U. Rehman (✉)
College of Computing and Informatic, Saudi Electronic University, Dammam, Saudi Arabia
e-mail: m.rehman@seu.edu.sa

S. K. Tiwari et al. (eds.), *AI-Driven Environmental Pollution Management*, Climate Risks and Solutions, https://doi.org/10.1007/978-3-031-96243-1_9

Numerous studies have shown a strong link to respiratory and cardiovascular diseases (Dominski et al., 2021). Water pollution is caused by industrial discharge, agricultural runoff, and untreated sewage, contaminates vital water sources and leads to severe health hazards and ecosystem disruption (Sharma et al. 2024). Noise pollution is mostly prevalent in urban areas due to transportation, construction, and industrial activities which has a detrimental impact on mental health and quality of life (Tiwari et al. 2023a, b; Chauhan et al. 2024). Soil pollution is driven by pesticide use, heavy metals, and industrial waste, compromises agricultural productivity and food security (Hossain et al. 2020). The urgency to address these challenges has been underscored by alarming trends like rise in global temperature, biodiversity loss, and increasing prevalence of pollution-related illnesses. Governments, researchers, and organizations worldwide are focusing on sustainable solutions to monitor, predict, and mitigate environmental pollution. One pivotal aspect of these efforts is leveraging advanced computational techniques to enhance the accuracy and efficiency of pollution prediction systems.

Nowadays, predictive modeling has emerged as a powerful tool in understanding and addressing environmental pollution. The predictive models analyze the trends and patterns from the pollution dataset for the prediction and formulate actionable strategies for mitigation. Traditional statistical approaches have laid the groundwork for understanding pollution dynamics (Rajula et al. 2020). However, their limitations are restricted to handle large-scale and high-dimensional data which gives the necessity for the adoption of machine learning (ML) and deep learning (DL) techniques (Tiwari et al. 2022a, b, c). Machine Learning techniques such as linear regression, K-Nearest Neighbor (KNN), decision trees, random forests, and support vector machines (SVMs) offer a robust solution for the classification, regression, and clustering problem statements (Liu et al. 2022a, b). These methods are particularly effective in situations where the relationships between variables are complex and nonlinear. On the other hand, DL techniques such as artificial neural network (ANN), convolutional neural networks (CNNs), recurrent neural networks (RNNs), auto encoders (AE), and transformers network are adept at processing high-dimensional data with spatial and temporal dependencies (Zhang et al. 2022). These capabilities make them invaluable for environmental pollution predict tasks.

The success of predictive models hinges on the availability and quality of data. Environmental monitoring generates vast amounts of data from satellite imagery, IoT-enabled sensors, and public health records (Hossain et al. 2024; Divine et al. 2024). The data poses significant challenges in terms of preprocessing, integration, and analysis. The advancement in data science have enabled the extraction of meaningful insights from these complex datasets. Real-time monitoring facilitated by IoT devices has revolutionized data collection processes. For example, air quality sensors deployed across urban areas provide continuous updates on pollutant concentrations and allows predictive models to deliver near-instantaneous forecasts (Karagulian et al. 2019). Similarly, water quality sensors measure parameters like pH, turbidity, and dissolved oxygen as a critical input for contamination prediction models (Cloete et al. 2016). For noise pollution analysis, the Central Pollution Control Board (CPCB) in India has developed several monitoring stations to monitor noise levels (Tiwari et al. 2023a, b). The integration of these technologies with predictive algorithms is driving significant progress in environmental management.

The present review study seeks to provide actionable insights for researchers, practitioners, and policymakers regarding environmental pollution. The findings of the study underscore the transformative potential of ML and DL algorithms in addressing the environmental pollution predictive challenges. Figure 9.1 shows the conceptual framework of the present paper. This review provides a comprehensive exploration of ML and DL techniques applied to environmental pollution prediction. It aims to:

Categorize various environmental pollution such as air, water, noise, and soil pollution.

Discuss the machine learning applications across different pollution domains.

Discuss the deep learning applications across different pollution domains.

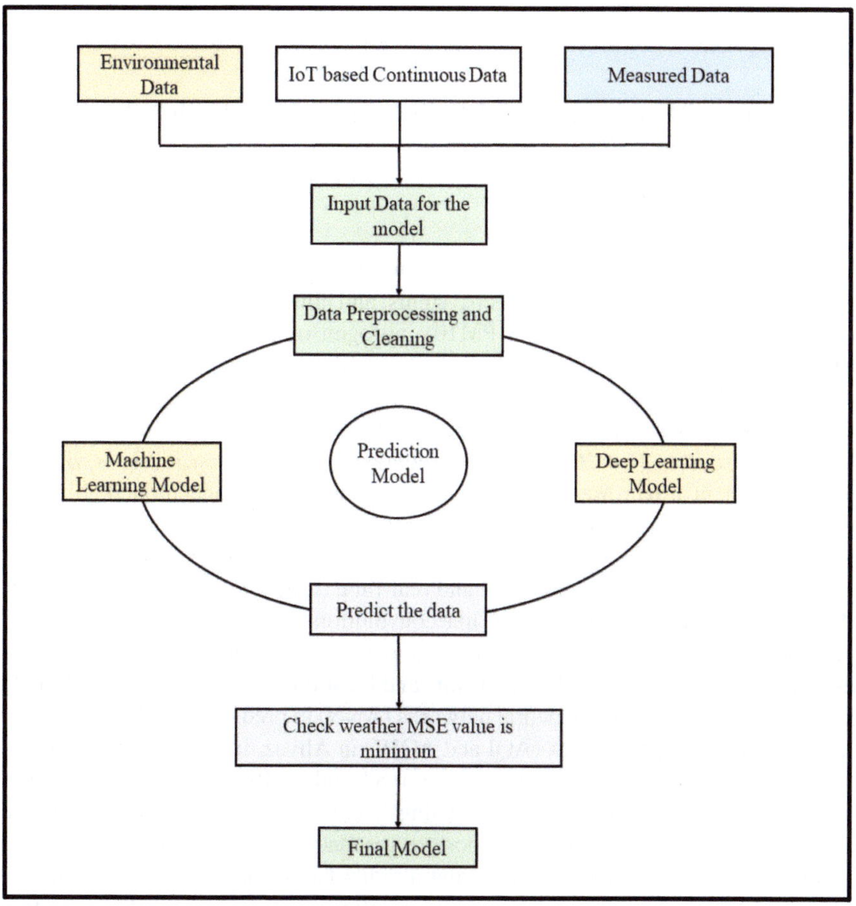

Fig. 9.1 Conceptual framework of the present study

Examine the role of emerging technologies such as IoT and explainable AI (XAI) in enhancing predictive modeling.

Identify existing challenges related to data quality issues, model interpretability, and scalability constraints.

Outline future directions for potential interdisciplinary collaboration.

The subsequent sections of this paper are organized as follows: Sect. 9.2 delves into the types of environmental pollution and the specific prediction needs associated with each domain. Section 9.3 provides an overview of ML techniques. Section 9.4 explores DL methodologies and emphasizes on their superiority in handling complex data. Section 9.5 offers a comparative analysis of ML and DL approaches. Section 9.6 discusses the integration of smart sensors and IoT in real-time pollution monitoring. Section 9.7 addresses the challenges and limitations in applying ML and DL to pollution prediction. Finally, Sects. 9.8 conclude the paper with key takeaways and future research directions.

9.2 Types of Environmental Pollution and Prediction Needs

9.2.1 Air Pollution

Air pollution is one of the most pervasive forms of environmental degradation which significantly impact public health, ecosystems, and climate systems. Pollutants such as particulate matter (PM2.5 and PM10), nitrogen oxides (NOx), sulfur dioxide (SO_2), carbon monoxide (CO), and ground-level ozone (O_3) are the primary contributors (Jyethi 2016; Akbarzadeh et al. 2018). These pollutants mostly originate from vehicular emissions in precedence with industrial activities, agricultural practices, and natural sources (Lu 2020). The prediction of air pollution levels is vital for public health in urban areas where pollution tends to be concentrated. Machine learning models have demonstrated considerable promise in predicting air pollution. Techniques such as support vector machines (SVM), random forests, and gradient boosting machines analyze historical and real-time data from sensors and satellites. More advanced deep learning models like convolutional neural networks (CNNs) and recurrent neural networks (RNNs) are particularly effective in capturing spatial and temporal dependencies, enabling real-time and location-specific predictions. Maleki et al. (2019) applied artificial neural networks (ANN) to predict air pollutant concentrations and air quality indices (AQI and AQHI) in Ahvaz, Iran. The model showed strong performance with an average $R = 0.87$ and RMSE of 59.9 across pollutants, providing a viable tool for mitigating health risks in highly polluted urban areas. Leong et al. 2020 utilized support vector machines (SVM) to predict the air pollution index (API) in Malaysia. Using a radial basis function kernel, the model achieved high accuracy with $R^2 = 0.98$, significantly improving API calculation efficiency while handling outlier removal and non-linear data dynamics effectively.

Open-source datasets like AirNow and OpenAQ provide comprehensive information for these models, ensuring high prediction accuracy and reliability.

9.2.2 Water Pollution

Water pollution arises from industrial discharge, agricultural runoff, and untreated sewage (Sharma et al. 2024; Singh et al. 2020). Contaminants such as heavy metals, pathogens, and chemical pollutants disrupt aquatic ecosystems and pose severe health risks to human populations. Critical parameters for water quality assessment include pH levels, turbidity, dissolved oxygen (DO), and chemical oxygen demand (COD) (Kothari et al. 2021). Accurate prediction of these parameters is essential for sustainable water resource management and pollution control. Predictive models for water pollution utilize sensor data and historical records to forecast contamination levels. Machine learning techniques are commonly employed for parameter prediction. Additionally, deep learning architectures such as long short-term memory (LSTM) networks are applied for time-series analysis of water quality data over extended periods. Ahmed et al., (2019) developed a WDT-ANFIS model combining wavelet de-noising with adaptive neuro-fuzzy inference systems for water quality prediction in Malaysia's Johor River. Their approach improved accuracy by 0.5–5% across stations, achieving $R^2 \geq 0.9$, outperforming traditional MLP and RBF networks. Liu et al. 2022a, b employed the Geodetector model to analyze heavy metal contamination in Xiangjiang River basin soils, focusing on interactions between industrial and natural factors. The study revealed that Cd, As, and Pb levels were strongly influenced by mining and smelting activities, while Hg and Cr were more dependent on pH and soil type. Spatial visualization identified high-risk areas, particularly in the mid-to-upper reaches, enabling targeted interventions. Wang et al. 2019 proposed an LSTM-based AI system for tracing water pollution sources in Shandong, China. By combining cross-correlation and Apriori algorithms, the system identified pollution trends and traced industrial sources with high accuracy. Results demonstrated that AI integration improves monitoring and prevention of aquatic pollution at large scales. Datasets from organizations like the Environmental Protection Agency (EPA) and World Bank serve as valuable resources for developing and validating these models.

9.2.3 Noise Pollution

Noise pollution primarily originates from transportation, industrial activities, and urban construction (Patel et al. 2024a, b). Long-exposure to noise pollution can lead to several health issues. Chronic exposure to high noise levels can lead to hearing loss, stress, and cardiovascular issues. Monitoring and predicting noise pollution patterns are crucial for designing effective mitigation strategies and urban planning (Tiwari et al. 2023a, b). IoT-enabled sensors combined with machine learning models provide robust solutions for noise level prediction. Algorithms such as k-nearest neighbors (KNN), SVMs, and ensemble methods analyze acoustic data to identify noise hotspots and predict temporal variations (Tiwari et al. 2022a, b, c; Tiwari et al. 2021; Tsalera et al. 2020). Advanced DL models like CNNs are employed for audio signal processing for the analysis of noise intensity and frequency patterns (Mustaqeem and Kwon 2019). The integration of these models with GIS (Geographic Information Systems) further enhances their utility in urban noise mapping (Patel et al. 2024a, b).

9.2.4 *Soil Pollution*

Soil pollution is caused by the accumulation of hazardous substances like pesticides, heavy metals, and industrial waste which poses a significant threat to agricultural productivity and ecological balance (Hossain et al. 2020). Key indicators of soil health include nutrient content, pH levels, moisture, and contaminant concentrations. Accurate prediction of soil pollution levels is essential for sustainable land management and agricultural planning. Li et al., 2019 developed an RNN-Dempster Shafer evidence theory model (RNN-DS) for water quality prediction in the Qiantang River, China. The hybrid model demonstrated superior accuracy and stability compared to traditional methods, reducing false alarms and enhancing predictions of indicators like DO and phosphorus, crucial for early risk management. Jia et al. 2021 proposed a novel approach for mapping arsenic soil contamination in agricultural fields by combining high-resolution aerial imaging (HRAI) with machine learning algorithms. Using 1068 samples from Zhongxiang, China, they demonstrated that the extreme random forest (ERF) model achieved superior accuracy (F1-score 0.8 for high-risk areas) compared to traditional kriging methods, offering a cost-effective solution for environmental monitoring. Wang et al. 2020 compared random forest (RF) and land use regression (LUR) models to predict heavy metal concentrations in Tianjin's agricultural soils. The RF model significantly outperformed the LUR model, achieving R^2 values above 0.9 for metals like Cd and as, revealing critical pollution sources such as sewage irrigation and atmospheric deposition.

9.3 Machine Learning Algorithms for Environmental Pollution Prediction

Machine learning algorithms have become indispensable for environmental pollution prediction due to their ability to handle high-dimensional complex data. Figure 9.2 shows the working principal of machine learning techniques. This section delves into some of the most widely used ML algorithms, including ARIMA, SVM, MLR,

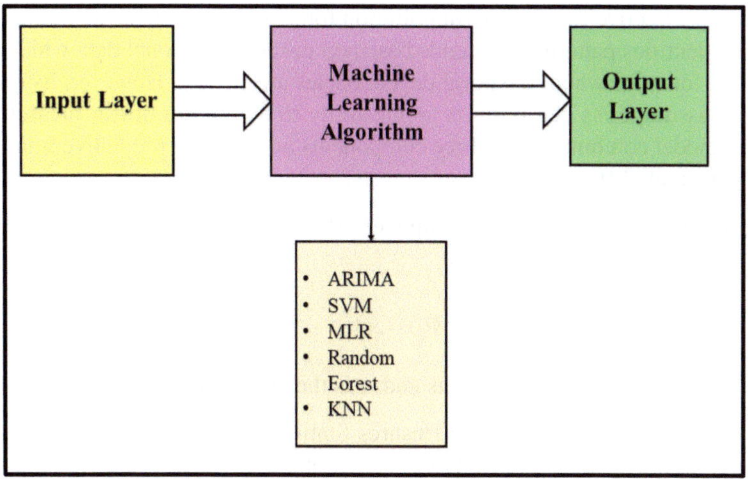

Fig. 9.2 Working principal of machine learning techniques

Table 9.1 Limitations and advantages of machine learning algorithms

Model	Advantages	Limitations
ARIMA	Effective for time-series data; interpretable results	Assumes linear relationships; not suitable for large or complex datasets
SVM	Works well with high-dimensional data; robust to overfitting	Computationally expensive; sensitive to kernel choice
MLR	Simple and interpretable; effective for linear relationships	Limited to linear problems; prone to multicollinearity issues
Random Forest	Handles non-linear data well; robust against overfitting	Requires large datasets; less interpretable compared to single decision trees
KNN	Simple implementation; effective for small datasets	Computationally expensive for large datasets; sensitive to choice of k

Random Forest, and KNN, outlining their methodologies, applications, and effectiveness in addressing environmental challenges. Table 9.1 represents the limitations and advantages of machine learning algorithms.

9.3.1 Auto-Regressive Integrated Moving Average (ARIMA)

ARIMA is a time-series forecasting method that combines autoregressive (AR) and moving average (MA) components to make the data stationary. ARIMA models are particularly effective for short-term pollution forecasting (Tiwari et al. 2022a, b, c). ARIMA identifies patterns and trends based on past observational data which make it suitable for tasks where temporal dependencies are critical. However, its reliance on linear assumptions can limit its applicability to highly non-linear datasets. The ARIMA model combines these three components into a single predictive framework (Shumway et al. 2017).

1. **Autoregression (AR)**: This Component Predicts the Current Value as a Linear Combination of Past Values:

$$y_t = \phi_1 y_{t-1} + \phi_2 y_{t-2} + \ldots + \phi_p y_{t-p} + \epsilon_t \tag{9.1}$$

where $\phi_1, \phi_2, \ldots, \phi_p$ are coefficients and ϵ_t is the error term.

2. **Differencing (I)**: Differencing Ensures Stationarity by Transforming the Data into Differences Between Consecutive Observations:

$$y_t' = y_t - y_{t-1} \tag{9.2}$$

3. **Moving Average (MA)**: This Component Models the Error Term as a Linear Combination of Past Error Terms:

$$y_t = \theta_1 \epsilon_{t-1} + \theta_2 \epsilon_{t-2} + \ldots + \theta_q \epsilon_{t-q} + \varepsilon_t \tag{9.3}$$

where $\theta_1, \theta_2, \ldots, \theta_q$ are coefficients.

9.3.2 Support Vector Machine (SVM)

SVM is a supervised learning algorithm that excels in classification and regression tasks. It works on the principal of finding the hyperplane that best separates data points into distinct classes (Stitson et al. 1996). SVM is widely used environmental pollution prediction problems. The kernel functions enhance the SVM's ability to handle non-linear relationships in data. Despite its robustness, SVM can be computationally intensive for large datasets. In classification problem statement SVM maximizes the margin between data points of different classes (Tiwari et al. 2022a, b, c):

$$Margin = \frac{2}{||w||}$$

where w is the weight vector is perpendicular to the hyperplane. The optimization problem is expressed as:

$$\min \frac{1}{2} ||w||^2 subjecttoy_i(w.x_i + b) \geq 1$$

for all $i = 1, \ldots, N_i$, where y_i represents class labels and x_i are feature vectors. For regression tasks, SVM uses Support Vector Regression (SVR), which minimizes the error within an ϵ margin:

$$\min \frac{1}{2}||w||^2 + C\sum_{i=1}^{N} \xi_i \tag{9.4}$$

where ξ_i are slack variables for error tolerance and C controls the trade-off between margin width and prediction error.

9.3.3 *Multiple Linear Regression (MLR)*

MLR is a statistical technique used to model the relationship between one dependent variable and multiple independent variables. The MLR is straightforward and interpretable model, and its performance diminishes when dealing with non-linear complex interactions among variables (Tiwari et al. 2024). Multiple Linear Regression (MLR) is a foundational algorithm that models the relationship between a dependent variable (y) and multiple independent variables ($x_1, x_2 \ldots, \ldots, x_p$). The model assumes a linear relationship expressed as (Keith 2019):

$$y = \beta_0 + \beta_1 x_1 + \beta_2 x_2 + \ldots + \beta_p x_p + \epsilon \tag{9.5}$$

where β_0 is the intercept, $\beta_1, \beta_2, \ldots, \beta_p$ are coefficients, and ϵ is the error term. MLR uses the least squares method to minimize the sum of squared residuals:

$$\min \sum_{i=1}^{n} (y_i - \hat{y})^2 \tag{9.6}$$

where $\hat{y}$ is the predicted value. MLR is commonly used for basic environmental predictions, such as correlating meteorological variables with air pollutant concentrations.

9.3.4 *Random Forest (RF)*

Random Forest is an ensemble learning method that constructs multiple decision trees during training and outputs the average prediction for regression tasks or the majority vote for classification tasks. RF is highly effective in handling large datasets with numerous features and is widely used in soil contamination prediction and air quality analysis (Wang et al. 2020; Zimmerman et al. 2018). Its ability to handle missing data and mitigate overfitting makes it a preferred choice for many environmental applications.

Tree Construction: Each decision tree splits the dataset based on the feature that minimizes impurity. For classification, Gini impurity is often used:

$$G = \sum_{k=1}^{K} p_k(1 - p_k) \tag{9.7}$$

where p_k is the proportion of samples belonging to class k.

Prediction Aggregation: For regression, the final prediction is the average of individual tree outputs:

$$\hat{y} = \frac{1}{T}\sum_{t=1}^{T} \hat{y}_t \tag{9.8}$$

where T is the number of trees and $\hat{y}_t$ is the prediction of the t-th tree.

9.3.5 *K-Nearest Neighbors (KNN)*

KNN is a simple yet powerful algorithm that classifies a data point based on the majority class of its nearest neighbors or predicts continuous values by averaging the values of the neighbors. KNN has been applied in tasks like noise level prediction and localized air quality assessments (Tsalera et al. 2020). Its simplicity and ease of implementation are advantages, but its performance can degrade with high-dimensional data or imbalanced datasets. For classification, KNN assigns a data point to the majority class among its neighbors:

$$\hat{y} = mode(y_1, y_2, \ldots, y_k$$

For regression, the prediction is the average of the neighbors' values:

$$\hat{y} = \frac{1}{k}\sum_{i=1}^{k} y_i$$

The distance between data points is typically calculated using the Euclidean distance:

$$d\left(x_i, x_j\right) = \sqrt{\sum_{m=1}^{M}\left(x_{im} - x_{jm}\right)^2} \tag{9.9}$$

Each of these algorithms offers unique strengths and limitations, making them suitable for specific pollution prediction tasks. The selection of an appropriate algorithm depends on the nature of the data, the desired outcome, and computational constraints.

9.4 Deep Learning Algorithms for Environmental Pollution Prediction

There are several limitations of machine learning algorithms such as low robustness and it is not able to handle big data which is represented in Fig. 9.3. To overcome this, deep learning techniques has been introduced. Table 9.2 represents the advantages and limitations of deep learning algorithms.

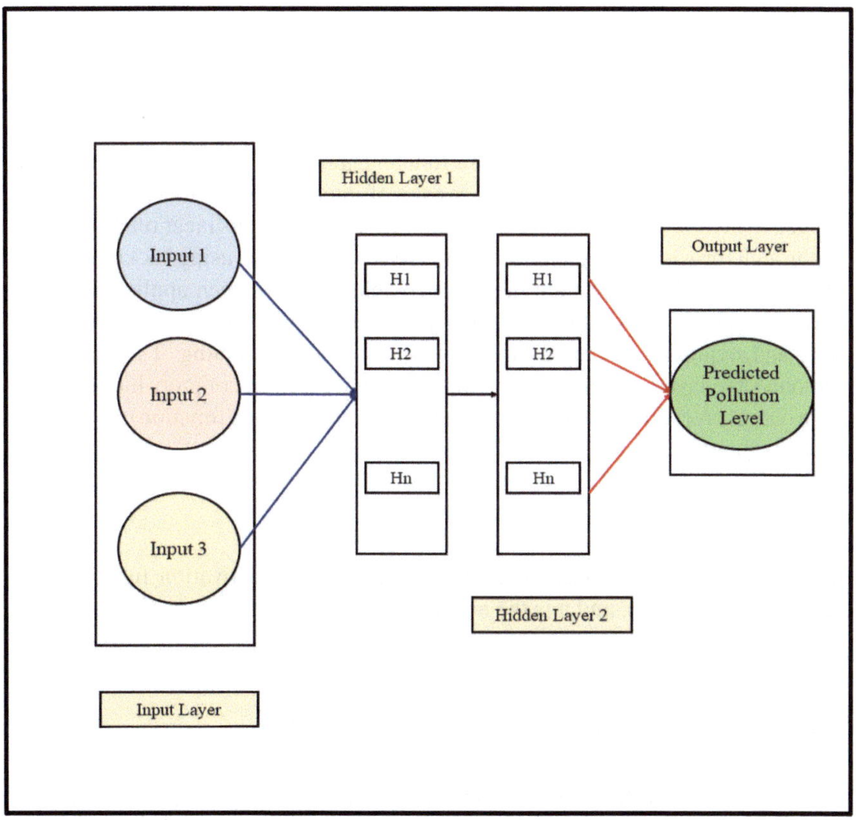

Fig. 9.3 Working principal of deep learning techniques

Table 9.2 Limitations and advantages of deep learning algorithms

Model	Advantages	Limitations
Artificial neural network (ANN)	Handles non-linear relationships and large datasets effectively	Prone to overfitting, requires careful tuning of hyperparameters
Convolutional neural network (CNN)	Excels in spatial data analysis such as images or spatial pollution maps	Computationally expensive, requires large datasets for training
Recurrent neural network (RNN)	Effective for sequential data, such as time-series pollution data	Suffers from vanishing gradient problems, challenging to train for long sequences
Autoencoder (AE)	Useful for dimensionality reduction and noise elimination in datasets	Limited interpretability may not capture all relevant features
Transformer network	Superior in handling long-term dependencies and parallel computation	High computational and memory requirements, complex architecture

9.4.1 Artificial Neural Network

Artificial Neural Networks (ANNs) are fundamental deep learning models inspired by the human brain's structure. They consist of interconnected nodes or "neurons" arranged in layers. The ANN networks can learn complex non-linear relationships in data through forward propagation and iterative weight updates using backpropagation (Tiwari et al., 2024). ANNs are highly adaptable and have been applied to various environmental domains. However, their performance depends on hyperparameter tuning and requires substantial labeled data for accurate training. They consist of interconnected layers of neurons, including an input layer, one or more hidden layers, and an output layer. Each neuron applies a non-linear activation function to a weighted sum of its inputs (Gupta and Raza 2019):

$$z = w^T x + b, a = \sigma(z) \tag{9.10}$$

where w represents weights,b is the bias term, σ is the activation function (e.g., ReLU, Sigmoid, Tanh), and a is the activated output.

9.4.2 Convolutional Neural Networks (CNNs)

Convolutional Neural Networks (CNNs) are particularly adept at handling grid-like data structures like images or spatial datasets. The convolutional layers extract hierarchical features by applying filters to the input data, while pooling layers reduce the spatial dimensions (Tiwari et al. 2022a, b, c). A convolution operation involves a filter f sliding over the input I:

$$S(i,j) = (I * f)(i,j) = \sum_m \sum_n I(m,n)f(i-m, j-n) \tag{9.11}$$

The layers typically include, (1) Convolutional Layers which extract spatial features using filters, (2) Pooling Layers that reduce dimensionality via max or average pooling, and (3) Fully Connected Layers which aggregate features for final predictions.

9.4.3 Recurrent Neural Networks (RNNs)

Recurrent Neural Networks (RNNs) are designed to process sequential data by maintaining a memory of past inputs through their hidden states. However, vanilla RNNs face issues with long-term dependencies due to vanishing gradients. To address this, advanced variants such as Long Short-Term Memory (LSTM) networks and Gated Recurrent Units (GRU) were developed (Tiwari et al., 2022a, b, c). Unlike ANNs, RNNs incorporate temporal dependencies by maintaining a hidden state h_t that evolves over time:

$$h_t = \sigma(W_h h_{t-1} + W_x x_t + b) \tag{9.12}$$

where W_h and W_x are weight matrices, and x_t is the input at time t.

9.4.4 Autoencoders (AEs)

Autoencoders are unsupervised learning models used for data compression and feature extraction. They consist of two parts (Li et al. 2023):

- **Encoder**: Compresses the input into a lower-dimensional representation (latent space).

$$h = \sigma(W_e x + b_e) \tag{9.13}$$

- **Decoder**: Reconstructs the input from the compressed representation.

$$\hat{x} = \sigma(W_d x + b_d) \tag{9.14}$$

Variants such as Denoising Autoencoders can remove noise from environmental datasets, improving data quality. Another extension, Variational Autoencoders

(VAEs), is used for generating synthetic environmental data, such as simulating pollutant dispersion patterns.

9.4.5 Transformer Networks

Transformers are sequence-to-sequence architectures that rely on self-attention mechanisms rather than recurrence. The attention mechanism computes the importance of each input token in predicting the output (Jaderberg et al. 2015):

$$Attention(Q, K, V) = softmax\left(\frac{QK^T}{\sqrt{d_k}}\right)V \tag{9.15}$$

where Q, K, and V represent queries, keys, and values derived from the input.

9.5 Comparative Analysis of ML and DL Models

Machine learning (ML) and deep learning (DL) have emerged as pivotal technologies in environmental pollution prediction. Each approach has its strengths and trade-offs. Table 9.3 shows a comparative analysis of machine learning and deep learning algorithms. This section explores the differences between ML and DL models based on factors such as data requirements, computational complexity, interpretability, and predictive accuracy.

9.5.1 Data Requirements and Scalability

ML algorithms are well-suited for smaller datasets with structured features. They excel in environments where feature engineering can enhance the quality of input data. For instance, random forests have demonstrated significant success in predicting air quality indices using meteorological and emission data. However, ML models often struggle with unstructured data such as images or audio signals which require extensive preprocessing. In contrast, DL models thrive on large and high-dimensional datasets. Their ability to learn hierarchical feature representations directly from raw data makes them ideal for tasks involving satellite imagery, audio signals, or spatiotemporal analysis. For example, CNNs have been widely applied in remote sensing to detect and predict pollution patterns from satellite images. The usage of the pre-trained models elevates the level of accuracy and scalability (Nallakaruppan et al. 2022).

Table 9.3 A comparative analysis of machine learning and deep learning algorithms

Aspect	Machine learning (ML)	Deep learning (DL)
Data requirements	Performs well with small to medium datasets	Requires large datasets to generalize well
Feature engineering	Manual feature selection and engineering required	Learns features automatically from raw data
Complexity	Lower computational complexity	Higher computational demands due to deep architectures
Accuracy	Moderate to high, depending on the model	Generally higher accuracy for large and complex data
Interpretability	Easier to interpret models like decision trees	Limited interpretability, often seen as a "black box"
Applications	Effective for structured data analysis	Excels in unstructured data like images, videos, and sequences
Training time	Faster training times	Slower training due to complex computations
Scalability	More scalable for simple tasks	Requires significant resources for scalability

9.5.2 Computational Complexity

ML models are typically less computationally intensive compared to DL models. Algorithms like multiple linear regression (MLR) and k-nearest neighbors (KNN) can be implemented with minimal hardware requirements which make them accessible for smaller-scale applications or in resource-constrained environments. This advantage is particularly relevant for real-time air or noise quality monitoring using IoT devices. The handling of the time-series data is still more challenging (Nallakaruppan and Iiango 2017). DL models on the other hand require significant computational resources due to their deep architecture and large parameter space. Training a CNN or a long short-term memory (LSTM) network often necessitates GPUs or TPUs along with a substantial amount of time. However, the trade-off is their unparalleled performance in handling complex, non-linear relationships in data, which ML models might not capture effectively.

9.5.3 Interpretability and Explainability

Interpretability is a crucial factor in environmental applications where insights from models influence public policies and health decisions. ML models are relatively interpretable which allows the stakeholders to understand the contribution of individual features for predictions. For example, in soil pollution studies, feature importance rankings derived from random forests provide actionable insights into key pollutants. DL models, while powerful, often act as "black boxes," making it challenging to interpret their inner workings. The emergence of explainable AI (XAI) techniques, such as SHAP (SHapley Additive exPlanations) values and attention mechanisms, has begun to address this limitation. These methods enable a better understanding of DL model outputs, enhancing their trustworthiness in critical applications such as water contamination forecasting. The financial forecasting for loan approvals is proposed in the work (Nallakaruppan et al. 2024).

9.5.4 *Predictive Accuracy and Flexibility*

DL models generally outperform ML models in predictive accuracy, particularly for tasks involving unstructured data or complex spatiotemporal relationships. For example, RNNs have shown superior performance in time-series forecasting of air pollution levels, capturing intricate temporal dependencies. However, ML models maintain an edge in scenarios with smaller datasets or where computational efficiency is paramount. For example, MLR and random forests provide reliable predictions in noise pollution studies with limited sensor data.

9.5.5 *Hybrid Approaches*

The integration of ML and DL techniques has given rise to hybrid models that leverage the strengths of both approaches. For instance, ML models can preprocess and structure data before feeding it into DL architectures for deeper analysis. Alternatively, DL models can extract high-level features, which are then used by ML algorithms for classification or regression tasks. Such hybrid models have shown promise in applications like water quality prediction and urban noise mapping.

9.6 Integration with Smart Sensors and IoT

The integration of smart sensors and Internet of Things (IoT) technologies with machine learning (ML) and deep learning (DL) models has significantly transformed the field of environmental pollution prediction. Smart sensors equipped with advanced sensing capabilities can continuously monitor various environmental parameters (Li et al. 2020). These sensors facilitate real-time data collection, transmission, and analysis when combined with IoT to create a seamless ecosystem for monitoring and predictive modeling. IoT-enabled systems operate through interconnected devices that transmit data to centralized or distributed databases via wireless communication protocols (Kishorebabu and Sravanthi 2020). These systems enable the deployment of numerous sensors across vast geographical regions and offers a comprehensive view of environmental conditions. For instance, air quality monitoring stations equipped with particulate matter (PM) sensors and gas analyzers can provide granular data on pollutants like carbon dioxide (CO_2), nitrogen dioxide (NO_2), and ozone (O_3) (Gupta and Singh 2023).

One of the critical advantages of IoT-based systems is their ability to support real-time analytics. Data collected by sensors is processed using ML and DL algorithms to generate actionable insights. The recurrent neural networks (RNNs) and long short-term memory (LSTM) networks are particularly effective in analyzing time-series data from sensors to predict future pollution levels (Tiwari et al. 2022a, b, c). Convolutional neural networks (CNNs) are useful for spatial data analysis and identifying pollution hotspots from satellite imagery (Bashardoost et al. 2024). Moreover, IoT systems enhance the accessibility and scalability of environmental monitoring. These systems can handle vast amounts of data efficiently by leveraging cloud computing and edge computing. Cloud-based platforms allow centralized storage and advanced analytics whereas edge devices enable localized processing (Stergiou et al. 2018). This dual approach ensures that pollution data is analyzed and acted upon promptly, even in remote or resource-constrained areas.

Table 9.4 shows the benefits and challenges in developing IoT system for pollution management. The integration of IoT with ML and DL also supports the development of predictive maintenance strategies for environmental sensors. These systems can identify sensor malfunctions or data inconsistencies by applying anomaly detection algorithms which further ensures the reliability of collected data (Cook et al. 2019). Additionally, IoT platforms facilitate seamless data sharing among policymakers, researchers, and the public for fostering collaborative efforts in pollution mitigation. Despite its transformative potential, the integration of IoT with smart sensors and predictive models faces several challenges. Data quality and reliability remain significant concerns. Ensuring data security and privacy is another critical aspect for the sensitive nature of environmental data. The high initial costs of deploying IoT infrastructure and maintaining it over time can be a barrier to widespread adoption.

Table 9.4 Benefits and challenges in developing IoT system for pollution management

Aspect	Benefits	Challenges
Air pollution	Real-time AQI updates, early warning systems	Sensor calibration, data drift issues
Water pollution	Accurate detection of chemical and physical contaminants	High cost of underwater sensors, maintenance
Noise pollution	Urban noise mapping, predictive trend analysis	Limited battery life of devices, network connectivity issues
Soil pollution	Monitoring agricultural soil health, contamination detection	Difficult deployment in large or rugged terrains

9.7 Limitations and Scope for Future Work

Despite significant advancements in the application of machine learning (ML) and deep learning (DL) techniques to environmental pollution prediction, there exists several limitations which are as follows:

1. One major challenge is the reliance on high-quality labelled datasets which are often unavailable or inconsistent across regions. Many existing datasets suffer from missing values, noise, or lack of representativeness, leading to suboptimal model performance.
2. Model interpretability remains another critical limitation. While complex DL models such as neural networks achieve high accuracy, but we could not understand about how the predictions are made as these models are black-box in nature. The lack of transparency can reduce trust among stakeholders and impede policy implementation. The scalability of these models vary significantly based on geographic, climatic, and socio-economic factors.

Future research should address these limitations by prioritizing the development of explainable AI (XAI) techniques. Enhanced data collection methods can improve the quality and diversity of datasets. The adoption of hybrid models such as combining the strengths of ML and DL with domain expertise can offer promising avenues for better generalization and scalability. Moreover, advancements in edge AI and energy-efficient algorithms can facilitate real-time predictions in resource-constrained environments.

9.8 Conclusion of the Study

The present study highlights the transformative potential of machine learning and deep learning techniques in predicting environmental pollution. The study examines various ML models such as ARIMA, SVM, MLR, Random Forest, and KNN along with advanced DL algorithms including CNNs, RNNs, Autoencoders, and Transformer Networks. As continuous monitoring of pollution level is as cumbersome process, the study also discussed about integration of predictive models with smart sensors and IoT devices. However, sudden challenges such as data quality, model interpretability, scalability, and computational requirements persist while addressing the integration. The present study concluded that ML and DL offer unprecedented opportunities to address the pressing challenges of environmental pollution prediction. This study serves as a comprehensive guide for researchers, practitioners, and policymakers, emphasizing the need for continued innovation and collaboration to safeguard environmental health for future generations.

References

Akbarzadeh MA, Khaheshi I, Sharifi A, Yousefi N, Naderian M, Namazi MH, Safi M, Vakili H, Saadat H, Parsa SA, Nickdoost N (2018) The association between exposure to air pollutants including PM10, PM2. 5, ozone, carbon monoxide, sulfur dioxide, and nitrogen dioxide concentration and the relative risk of developing STEMI: a case-crossover design. Environ Res 161:299–303

Bashardoost A, Mesgari MS, Karimi M (2024) Enhancing spatial modeling and risk mapping of six air pollutants using synthetic data integration with convolutional neural networks. Front Environ Sci 12:1399339

Chauhan BS, Garg N, Tiwari S (2024) Predictive noise annoyance and noise-induced health effects models for road traffic noise in NCT of Delhi India. Noise Mapp 11(1):20240008

Cloete NA, Malekian R, Nair L (2016) Design of smart sensors for real-time water quality monitoring. IEEE Access 4:3975–3990

Cook AA, Mısırlı G, Fan Z (2019) Anomaly detection for IoT time-series data: a survey. IEEE Internet Things J 7(7):6481–6494

Divine ILOH, Aguma CP, Olagunju AO (2024) Integrating AI enhanced remote sensing technologies with IOT networks for precision environmental monitoring and predicative ecosystem management

Dominski FH, Branco JHL, Buonanno G, Stabile L, da Silva MG, Andrade A (2021) Effects of air pollution on health: a mapping review of systematic reviews and meta-analyses. Environ Res 201:111487

Gupta TK, Raza K (2019) Optimization of ANN architecture: a review on nature-inspired techniques. In: Machine learning in bio-signal analysis and diagnostic imaging, pp 159–182

Gupta SK, Singh B (2023) Instruments for monitoring air pollution and air quality. In: Handbook of metrology and applications. Springer Nature Singapore, Singapore, pp 1657–1705

Hossain MS, Rahman MK, Dalim HM (2024) Leveraging AI for real-time monitoring and prediction of environmental health hazards: protecting public health in the USA. Revista De Inteligencia Artificial En Medicina 15(1):1117–1145

Hossain A, Krupnik TJ, Timsina J, Mahboob MG, Chaki AK, Farooq M, Bhatt R, Fahad S, Hasanuzzaman M (2020) Agricultural land degradation: processes and problems undermining future food security. In: Environment, climate, plant and vegetation growth. Springer International Publishing, Cham, pp 17–61

Jaderberg M, Simonyan K, Zisserman A (2015) Spatial transformer networks. In: Advances in neural information processing systems, p 28

Jia X, Cao Y, O'Connor D, Zhu J, Tsang DC, Zou B, Hou D (2021) Mapping soil pollution by using drone image recognition and machine learning at an arsenic-contaminated agricultural field. Environ Pollut 270:116281

Jyethi DS (2016) Air quality: global and regional emissions of particulate matter, SOx, and NOx. In: Plant responses to air pollution, pp 5–19

Karagulian F, Barbiere M, Kotsev A, Spinelle L, Gerboles M, Lagler F, Redon N, Crunaire S, Borowiak A (2019) Review of the performance of low-cost sensors for air quality monitoring. Atmosphere 10(9):506

Keith TZ (2019) Multiple regression and beyond: an introduction to multiple regression and structural equation modeling. Routledge

Kishorebabu V, Sravanthi R (2020) Real time monitoring of environmental parameters using IOT. Wireless Pers Commun 112(2):785–808

Kothari V, Vij S, Sharma S, Gupta N (2021) Correlation of various water quality parameters and water quality index of districts of Uttarakhand. Environ Sustain Indic 9:100093

Leong WC, Kelani RO, Ahmad ZJJOECE (2020) Prediction of air pollution index (API) using support vector machine (SVM). J Environ Chem Eng 8(3):103208

Li L, Jiang P, Xu H, Lin G, Guo D, Wu H (2019) Water quality prediction based on recurrent neural network and improved evidence theory: a case study of Qiantang river, China. Environ Sci Pollut Res 26:19879–19896

Li J, Pei Y, Zhao S, Xiao R, Sang X, Zhang C (2020) A review of remote sensing for environmental monitoring in China. Remote Sens 12(7):1130

Li P, Pei Y, Li J (2023) A comprehensive survey on design and application of autoencoder in deep learning. Appl Soft Comput 138:110176

Liu X, Lu D, Zhang A, Liu Q, Jiang G (2022a) Data-driven machine learning in environmental pollution: gains and problems. Environ Sci Technol 56(4):2124–2133

Liu Z, Fei Y, Shi H, Mo L, Qi J (2022b) Prediction of high-risk areas of soil heavy metal pollution with multiple factors on a large scale in industrial agglomeration areas. Sci Total Environ 808:151874

Lu JG (2020) Air pollution: a systematic review of its psychological, economic, and social effects. Curr Opin Psychol 32:52–65

Maleki H, Sorooshian A, Goudarzi G, Baboli Z, Tahmasebi Birgani Y, Rahmati M (2019) Air pollution prediction by using an artificial neural network model. Clean Technol Environ Policy 21:1341–1352

Mustaqeem N, Kwon S (2019) A CNN-assisted enhanced audio signal processing for speech emotion recognition. Sensors 20(1):183

Nallakaruppan MK, Ramalingam S, Somayaji SRK, Prathiba SB (2022) Comparative analysis of deep learning models used in impact analysis of coronavirus chest X-ray imaging. Biomedicines 10(11):2791. https://doi.org/10.3390/biomedicines10112791

Nallakaruppan MK, Ilango HS (2017) Location aware climate sensing and real time data analysis, pp 73–79. https://doi.org/10.1109/WCCCT.2016.27

Nallakaruppan MK, Balusamy B, Shri ML, Malathi V, Bhattacharyya S (2024) An explainable AI framework for credit evaluation and analysis. Appl Soft Comput 153. https://doi.org/10.1016/j.asoc.2024.111307

Patel R, Tiwari SK, Kumar S, Garg N (2024) Noise modeling, mapping, and control: promotion of sustainable and quieter environment. In: Handbook of Vibroacoustics, noise and harshness. Springer Nature Singapore, Singapore, pp 235–254

Patel R, Singh PK, Saw S, Tiwari SK, Kumar R (2024) Assessment and mapping of traffic noise levels in coal capital of India, Dhanbad: implications for noise pollution management. In: Dynamics of transportation ecosystem, modeling, and control. Springer Nature Singapore, Singapore, pp 317–332

Rajula HSR, Verlato G, Manchia M, Antonucci N, Fanos V (2020) Comparison of conventional statistical methods with machine learning in medicine: diagnosis, drug development, and treatment. Medicina 56(9):455

Sharma K, Rajan S, Nayak SK (2024) Water pollution: primary sources and associated human health hazards with special emphasis on rural areas. In: Water resources management for rural development. Elsevier, pp 3–14

Shumway RH, Stoffer DS, Shumway RH, Stoffer DS (2017) ARIMA models. Time series analysis and its applications: with R examples, pp 75–163

Singh J, Yadav P, Pal AK, Mishra V (2020) Water pollutants: origin and status. In: Sensors in water pollutants monitoring: role of material, pp 5–20

Stergiou C, Psannis KE, Kim BG, Gupta B (2018) Secure integration of IoT and cloud computing. Futur Gener Comput Syst 78:964–975

Stitson MO, Weston JAE, Gammerman A, Vovk V, Vapnik V (1996) Theory of support vector machines. Univ Lond 117(827):188–191

Tiwari SK, Kumaraswamidhas LA, Garg N (2022b) Time-series prediction and forecasting of ambient noise levels using deep learning and machine learning techniques. Noise Control Eng J 70(5):456–471

Tiwari SK, Kumaraswamidhas LA, Gautam C, Garg N (2022c) An auto-encoder based LSTM model for prediction of ambient noise levels. Appl Acoust 195:108849

Tiwari SK, Kumaraswamidhas LA, Garg N (2023a) Accuracy of temporal sampling strategies for the estimation of the annual average equivalent noise levels in urban scenario. Mapan 38(2):343–362

Tiwari SK, Kumaraswamidhas LA, Garg N (2023b) Assessment of noise pollution and associated subjective health complaints in Jharia Coalfield, India: a structural equation model analysis. Noise Mapping 10(1):20220172

Tiwari SK, Kumaraswamidhas LA, Garg N (2021) Accuracy of short-term noise monitoring strategy in comparison to long-term noise monitoring strategy

Tiwari SK, Kumaraswamidhas LA, Garg N (2022) Comparison of SVM and ARIMA model in time-series forecasting of ambient noise levels. In: Advances in energy technology: select proceedings of EMSME 2020. Springer Singapore, pp 777–786

Tiwari SK, Kumaraswamidhas LA, Patel R, Garg N, Vallisree S (2024) Traffic noise measurement, mapping, and modeling using soft computing techniques for mid-sized smart Indian city. Meas Sens 33:101203

Tsalera E, Papadakis A, Samarakou M (2020) Monitoring, profiling and classification of urban environmental noise using sound characteristics and the KNN algorithm. Energy Rep 6:223–230

Wang H, Yilihamu Q, Yuan M, Bai H, Xu H, Wu J (2020) Prediction models of soil heavy metal (loid) s concentration for agricultural land in Dongli: a comparison of regression and random forest. Ecol Ind 119:106801

Wang P, Yao J, Wang G, Hao F, Shrestha S, Xue B, Xie G, Peng Y (2019) Exploring the application of artificial intelligence technology for identification of water pollution characteristics and tracing the source of water quality pollutants. Sci Total Environ 693:133440

Zhang B, Rong Y, Yong R, Qin D, Li M, Zou G, Pan J (2022) Deep learning for air pollutant concentration prediction: a review. Atmos Environ 290:119347

Zimmerman N, Presto AA, Kumar SP, Gu J, Hauryliuk A, Robinson ES, Robinson AL, Subramanian R (2018) A machine learning calibration model using random forests to improve sensor performance for lower-cost air quality monitoring. Atmos Meas Tech 11(1):291–313

Chapter 10
Global Cooperation in AI-Driven Environmental Management

Varun Tripathi, Manish Ghalwan, and Sivarao Subramonian

Abstract Ecosystem sustainability is a prevalent topic in the current worldwide scenario. It challenges the human being for long-term healthy survival considering several factors, including adaptability, applicability, survivability, and immunity. Global cooperation uses current emerging trends to protect worldwide climate and biological diversity. The present chapter reviewed the challenges of global cooperation in tackling environmental management. The extensive brainstorming seeks to support the natural-based solution with artificial intelligence-driven environment management. The chapter develops a novel artificial intelligence-based approach for environmental sustainability by targeting the severity source. The developed approach helps make decisions to avoid issues by monitoring living beings within the constraints of the ecosystem. The chapter reveals that promoting natural sources boosts global cooperation efforts in the field of peace protection of humanitarian. The present review is aware of the advantages of Artificial intelligence-based global efforts to strengthen immunity with survivability in the current ecosystem. Artificial intelligence-driven environment management can control the feasibility of the environment within the protocol.

Keywords Artificial intelligence · Environmental management · Global cooperation · Ecosystem · Renewable energy

V. Tripathi
Uttaranchal Institute of Management, Uttaranchal University, Dehradun, Uttarakhand 248007, India

M. Ghalwan (✉)
Uttaranchal Institute of Technology, Uttaranchal University, Dehradun, Uttarakhand 248007, India
e-mail: imanishghalwan@gmail.com

S. Subramonian
Faculty of Manufacturing Engineering, University Teknikal Malaysia Melaka, Durian Tunggal, Melaka, Malaysia

S. K. Tiwari et al. (eds.), *AI-Driven Environmental Pollution Management*, Climate Risks and Solutions, https://doi.org/10.1007/978-3-031-96243-1_10

10.1 Introduction

Global population growth, urbanization, economic development, and the ensuing increase in consumption habits have all combined to cause an unprecedented increase in garbage production worldwide (Doron and Jeffrey2018; Brown 2015; Noor et al. 2020). Based on projections, the amount of garbage generated will rise by an astounding 73% by the year 2050 in comparison with 2020 levels. This trend emphasizes how critical it is to take proactive, sustainable action to solve the growing trash challenge. The SDGs can be more easily attained because to recent developments in computer vision and artificial intelligence (AI). Three drawbacks prevent AI networks from being effectively used in application fields: poor generalizability and overfitting, which is common, expensive computation, and unclear prediction causes (Arashpour 2023). Applications of artificial intelligence (AI) may be found in a wide range of fields, including engineering, mathematics, physics, astronomy, robotics, medicine, chemistry, transportation, molecular biology and genetics, digital marketing, social sciences, forensics, and precision agriculture. AI tools' quick development and effective solutions have made them more and more popular in a variety of sectors and research areas (Issa et al. 2024; Ye et al. 2020; Bibri et al. 2024). Numerous fields, including environmental science and engineering, can profit from the use of AI technologies on a professional and personal level (Tiwari et al. 2023, Tiwari et al. 2024, Rehman et al. 2025). Even though the environmental area has been stagnant for decades, new research indicates that environmental science and engineering are becoming more interested in using AI techniques (Konya & Nematzadeh 2024).

In a period of unparalleled environmental concerns, the need to promote sustainability has emerged as a worldwide call to action. The world is challenged with issues such as depletion of resources, degradation of the environment and global warming which is where information technology (IT) plays a very important role in the development towards sustainable future (Abatan et al. 2024). The earth is suffering badly from a whole range of interconnected environmental disasters, some of which are also crossing national boundaries. But as music of sorts, it is vital—more than ever before—to deal with such matters, ranging from powerful climate changes such as extreme weather events, rising sea levels and damaged ecosystems; depletion of essential resources; astonishing loss of species (Dwivedi et al. 2021). Figure 10.1 illustrates the observed sources for global disturbance in environment management.

10.1.1 Background

Without question, the spreading climate crisis, pollution, and resource scarcity are defining phenomena of our time, which are all intertwined and equally how powerful

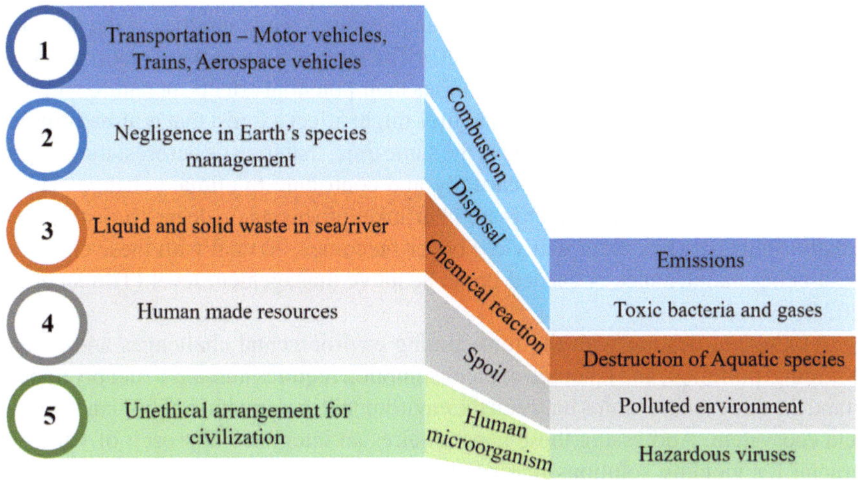

Fig. 10.1 Illustration of sources for global environment disturbance (Schwarzenbach et al. 2010; Appannagari et al. 2017; Ukaogo et al. 2020; Jhariya et al. 2020; Bose et al. 2010)

threats to both humans and the environment. As a result of vast greenhouse gas emissions, climate change continues to evolve, causing increases in global average temperature, modification of precipitation patterns, and intensification of extreme climatic events. These events wholly modify the ecosystem which eventually results in loss of biodiversity and undermining the ecosystem services on which ordinary human societies are reliant. Altered rainfall patterns can significantly affect the agricultural sector, impacting the food security for upwards of millions. Rigorous research showed evidences on how the unpredictable elements of climate change are progressively deteriorating the viability of various sectors across the globe, which calls for adaptive mechanisms to mitigate impacts (Abbass et al. 2022; Thornton et al. 2014; Raihan et al. 2023). Currently, both natural and anthropogenic pollution, like air, water and soil pollution, continues to compromise the health of ecosystems and pose serious threats to the health of people. Industrial processes introduce waste substances in the water bodies which affects the flora and fauna in the water to the community drinking water. Inhalation of polluted air, for example from burning of fossil fuels, causes respiratory disorders and also worsens global warming by spewing greenhouse gases into the air. These problems are interlinked. It is not only nature, but also mankind and even the economy that suffers because of pollution.

The problems are exacerbated by overconsumption of resources. Habitat destruction, loss of biodiversity, and decreased ecosystem integrity often results from resource depletion, including from deforestation, overfishing, and excessive extraction of ground water. Because of social and economic reasons, this resource exploitation is likely to stem from impermissible patterns of consumption. One research article focusing on the environmental impacts of globalization anticipates that resource consumption and production will be done in a manner that is chewing the environment, and furthermore, it will expunge habitats while modifying the land use and endanger biodiversity (Bal et al. 2020). The interrelation of these environmental issues creates a vicious cycle that multiplies their potential effects on nature within a specific region and the world. Pests and fires might affect a forest that is already under the influence of climate change, and at the same time, that forest's deforestation might reduce the carbon sequestration capacity and contribute to climate change. These social and environmental problems are interlinked; focusing on deeper holistic interrelationships between them will yield better outcomes. To deal with these changes, well-being maintenance is needed on an adaptive management level (Tilman et al. 2024).

As a result, dealing with the forthcoming environmental challenges and global warming, pollution, and resource overconsumption requires humans to deeply understand the three relationships between the environment, polarization of human activity, and ecosystem. Addressing these gaps requires an integrated approach of which is crucial for yielding solutions that enhance environmental sustainability alongside promoting human well-being.

10.1.2 Importance of Environmental Management

It is clear that environmental management is required within the organizations which want to achieve goals of sustainability and minimize the consequences on the environment. The use of Environmental Management Systems enables organizations to deal with environmental issues in a more structured manner, which translates to improved corporate social responsibility, and reduced environmental impacts. One research on the use of EMS in the company found out that these systems not only ensure compliance with the various environmental obligations, but also improve the management of information, which in turn improves the decision-making processes (Walker et al. 2007). The implementation of an EMS is associated with improving the cost effectiveness of processes and resource usage. Organizations can identify areas in which resources are being over utilized or wasted and seek to optimize resource usage which brings in cost benefits. Studies show that firms that are more committed to implementing environmental management tend to have better performance and competitiveness (Hamdoun 2020).

Diligent management of the environment creates a favorable impression about an organization and also its stakeholder relations. It is more probable that these organizations will enjoy the confidence and support of their consumers, investors, and the wider community and this can improve their standing in the eyes of the public. Such perception may lead to an increase in loyalty as well as market share. Research indicates that environmental performance is increasingly emerging as an important criterion for consumers, with many preferring to do business with environmentally friendly companies (Shobande and Asongu 2021).

To summarize, the benefits of managing the environment are multi-dimensional in that it enhances a company's reputation, strengthens engagement with the stakeholders, and promotes the protection and preservation of the environment. As shown in the report, the implementation of effective environmental management systems in the corporate world is not just a legal obligation, but a strategy for sustainable economic growth and a clean environment.

10.2 The Role of AI in Environmental Management

Global climate monitoring centers supports the advancement in environmental management. AI attracts the intellectual individuals for tackling the issues faced in controlling obstacles faced in sustaining health environment (Jones et al. 2024; Kaginalkar et al. 2021). The previous studies showed that the human civilization could achieve long term commitment using AI based systems (Mehmood et al. 2019; Som 2021; Li 2020; Ly et al. 2019; Felici-Castell et al. 2023; Popkova et al. 2021). Figure 10.2 describes the role of AI considering environment management in the current scenario.

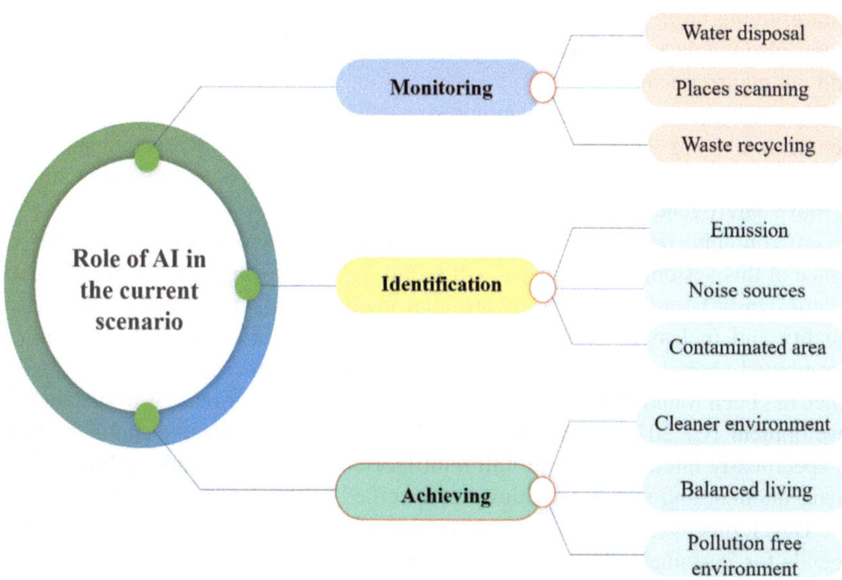

Fig. 10.2 Role of AI in the current scenario in terms of environment management

10.2.1 AI Technologies in Environmental Monitoring

The world needs a lot of things to grow sustainably like economy, good education and many more but one of the most important requirements is Environment. Civilization and the progress of any country will be hampered without measures that favour health, which is impossible if there is a hostile and contaminated environment (Danilov-Danil'yan et al. 2009; Raimi et al. 2021). Hence, monitoring tribe health becomes as necessary to make sure this country can all have healthy lives. This module of Environment monitoring (EM) which deals with preparing and handling catastrophes, limiting the contaminants and provides measures to redress problems resulting from unfit environmental conditions. Some significant issues that really cause problems for the environment include radiation, water pollution, and air quality. Sufficient oversight is required in order to ensure a healthy society and enable sustainable global growth. In past five years, environment monitoring has been transformed into a smart environment monitoring (SEM) system due to improvement in the internet of things (IoT), and modern sensor technology (Haji and Sallow 2021; Vo et al. 2021; Pawar 2019; Ghazal et al. 2023; Ullo and Sinha 2020). Some of the main applications of environment monitoring that you can see in everyday life are weather forecasting, air pollution management, water quality control and monitoring, crop damage assessment. Our aim is to produce an environment suitable for either agriculture, living space or a combination of the two. After the wireless network and things, also thanks to artificial intelligence, it is easier to monitor what is happening with our environment.

AI-based smart environmental monitoring (SEM) is an advanced general environmental monitoring which has also been considered. It utilizes W3C specifications for handling sensor data but might struggle ensuring a seamless communication among all those heterogeneous sensors. These sensors are effective in collecting data from the environment, but their compatibility is a problem, and this limits the performance of this system (Tadejko 2017). It seems as if a growing number of industries are constantly being influenced by artificial intelligence (AI). Global productivity, equality and inclusivity, ecological impacts and many other fields are expecting a combined short- and more long-term effects of AI to be seen. Artificial intelligence has been found to bring both positive and negative contributions to sustainable development (Cf 2015). Previous research has not systematically appraised how AI specifically interacts with the full multifaceted set of sustainable technological, environmental, and social priorities—as specified in this study by the 17 Sustainable Development Goals (SDGs) and 169 internationally agreed targets in the 2030 Agenda for Sustainable Development. We find a considerable research gap on how AI might affect the ability to achieve the entire set of SDGs (Vinuesa et al. 2020).

10.2.2 Role of AI in IoT for Sustainable Environmental Monitoring

Over the past few years air pollution monitoring through IoT has gained popularity. Third research examined a mobile system for air pollution prediction known as "IoT-Mobair" with low accuracy, sensitivity, and computational load capability, however it is innovative. Although the use of highly sensitive gas sensors with IoT can give real-time data, more research is needed to improve their accuracy and efficiency. The large amount of data generated for example—almost 1 M samples during model training are helping perform better but needs to solve computational challenges (Dhingra et al. 2019). Air Quality Monitoring IoT use across the large areas has some important results. Data fragmentation, noise, collision are still the bottlenecks hindering its large-scale practice as demonstrated in some research. Geomatics sensors with IoT have already been employed to improve data quality but face several challenges (Gaglio et al. 2014).

One of the key areas to study is Oceanic Environment Monitoring (OEM). This exercise maintains the primary objective of keeping an eye on the oceanic environments in search of different ecological and atmospheric observables. The results underscore the difficulty of deploying inexpensive non-invasive lightweight sensor networks, with these sensors nearly always expensive to operate and invasive, disrupting marine environments. It is because of their ability to sense and measure the data from an area, regardless of the vastness of that oceanic location over which they write to a remote terminal out in deployed at any time (Shaikh and Hussain 2019). Another important one is the field of Marine environment acoustic monitoring (MEM) where IoT protocols have been used to monitor acoustic signals in marine ecosystems. Although helpful to lower latency and power consumption, these protocols suffer from poor replication performance owing to the deployment complexity and limitation of sensor coverage. For such problems WSNs and IoT systems are used but they need improvement for better coverage (Durante et al. 2018). The AI-based systems provide wide coverage and efficiency at a very low cost to the dust and humidity monitoring. These sensors are tiny and can monitor the environment efficiently, but still, it is necessary that problems in these technologies would be solved to collect robust information. While high-pressure xenon (HPXe) chambers are used in radiation monitoring, these methods are expensive and do not function reliably in temperature variation (Beebi 2018).

IoT based soil monitoring (SM) is another innovation that has ventured in the field of agriculture to promote the smart vegetable farming. This system is good at monitoring green crops, especially for vegetable farmers, but as greenhouse gases start building up the effect of this on tomatoes are quite challenging. Continuous data recording is another advantage of these systems, which use wireless sensors (Shinde and Siddiqui 2018). Another area is the implementation of pollution monitoring systems, environmentally friendly projects such as a real-time air pollution monitoring system has help in containing pollutants but remains inaccurate. The sensors in combination with the low-cost of MQ3 Analog Sensor, Raspberry Pi and IoT detects real-time information but with a lack of accuracy. This is followed by investigating some sensor-based air quality monitoring (AQM) systems, often targeting small scale areas. It have these systems that require low cost and easy installation, but it has a limitation of monitoring number of pollutions feed which are collect date by the gas and LASER sensors (Dhas and Jeyanthi 2017).

The research showed challenges for water quality and quantity control which is especially difficult at those higher carbon related emissions and energy output layers. Odor, pH, and electro-conductance sensors are employed in conjunction with temperature as the indicators to monitor wastewater levels when it is necessary detect odor detection status; however, more works on eco-toxicity reduction/prevention should be done (Chen et al. 2013). The research in the multi-agent supervising systems can be seen in other areas like for e-health monitoring in concern to environmental parameter such as temperature and radiation. These systems efficiently pick up on emergency situations due to changing environmental conditions, especially those including AI to improve predictive software (Mocanu and Florea 2012).

Environmental conditions are monitored using smart environmental monitoring (SEM) systems during the colder seasons. Reliability is a big problem with battery performance and sensor effectiveness on low temperature, and more research studies have to be made in each outdoor sites for wireless sensor networks (Yamashita et al. 2016). LoRa technology is not only used for climate and ecological studies like emissions in the atmosphere. While it works great, however the problems related to installation and sensor coverage remain a concern for its effective use. In smart city environmental monitoring, IoT-based systems are deployed to monitor temperature and humidity in data centers or to track energy consumption; however, management of energy and scale deployment remain problematic (Okafor and Delaney 2019). In smart industries, they apply ZigBee-based environmental monitoring for examining hazardous conditions. This system, based on ZigBee and WSN technology, is intended to determine harmful industrial effects, however the whole coverage of this approach remains unexploited and needs depth research to enhance its performance (Chehri and Saadane 2019).

10.3 AI Based Environment Management Approach

Environment management needs a revolutionary approach to control the severe issues in the current worldwide Earth scenario. Scientists need help to maintain a sustainable climate with the availability of long-lived resources (Giordano 2012; Clausen et al. 2024; Mathew et al. 2024). It has been found that several human-made facilities strongly hamper long-lived dreams and continuously deteriorate natural resources. The present chapter proposed an AI-based approach to environment management. The proposed approach enables tackling the issues faced in the current global living environment. The approach consists of six steps. The first step reviews the current global environment scenario by collecting environmental reports. The second step categorizes the reports by considering sources, sectors, and areas. The third step

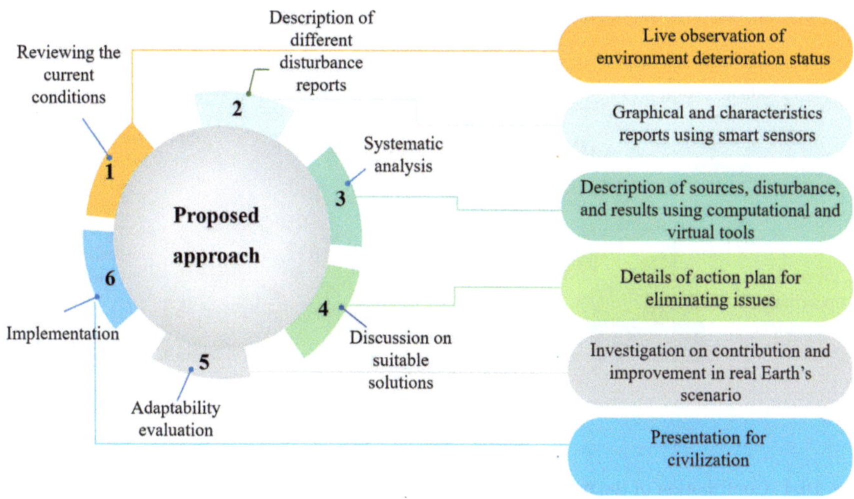

Fig. 10.3 Proposed approach

analyses the different parameters using suitable computational and virtual tools. The fourth step includes a discussion on collecting suitable plans using the Internet of Things inbuilt equipment. The fifth step evaluates the response of the action plan in the natural Erath environment using intelligent sensors. Finally, the sixth step implements the action plan in the concerned civilization. Figure 10.3 describes the proposed approach.

10.4 Summary

The intellectual personnel are always worried for healthy environment in the current earth's living scenario. The government and non-government agencies emphasizing novel approaches for tackling the challenges faced in eliminating the disturbance sources using suitable technologies. Emerging technologies helps in targeting the sources effectively with maintaining environment sustainability. There are several techniques used in controlling environment threatening sources, AI is one of preferred technique for Environment management in global scenario. The present chapter proposed an AI based approach for environment management considering the imbalance sources. The proposed approach can provide a cleaner, safer, and healthy Earth's species living environment in long-term. Figure 10.4 discussed the contribution of this chapter in the current and upcoming Global Cooperation in AI-Driven Environmental Management.

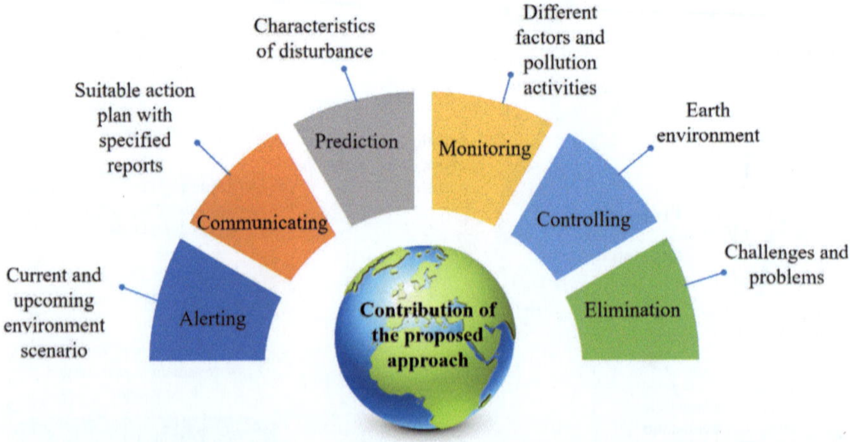

Fig. 10.4 Contribution of the present chapter

10.5 Conclusion

This chapter reviews the current global scenario regarding environmental deterioration effects on the Earth's living species. It helps us understand the sources that shatter the dreams of environmental sustainability. The environment report alerts for upcoming dreadful consequences on Earth's scenario. This chapter proposed an AI-based approach to tackle severe disturbance sources by real-time identification and information dispersion. Concerned civilizations get the opportunities to tackle environmental issues within available constraints, including different age and patient groups. The Intellectual personnels can identify the source of disturbance before deterioration take place. Using a suitable action plan within standard ethical guidelines, they can eliminate the sources. The action plan can provide long-lived and environmental sustainability dreams within available resources. The approach can predict future environmental problems using intelligent sensors. The chapter reveals that promoting natural resources improves global cooperation efforts to protect civilization. The present boosts the awareness of AI-based efforts to strengthen the Earth's species immunity.

References

Abatan A, Adeyinka MA, Sodiya EO, Jacks BS, Ugwuanyi ED, Daraojimba OH, Lottu OA (2024) The role of IT in sustainable environmental management: a global perspective review. Int J Sci Res Arch 11(1):1874–1886

Abbass K, Qasim MZ, Song H, Murshed M, Mahmood H, Younis I (2022) A review of the global climate change impacts, adaptation, and sustainable mitigation measures. Environ Sci Pollut Res 29(28):42539–42559

Appannagari RR (2017) Environmental pollution causes and consequences: a study. North Asian Int Res J Soc Sci Humanities 3(8):151–161
Arashpour M (2023) AI explainability framework for environmental management research. J Environ Manage 342:118149
Bal P, Kapitza S, Cadenhead N, Kompas T, Van Ha P, Wintle B (2020) Predicting the ecological outcomes of global consumption. arXiv preprint arXiv:2003.04231
Beebi F (2018) Environmental monitoring system using IoT. India Res Tech Organiz 5:64–68
Bibri SE, Krogstie J, Kaboli A, Alahi A (2024) Smarter eco-cities and their leading-edge artificial intelligence of things solutions for environmental sustainability: a comprehensive systematic review. Environ Sci Ecotechnology 19:100330
Bose BK (2010) Global warming: energy, environmental pollution, and the impact of power electronics. IEEE Ind Electron Mag 4(1):6–17
Brown DP (2015) Garbage: how population, landmass, and development interact with culture in the production of waste. Resour Conserv Recycl 98:41–54
Cf ODDS (2015) Transforming our world: the 2030 agenda for sustainable development. United Nations, New York, NY, USA
Chehri A, Saadane R (2019) Zigbee-based remote environmental monitoring for smart industrial mining. In: Proceedings of the 4th international conference on smart city applications, pp 1–6
Chen WP, Wang LK, Wang TT, Chen YT (2013) An intelligent management system for aquaculture's environmental monitoring and energy conservation. In: 2013 International workshop on computer science in sports. Atlantis Press, pp 194–198
Clausen CA, Bjørn A, Sanyé-Mengual E, Ryberg M (2024) Applying environmental sustainability boundaries for climate change in life cycle assessment: a review of approaches and implications for policymaking. J Ind Ecol
Danilov-Danil'yan VI, Losev KS, Reyf IE (2009) Sustainable development and the limitation of growth: future prospects for world civilization. Springer Science & Business Media
Dhas YJ, Jeyanthi P (2017) Environmental pollution monitoring system using internet of things (IoT). J Chem Pharm Sci 10(3):1391–1395
Dhingra S, Madda RB, Gandomi AH, Patan R, Daneshmand M (2019) Internet of Things mobile–air pollution monitoring system (IoT-Mobair). IEEE Internet Things J 6(3):5577–5584
Doron A, Jeffrey R (2018) Waste of a nation: garbage and growth in India. Harvard University Press
Durante G, Beccaro W, Peres H (2018) IoT protocols comparison for wireless sensors network applied to marine environment acoustic monitoring. IEEE Lat Am Trans 16(11):2673–2679
Dwivedi YK, Hughes L, Ismagilova E, Aarts G, Coombs C, Crick T, Duan Y, Dwivedi R, Edwards J, Eirug A, Galanos V (2021) Artificial Intelligence (AI): multidisciplinary perspectives on emerging challenges, opportunities, and agenda for research, practice and policy. Int J Inf Manage 57:101994
Felici-Castell S, Segura-Garcia J, Perez-Solano JJ, Fayos-Jordan R, Soriano-Asensi A, Alcaraz-Calero JM (2023) AI-IoT low-cost pollution-monitoring sensor network to assist citizens with respiratory problems. Sensors 23(23):9585
Gaglio S, Re GL, Martorella G, Peri D, Vassallo SD (2014) Development of an IoT environmental monitoring application with a novel middleware for resource constrained devices. In: Proceedings of the 2nd conference on mobile and information technologies in medicine (MobileMed 2014)
Ghazal TM, Hasan MK, Alzoubi HM, Alshurideh M, Ahmad M, Akbar SS (2023) Internet of things connected wireless sensor networks for smart cities. The effect of information technology on business and marketing intelligence systems. Springer International Publishing, Cham, pp 1953–1968
Giordano T (2012) Adaptive planning for climate resilient long-lived infrastructures. Utilities Policy 23:80–89
Haji SH, Sallow AB (2021) IoT for smart environment monitoring based on Python: a review. Asian J Res Comput Sci 9(1):57–70

Hamdoun M (2020) The antecedents and outcomes of environmental management based on the resource-based view: a systematic literature review. Manag Environ Qual: Int J 31(2):451–469

Issa J, Abdulrahman LM, Abdullah RM, Sami TMG, Wasfi B (2024) AI-powered sustainability management in enterprise systems based on cloud and web technology: integrating IoT data for environmental impact reduction. J Inf Technol Inform 3(1):154

Jhariya DC, Tiwari AK (2020) Ganga River: a paradox of purity and pollution in India due to unethical practice. In: IOP conference series: earth and environmental science, vol 597, no. 1. IOP Publishing, p 012023

Jones J, Harris E, Febriansah Y, Adiwijaya A, Hikam IN (2024) AI for sustainable development: applications in natural resource management, agriculture, and waste management. Int Trans Artif Intell 2(2):143–149

Kaginalkar A, Kumar S, Gargava P, Niyogi D (2021) Review of urban computing in air quality management as smart city service: an integrated IoT, AI, and cloud technology perspective. Urban Climate 39:100972

Konya A, Nematzadeh P (2024) Recent applications of AI to environmental disciplines: a review. Sci Total Environ 906:167705

Li R (2020) Artificial intelligence revolution: how AI will change our society, economy, and culture. Simon and Schuster

Ly HB, Le LM, Phi LV, Phan VH, Tran VQ, Pham BT, Le TT, Derrible S (2019) Development of an AI model to measure traffic air pollution from multisensor and weather data. Sensors 19(22):4941

Mathew N, Somanathan A, Tirpude A, Arfin T (2024) The impact of short-lived climate pollutants on the human health. Environ Pollut Manag 1:1–14

Mehmood MU, Chun D, Han H, Jeon G, Chen K (2019) A review of the applications of artificial intelligence and big data to buildings for energy-efficiency and a comfortable indoor living environment. Energy Build 202:109383

Mocanu I, Florea AM (2012) A multi-agent supervising system for smart environments. In: Proceedings of the 2nd international conference on web intelligence, mining and semantics, pp 1–4

Noor T, Javid A, Hussain A, Bukhari SM, Ali W, Akmal M, Hussain SM (2020) Types, sources and management of urban wastes. In: Urban ecology, pp 239–263. Elsevier

Okafor NU, Delaney D (2019) Considerations for system design in IoT-based autonomous ecological sensing. Procedia Comput Sci 155:258–267

Pawar P (2019) Design and development of advanced smart energy management system integrated with IoT framework in smart grid environment. J Energy Storage 25:100846

Popkova EG, Inshakova AO, Bogoviz AV, Lobova SV (2021) Energy efficiency and pollution control through ICTs for sustainable development. Front Energy Res 9:735551

Raihan A (2023) A review of the global climate change impacts, adaptation strategies, and mitigation options in the socio-economic and environmental sectors. J Environ Sci Econ 2(3):36–58

Raimi MO, Vivien OT, Oluwatoyin, O.A., 2021. Creating the healthiest nation: Climate change and environmental health impacts in Nigeria: a narrative review. Raimi MO, Odubo TV, Omidiji AO (2021) Creating the healthiest nation: climate change and environmental health impacts in Nigeria: a narrative review. Scholink Sustainability in Environment. ISSN

Rehman MU, Kamal M, Tiwari SK (2025) Statistical analysis based on a two-staged SEM-SVM approach for environmental noise annoyance prediction and identification of influencing factors. Int J Hum Cap Urban Manag 10(1)

Schwarzenbach RP, Egli T, Hofstetter TB, Von Gunten U, Wehrli B (2010) Global water pollution and human health. Annu Rev Environ Resour 35(1):109–136

Shaikh SF, Hussain MM (2019) Marine IoT: non-invasive wearable multisensory platform for oceanic environment monitoring. In: 2019 IEEE 5th World Forum on Internet of Things (WF-IoT), pp 309–312. IEEE

Shinde D, Siddiqui N (2018) IOT Based environment change monitoring and controlling in greenhouse using WSN. In: 2018 International conference on information, communication, engineering and technology (ICICET), pp 1–5. IEEE
Shobande O, Asongu AS (2021) Management of environmental quality: an international journal
Som T (2021) Sustainability in energy economy and environment: role of AI based techniques. Computational management: applications of computational intelligence in business management. Springer International Publishing, Cham, pp 647–682
Tadejko P (2017) Environmental monitoring systems using internet of things–standards and protocols. Ekonomia i Środowisko
Thornton PK, Ericksen PJ, Herrero M, Challinor AJ (2014) Climate variability and vulnerability to climate change: a review. Glob Change Biol 20(11):3313–3328
Tilman AR, Krueger EH, McManus LC, Watson JR (2024) Maintaining human wellbeing as socio-environmental systems undergo regime shifts. Ecol Econ 221:108194
Tiwari SK, Kumaraswamidhas LA, Prince Kamal M, Rehman MU (2023) A hybrid deep leaning model for prediction and parametric sensitivity analysis of noise annoyance. Environ Sci Pollut Res 30(17):49666–49684
Tiwari SK, Kumaraswamidhas LA, Patel R, Garg N, Vallisree, S., 2024. Traffic noise measurement, mapping, and modeling using soft computing techniques for mid-sized smart Indian city. Measurement: Sensors 33:101203
Ukaogo PO, Ewuzie U, Onwuka CV (2020) Environmental pollution: causes, effects, and the remedies. In: Microorganisms for sustainable environment and health, pp 419–429. Elsevier
Ullo SL, Sinha GR (2020) Advances in smart environment monitoring systems using IoT and sensors. Sensors 20(11):3113
Vinuesa R, Azizpour H, Leite I, Balaam M, Dignum V, Domisch S, Felländer A, Langhans SD, Tegmark M, Fuso Nerini F (2020) The role of artificial intelligence in achieving the sustainable development goals. Nat Commun 11(1):1–10
Vo DT, Nguyen XP, Nguyen TD, Hidayat R, Huynh TT, Nguyen DT (2021) A review on the internet of thing (IoT) technologies in controlling ocean environment. In: Energy sources, part a: recovery, utilization, and environmental effects, pp 1–19
Walker D, Pitt M, Jha Thakur U (2007) Environmental management systems: information management and corporate responsibility. J Facil Manag 5(1):49–61
Yamashita K, Suzuki T, Li H, Ao C, Xu Y, Tian J, Kimura K, Kasahara H (2016) Architecture design for the environmental monitoring system over the winter season. In: Proceedings of the 14th ACM international symposium on mobility management and wireless access, pp 27–34
Ye Z, Yang J, Zhong N, Tu X, Jia J, Wang J (2020) Tackling environmental challenges in pollution controls using artificial intelligence: a review. Sci Total Environ 699:134279

Part IV
Economic and Sustainable Solutions

Chapter 11
Driving Sustainable Innovation: Jendamark India's Journey to Environmental Excellence

Supriya Paigude, Gautam S. Bapat, and Deependra Sharma

Abstract This case study has focused on sustainability practices and operational efficiency within a subsidiary of South Africa's Jendamark Automation, namely JMKI, based in Pune. The theme of Industry 4.0 offers local expertise and industry-specific technical know-how together to the Indian automotive and aerospace industries. With the challenge of maintaining stringent manufacturing standards to meet quality standards while reducing environmental degradation, JMKI adopted a holistic systemic approach incorporating pollution prevention, resource conservation, lean manufacturing practices, and alternative sources of energy. The major programs consist of a 200 KW rooftop solar power facility, water-conserving efforts, and eco-friendly material usage. The benefits included significant reductions in waste, emissions, and operational costs, well-balanced with the environment. This reflects their efforts to continue improvement that the organization carry in its ISO certifications and active initiatives towards sustainability. This study also highlights how JMKI, in a way, combines emerging manufacturing techniques with strong environmental orientation, thereby making it a market leader in sustainable manufacturing.

Keywords Sustainable development goals (SDGs) · Industry 4.0 · Environmental sustainability · Renewable energy · Lean manufacturing

11.1 Introduction

11.1.1 Jendamark India

JMKI is Jendamark Automation, the global leader in next-generation specialized manufacturing solutions and an Indian subsidiary of Pune, which pushes Jendamark Automation to support the rapidly growing Indian automobile industry, which it supports automobile value chain. The foundation of JMKI is based on the integration

S. Paigude · G. S. Bapat (✉) · D. Sharma
Dr. Vishwanath Karad MIT World Peace University, Pune, India
e-mail: gautam.bapat@mitwpu.edu.in

S. K. Tiwari et al. (eds.), *AI-Driven Environmental Pollution Management*, Climate Risks and Solutions, https://doi.org/10.1007/978-3-031-96243-1_11

of the industry 4.0 practices with deep local knowledge of its industries, wherein it enables the company to produce products that have less substitute available in the global market. This strategy makes the Jendamark a market leader in the automotive domain.

Background and Industry Context

Jendamark Automation, South Africa, has a long history of partnership with European automotive companies, offering extensive expertise in automation and manufacturing technologies to various international markets. Due to this rich heritage, Jendamark India was established to address the requirement of the Indian automotive market and provide cutting-edge solutions conducive to local requirements. The establishment of JMKI was a strategic move to focus on India's rapidly growing industrial landscape, particularly in the automotive sector, which has seen exponential growth due to increased domestic demand and foreign investment in terms if joint collaboration and FII.

India's automotive sector considered the world's largest and has traditionally been a key driver of economic growth and employment. The sector contributes high manufacturing output as part of the index of industrial production (IIP). It is also increasing the per capita income of the country by employing a larger number of people. While electric vehicles, and other related sustainable vehicle (EV) solutions, are not yet accepted, although, their use grows gradually and provides a competitive opening this sector. JMKI is leading this revolution with cutting edge powertrain and assembly system solutions apart from being a pioneer of Industry 4.0-enabled assembly solutions for the automotive industries.

11.1.2 Adoption of Industry 4.0 Technologies

Jendamark India has adopted Industry 4.0 practices which comprised high automation, real-time exchange of data, and advanced state-of-the-art manufacturing. This further improves production output and quality and enables JMKI to offer highly customized product and services to the customers (Althabatah et al. 2023; Martín-Gómez et al. 2024; Pansare et al. 2024). The company integrates Industry 4.0 practices into the traditional manufacturing, which therefore also means combining the internet of things (IoT), artificial intelligence, and machine learning in processes of production to improve and enhance operational efficiency (Srhir et al. 2023; Piron et al. 2024). JMKI has emphasize the operational efficiency through implementation of innovative systems to increase the reliability and performance as reliability is important consideration to mitigate the environmental damages (Jyotish et al. 2023a, b). The major contribution of JMKI in the market is to design the first completely automated assembly line for battery packs and inverters for electric vehicle (EV) vehicles (Jyotish et al. 2023a, b). This shows the operational efficiency through which JMKI offers precise engineering solutions fit for the stringent requirements of the growing electric vehicle industry. By using real time digital technologies and high-precision manufacturing, JMKI ensures that its solutions are innovative and cost-effective and add unique value to its customers.

11.1.3 Connecting Local and Global Needs

Core philosophy of Jendamark lies in transforming the local automotive requirement into utility products by using advanced cutting-edge technologies (Mattsson 1992). It makes the products and services competitively fit on an international level with relevance to specific local requirements by combining local market knowledge and international standards. A case in point is the global machinery and advanced technology set up at the JMKI manufacturing facility in Pune, which is enough to create high quality products as per the Indian market as well as for the global customers base of Jendamark. It also serves in the automotive vendor industries where massive shifts to electric vehicles are occurring as well as their best manufacturing practices changes (Flint et al. 2019). It also penetrates into the aeronautical industries, for instance, customized solutions to specific operations. This expansion into complex manufacturing fields would ensure the market capitalization of JMKI and enhance flexibility in providing service to changing demands.

11.1.4 Organization Vision Towards Environmental Sustainability

As sustainability is an essential part of the current industrial landscape. Jendamark India is aware of environmental sustainability and has depicted in its operations (Gupta et al. 2018; Karedla et al. 2021; Piron et al. 2024). The concept of sustainability is evident in ongoing company initiatives that control negative externalities and enhance the resources efficiency. In this regard, JMKI has prepared an outstanding environmental policy and sustainable procurement policy that will guide all processes aimed at reducing the environmental impact of its operations. Therefore, the policies are primarily focused on natural resource conservation, pollution prevention, and waste reduction so that the manufacturer is well aligned with the developmental notion of other countries around the globe toward sustainability initiatives (Gupta et al. 2018; Martín-Gómez et al. 2024). The company is trying to redesign the process of producing the product, making the products use eco-friendly material and ways of preserving them so that it ensures less wastage and emissions-free production. JMKI adopted eco-friendly packaging where reuse and recycle the packaging is being encouraged while opting out of once used packets in final delivery to the customers. Optimum resource management is the most critical practice of JMKI to address operational pitfalls to ensure that water, energy, and raw materials are utilized with maximum benefits and proper management. Some examples such as automatic deactivation of lights, energy-saving lighting, and installation of rainwater harvesting systems. These reduce operational costs and promotes circular economy and protects the natural resources.

11.1.5 Adoption of Lean Manufacturing Practices

Lean manufacturing ensures Jendamark India with zero wastage and also add value to the customer in terms of cost reduction. It is adding value to the production system, elimination of redundant process and bottleneck operations to optimize the overall production process (Gupta et al. 2018; Ahmad et al. 2021). JMKI operationalize on lean manufacturing principles, including Just in Time (JIT) production and continuous improvement. JIT ensures that the materials man and machines need to be coordinated in such a way to avoid over storage of inventories. Continuous improvement, also known as KAIZAN, ensures that all the processes in operation are regularly checked and reported so that the operational efficiency may reaches to maximum level. This is not possible without the top leadership commitment towards operational excellence (Gandhi et al. 2021).

11.1.6 Certifications and Regulatory Compliance

Jendamark India has attained all necessary certifications and maintained its operations as per regulatory compliance to ensure quality and sustainability. The company has been certified with ISO 14001:2015 which focuses on its environmental management system and proves the commitment to robust environmental ownership. Furthermore, JMKI is also pursuing to acquire ISO 50001 certification for their Energy Management System, which further enhances their commitment to energy efficiency and sustainability. These certifications show commitment by JMKI to become an environmentally responsible organization, and this makes the company reputation an assurance as a sustainable manufacturer. JMKI operations ensures the best practices globally for quality and sustainability.

11.2 Background and Industry Context

11.2.1 Jendamark Automation: A Journey of Innovation

Jendamark is has high expertise in the providing solutions to a heterogenous industries as Jendamark India has been offering high value added services in several industries in Europe and South Africa, and now they have emerged as the market pioneer of advanced automation that yield optimum productivity and high precision in operations. Jendamark Automation came up with the idea of starting the company called JMKI due to growth prospect of Indian market.

11.2.2 The Indian Market: Opportunities and Growth

The Indian economy has witnessed the rapid emergence of the industrial sector due to 1991 LPG (Liberalization, privatization, and Globalization), FDI, and vast middle-class income group. The automobile industry has also attained its economies of scale in the country's economy. In present days, India is considered among the largest automobile markets globally in terms of market capitalization. Increasing per capita income, urbanization, and conducive government policies related to manufacturing also influence the growth trajectory. Heavy investment has been initiated in the Indian automobile sector with a vision of making India the global leader in manufacturing as the increasing aggregate demands from the Indian government have been supplied by policies such as "Make in India" (Gandhi et al. 2021; Pulicherla et al. 2022). New manufacturing facilities and plants were opened, and new emerging technologies have been adopted. This provided Jendamark Automation an opportunity to get its place of its Indian sister, JMKI, and all the technical expertise and experience it needed to be able to offer solutions that did not only meet the market standards but also meet the local market demand.

11.2.3 Industry 4.0: Transforming Manufacturing Landscape

Industry 4.0 is characterized by emerging ways of manufacturing practices. It is characterized by digital technologies like IoT, AI, and intelligent automation. Industry 4.0 enables manufacturers to achieve greater levels of efficiency, accuracy, and customization in their products and services (Althabatah et al. 2023). These technologies are implemented by Jendamark India which transforms the industry and Industry 4.0 technologies applied in smart manufacturing practice by JMKI allow improving the production processes of the organization through having real-time analytics and machine learning. This technological upgradation enables JMKI to deliver customized services based on customer requirements in minimum cost.

11.2.4 The Automotive Sector: A Shift Towards Electric Vehicle (EV)

The market shift to electric vehicles (EV) is the significant trend in the automotive industry and that will continue to be trending by increasing environmental pressure and stringent emissions standards. There is high government impetus to include electric vehicles in public transport lifeline of the country. India has huge market and is continuing to rise. The Indian government has set the vision of adoption of electric vehicles along with its ambitious Make in India (Kwag et al. 2021; Turoń et al. 2021; Vanus and Bilik 2022).

Jendamark India is the first organization that employs cutting edge technologies available for electric vehicle manufacturing. Therefore, it was JMKI that built the first assembly line of batteries and inverters in the domain of the Indian e-vehicle (EV) segment. By this move, the assembly line project puts the expertise of JMKI in delivering precise automation solutions in the EV market.

11.2.5 Aerospace Industry

JMKI is not just a solution provider for the tooling, but it is an expertise in providing special solutions to the aerospace-specific solution. The requirement for the accuracy and innovation of processes in aerospace industry leads the company's involvement as an integral element for its advanced manufacturing capabilities in the field of expertise. The company develops the next generation of aerospace technologies with special tooling and assembly solutions.

Jendamark India's foundation marks a strategic step toward an evolving and fast-changing industrial landscape in India. Industry 4.0 technologies are integrated into a deep understanding of local market demands for the ever-increasing complex requirements of the automotive and aerospace industries. Electric mobility and precision tooling solutions put JMKI in a vanguard position regarding innovation and sustainable development for the industrial sector of India.

11.3 Challenges

11.3.1 Environmental Challenges

Pollution Control and Waste Management: One of the important environmental issues for JMKI is pollution control and waste management. The manufacturing process involves a huge amount of by-products, which may include poisonous chemicals, metallic scraps, and emissions that can greatly affect the environment. Proper disposal and treatment of these products help avoid pollution in soil, water, and air. The problem goes further with the reduction of greenhouse gases and other pollutants. Environment regulations, since now strictly being implemented across the globe including India, would demand regulatory compliance through JMKI in avoidance of legal implications and positive contributions to environmental conservancy.

Resource Conservation: This is because the manufacturing industry consumes high volumes of water, energy, and raw materials. Depletion of these resources happens through inefficient use, resulting in higher costs of operations. JMKI, therefore, needs to implement practices that ensure improved resource efficiency without lowering productivity and quality.

Sustainable sourcing and supply chain management: The whole procedure of ensuring that materials and components are sourced responsibly and sustainably requires extensive audit work and coordination with the suppliers, which raises complexity factors. There is also a need to verify the environmental practices of a number of suppliers and confirm that those are in line with what JMKI has set on its sustainability goals.

11.3.2 Operational Challenges

Ensuring Higher Level of Production: Jendamark India is generally synonymous with higher standards of production and is known more within automotive and aerospace sectors. Yet competition is getting fierce and the growing demand to sustain these higher levels in production becomes an operational burden on the company. Sustaining such levels requires on-going investment in leading technology, skilled people, and quality control systems.

Although the benefits of using Industry 4.0 technologies are numerous, their integration is highly operational challenging. IoT, AI, and advanced automation require a high initial investment and require a well-trained workforce who can handle and maintain the systems. Training employees to ensure they are equipped with these new technologies is very difficult in today's environment because technology changes rapidly.

Cost Management: Innovation and sustainability, however, still integrate cost management into the operational process. Introducing sustainable practices through new technologies requires high investment costs. Budgeting and financial planning is the delicate task entrusted to JMKI, which ensures that costs incurred by the company in sustainability and innovation do not eat into profitability.

11.3.3 Strategic Responses to Challenges

Pollution Control and Waste Management Initiatives: JMKI has taken several proactive mandates towards pollution control and waste management. The company has optimized the production process to reduce the generation of waste, used eco-friendly materials, and encouraged recycling of materials. The company has also invested in the globally best waste treatment facilities so that if any waste is generated then it is disposed of in an eco-friendly manner so that it reduces the carbon footprint.

Zero waste to landfill: Zero waste to landfill is an important initiative on the part of JMKI. It establishes brief recycling and composting checklist while trying to eliminate landfills completely. This brings two significant contributions: it minimizes the environmental impacts, but also the company's operational effectiveness increases through reduction in cost incurred in waste management.

Resource Conservation Strategies: JMKI has implemented many resource conservation strategies. Strategies such as provision of energy-efficient lighting, low-flow water fittings, and rainwater harvesting systems implemented to save the overuse of resources. It also promotes the use of renewable energy sources such as solar energy, thereby decreasing its reliance on non-renewable sources of energy so that concentration of carbon emissions in the air may be reduced.

Lean Manufacturing and Cost Management: All the principles of lean manufacturing are incorporated into the approach of JMKI in the area of cost management and improvement in operational efficiency. Lean manufacturing increases productivity and reduces costs through waste reduction, streamlined processes through removing bottleneck operations, and continuous improvement-based production techniques such as implementation of KAIZAN. JIT and other lean manufacturing-based techniques ensure optimum resource utilization by maintaining optimum inventory levels with reduction in waste. JMKI address this issue by auditing its suppliers regularly and adopting a collaborative approach so as to develop synergy with supplier network. This enables the organization to team up with suppliers, ensuring that all the business functions are sustainable. The approaches also help JMKI to achieve sustainable supply chain by improving sustainability in operations.

Training and Workforce Development: JMKI invests in training and development programs of employees and workers which aimed to create a competent workforce to overcome the challenges posed by the integration of Industry 4.0 technologies with traditional technology. The types of programs include upgrading employees with the necessary skills and knowledge they would need to work with advanced manufacturing technologies. Human resource development (HRD) functions which are Continuous learning and development will keep the JMKI workforce competent and responsive to changes in technology.

11.4 Need for Sustainable Practices

The increasing negative externalities of economic growth and the evolving regulatory requirement have created a demand base for the adoption of sustainable practices in the manufacturing industry. The manufacturing industry is one of the major contributors of environmental degradation as majority of manufacturers have not adopted sustainable manufacturing practices and bypassing the regulatory compliance as well, hence it is aimed for adopting sustainable practices to reduce its negative impacts on environment.

11.4.1 Environmental Impact

The manufacturing processes use raw material which may be extracted from natural resources which is constantly depleting, and it further produce hazardous waste that causes environmental degradation and deplete natural resources. Standard sustainable practices like pollution control, management of waste, drip irrigation, and resource saving are essential to reduce the ecological footprint of manufacturing and save the environment for future generations.

11.4.2 Regulatory Compliance

All developed and emerging economies are developing stricter environmental policies to combat climate change and foster sustainability. Compliance with the rules is not an option but a necessity, which needs to be implemented to maintain operating licenses and avoiding legal notices. Sustainability ensures compliance with these policies while simultaneously enhancing the image of the corporation.

11.4.3 Economic Benefits

Sustainable practices significantly reduce costs by enhancing efficiency and optimum utilization of resources. Some examples of best industrial practices include energy-efficient technologies and measures for waste reduction that reduce the overall cost of operation. This will also attract more customers and investors concerned about sustainability, further it has fouled the competitive spirit in the market.

11.4.4 Social Responsibility

Organizations must possess the moral obligation to protect the environment and benefit society in the long run. Sustainability practices demonstrate their commitment to society by instilling confidence and loyalty in all associated stakeholders, which may include employees and customers. Jendamark India became a pioneering example of sustainable practices adoption due to their policies and their commitment towards society.

11.5 Solution Implemented

11.5.1 Environmental Sustainability Initiatives

Pollution Prevention and Resource Conservation: The approach of sustainable manufacturing comprises a reduction in the negative environmental impacts and also enhances the optimum usage of resources. Pollution prevention through the reduction or elimination at its source of creation, change in production methods, use of eco-friendly materials, and conservation is incorporated into the production process at Jendamark India. For instance, JMKI has replaced hazardous chemicals with less harmful chemicals and used environment-friendly packaging that reduces negative ecological impacts. In addition, the company is also concerned with transformation of by-products arising out of production and defective raw materials and work in process that have been generated into raw material for other product productions using the concept of circular economy. Resource conservation is the first step towards pollution prevention by prudent use of water, energy and raw materials effectively. These steps include installing low-flow water fixtures, installing rainwater harvesting systems, and energy-efficient lighting among others are proven as the most effective practices of pollution prevention. The measures such as reducing the operating costs result in a remarkable reduction in the consumption of natural resources. Optimizing resource utilization helps JMKI to promote the sustainability practices of its operation and the long-term goal of environment protection. Furthermore, renewable energy sources such as solar energy reduce their carbon footprint and maintain environmentally friendly manufacturing operations. Therefore, JMKI gives a model of pursuing both objectives such as environmental harmony and the profit maximization with operational excellence as mentioned in Fig. 11.1.

Fig. 11.1 Recycled plant bin made up of waste material. *Source* Jendamark official document

11.5.2 Zero Waste to Landfill

One of the most ambitious environmental protection goals of JMKI was Zero Waste to Landfill in the process of minimizing the carbon footprint, which is essential in designing a process from source of material procurement to landfill without waste. State-of-the-art production methods such as lean production systems that minimize waste have been implemented. They include using less hazardous materials, thereby increasing the efficient utilization of resources. The company has initiated recycling programs to ensure the systematic storing and further recycling of metal scraps, defective products packaging, and production waste. Composting organic waste also reduces landfills and puts minerals back into the soil and enhances the fertility of the soil and prevents soil erosion. These measures are making JMKI a more sustainable organization. JMKI has also developed a sustainable culture at work by discouraging the use of single-use plastics by embracing the circular principles of reuse, and recycling principles. It further expands its zero-waste philosophy through strategic partnerships with suppliers and customers to explore alternative sustainable packaging and return systems. This commitment proven more beneficial than just simply conserving natural resources and reducing environmental pollutants.

11.5.3 Zero Discharge and Emission

Zero discharge and emission are ambitious goals for Jendamark India as far as environmental sustainability is concerned. Zero discharge ensures pollutants emitted by industries in the surrounding water resources are removed to protect the aquatic ecosystem and guarantee clean water supply for society. Jendamark India achieves this by adopting cutting edge technology on water treatment such as recycling industrial wastewater in a manner that minimizes the water used. The organization ensures that water utilization in the manufacturing process is treated and recycled. On the contrary, zero emissions concentrate more on the elimination of greenhouse emissions and other pollutants emitted within the environment. Jendamark India addresses this by ensuring the inclusion of renewable energy sources, such as sunlight, within its operations, hence relying less on fossil fuel and reducing its carbon footprint.

Investments have been made in energy-efficient technologies that minimize emissions related to the production process. Due to landmark initiatives such as zero-discharge and zero-emissions initiatives, Jendamark India is contributing significantly to the global initiatives to combat climate change and improving operational efficiencies with sustainability.

11.5.4 Lean Manufacturing Practices

Lean production practices bring operational excellence and sustainability to Jendamark India by attaining economies of scale in the production output. It adds to customer value, productivity, and overall efficiency while reducing waste.

Materials and components are delivered at exactly the right time they are needed at their definite place and further, it is received by the right authorized people, which is the core principle of JIT production, which reduces inventory carrying cost by maintaining optimum level of inventory and waste. The main characteristics of lean manufacturing include continuous improvement processes, focused on regular assessment and further improvement in the production process. Major assessment method employed by organization is total productive maintenance (TPM) where the critical aspect of the operations listed in the checklist and employees regular assess these critical checkpoints during the production processes. It ensures the process speed but also reduces the chance of producing defective products. These practices engage employees in identification and elimination of bottlenecks and redundant processes to reduce optimize the utilization of resources. Lean manufacturing also reduces throughput time and enhances product quality, which assists in meeting the customer demand regularly. Lean manufacturing brings cost efficiency and maintaining productivity and reduction in resource wastage, therefore, represent the commitment of the company to innovative and sustainable manufacturing practices.

11.5.5 Employing Alternating and Renewable Energy Sources

The company has reduced its dependence on fossil fuel sources while using solar power based renewable energy in their operations, this move has reduced carbon emissions. The installation of a 200 KW rooftop solar system was the first initiative in the adoption of sustainable energy. It saves electricity costs and ensures decreased greenhouse gas emissions. Apart from solar, Jendamark India is also exploring alternative renewable energy including wind and biomass, which forms a diversified and robust portfolio of energy. Such practices not only support the Global reduction in carbon emission goal but also ensure energy saving along with improved efficiency in existing operations. Thus, investments in renewable energy by Jendamark India reflect its leadership to combat futuristic challenges in sustainable manufacturing as the advanced economies now shifts toward more renewable energy. Therefore, adoption of renewable and non-renewable energy sources simultaneously is becoming a new benchmark in the industry as mentioned in Fig. 11.2.

Figure 11.2 depicts a stock-flow diagram of integration of renewable energy and traditional energy sources. This diagram also shows the inherent advantage that arising out from integrating renewable energies with traditional energy in industrial practice. The integration of traditional and renewable energy is coming to operation 1 as input within the system. The residual energy form utilized in operation 2 is depicted by module 2. Further, total cost saving from installation of a 200 KW rooftop solar system has drastically improved from 66,227 ₹ in September 24 to 133,912 ₹. It depicts the effectiveness of installing 200 KW rooftop solar in jendamark facilities.

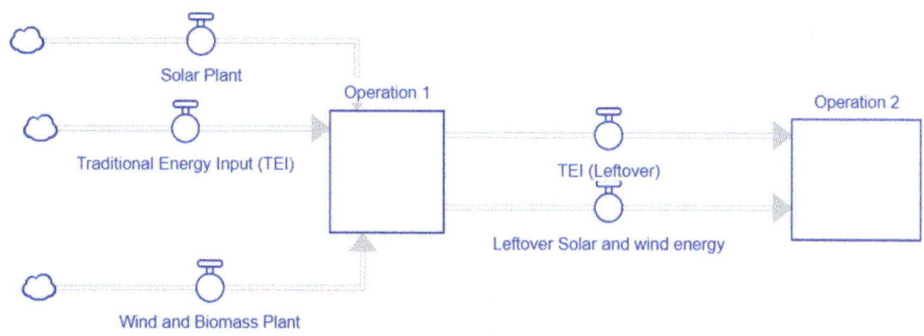

Fig. 11.2 System dynamic model: stocks and flows diagram integration of renewable energy and traditional energy source. *Sources* Author's construction using Stella

11.5.6 Certifications and Compliance

Jendamrk India has never compromised the standards of already attained certifications and compliance and maintain the higher standards of environmental and operational excellence. The company has attained ISO 14001:2015 for its Environmental Management System, displaying its commitment to organized and sustainable environmental practices. Also, Jendamark India is also moving on the path to achieve ISO 50001 registration of its another energy management system certification. This tells and showcases the story of efficiency and continuous improvement at energy front. Achieving certifications ensures the company's commitment towards the environment. It creates an organization image and poses as sound organization as far as legal and environmental issues in concerned and it further enhances the organization position in society as a responsible and environment-friendly manufacturer. Jendamark India reiterates commitment towards sustainability by following certification and compliance protocol, excellence in operations, and a global leadership position in manufacturing. This is represented in the flow chart which shows the sustainability activities performed at Jendamark India. Jendamark operations are following United Nations four SDG goals; renewable energy, SDG 17, conservation, SDG 15, operations and production, SDG 12, and waste management, SDG 12.3–12.5 as mentioned in Fig. 11.3.

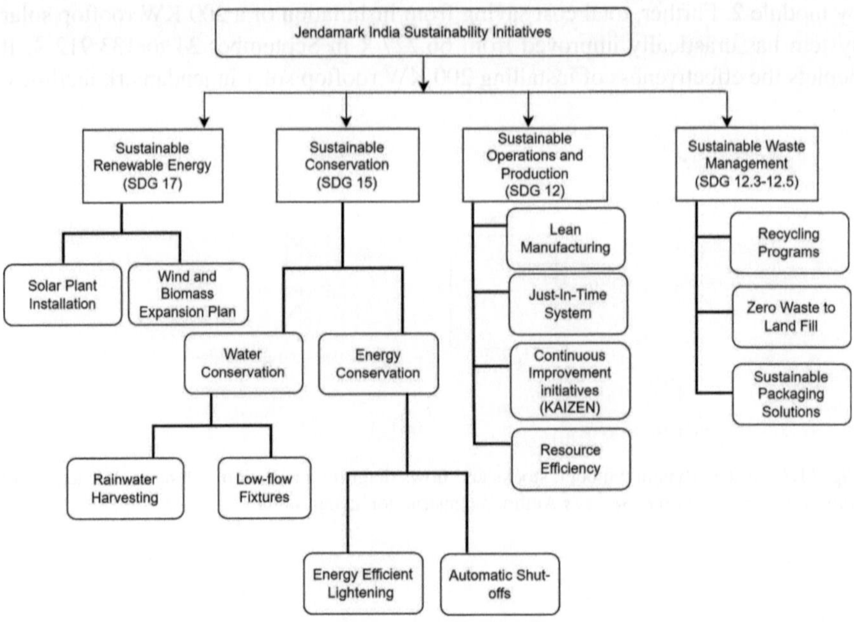

Fig. 11.3 Flow chart depiction of sustainability initiatives at Jendamark India. *Sources* author's construction

11.6 Outcome of Sustainability Initiatives

11.6.1 Environmental Impact and Benefits

Environmental sustainability provides many advantages to Jendamark India. Their initiatives have enabled them to preserve water for reuse by controlling the industrial pollutant waste into water bodies through advanced water treatment recycling system. The installation of 200 KW rooftop solar energy system has reduced emission of greenhouse gases and improves air quality, and it eventually reduces carbon footprints.

These activities have solely focused towards combating global warming but also help to enhance the growth of biodiversity. Due to the adoption of sustainable processes, such as sustainable packing, Jendamark India is rising as true example of CSR due to its commitment to balance industrial development and maintaining the exiting ecological system.

11.6.2 Improvements in Resource Productivity

These brought forth acceptable progress from dedicated efforts to exploit every resource on energy, water, and materials as much as possible. The energy saved considerably has been harnessed thanks to energy-saving lights, turn-off devices of light switches, and the very modern high-tech gadgetry which uses less electricity. Not even low flow fittings and catching rainwater managed to lower its own consumptions and promote careful utilization of water.

Jendamark India aims at improving the productivity of production and promotes the use and reuse of materials to reduce the amount of waste generated from using raw material. This reduces the cost of operations. It also reduces the environmental impacts related to the manufacturing process. Jendamark India hence commits itself to a green strategy focused on saving of critical resources for tomorrow and following amplified production standards basing their importance on resource efficiency. The installation of aerators instead of regular taps has significantly reduced water consumption from 287 m^3 in April 23 to 248 m^3 in March 24 which is 13% reduction. Further, 4800 L of water is saved daily which leads to saving 9900 ₹ per month.

11.6.3 Success Stories in Waste Reduction

This was also made possible in reducing waste through sustainability and operational efficiency by ensuring that a huge portion of scrap metal, plastics, and packages may be recycled or used for another purpose with absolute efficiency to divert a very big chunk of material not ending up in landfills. Also, it contributed through sustainable packaging solutions as being used as a means towards minimizing the generation of the same waste. The use of environmentally friendly cutting oils, and other biodegradable materials in the manufacturing has reduced hazardous waste production tremendously. Initiatives by Jendamark India to get zero waste to landfill best illustrate a good strategy that prevents waste generation. Apart from reducing the cost associated with disposal and minimizing harmful effects on the environment, this approach also positions the company as a leader in sustainable production methods. The waste minimization initiative forms a benchmark for responsible industrial practices that the company demonstrates its commitment towards Jendamark India.

11.6.4 Economic and Cost Savings

Jendamark India initiatives towards sustainability have saved much on finances and costs thereby proving for a financially viable relationship in terms of environmental protection. The saving in terms of electricity charges is significant owing to installations of 200 KW solar rooftop systems which availed approximately 2.96 million INR totalized per annum. Compulsive needs towards water conservation added with energy efficient technologies displayed a more prominent percentage as far as the saving concerning cost mainly because of resources consumption.

The lean manufacturing methodologies undertaken by the organization ensure efficiency and waste elimination, making the organization cost-efficient in the process. Further, eco-friendly material and sustainable packaging help in saving the cost of waste disposal and material utilization maximization. These cost benefits result in enhanced profitability besides giving a competitive edge in the marketplace. The integration of economic objectives with sustainable practices at Jendamark India clearly demonstrates that environmental accountability and financial achievement are supportive of each other.

11.6.5 Other Related Programs and Initiated

Jendamark India is implementing auxiliary programs and projects. Mass tree plantation programs are conducted wherein the company has planted 10,000 saplings just to prevent the increase in the rate of deforestation and the level of biodiversity. The company has used non-toxic cleaning products in the office and manufacturing area along with energy-saving appliances. Sustainable waste management through color-coded recyclable containers and use of reusable materials is also attained on its premises. The company takes in electric transport, which cuts the carbon footprint and decreases its operational cost. Jendamark India has taken up a new type of packaging, which is recyclable. It is doing this so that negative environmental impacts will be decreased. These developments are again proof of this company's whole approach toward sustainability by which it not only contributes toward manufacturing but also the problems of the environment and welfare of the society. All these efforts ensure that Jendamark India is committed to a sustainable future, apart from raising new standards in the manufacturing industry.

11.7 Continuous Improvement

11.7.1 Sustainability Activity on Continual Basis

Jendamark India has been dedicated to continuing activities that go beyond those established originally. The company periodically updates its environmental policies and adopts new technologies that minimize its ecological footprint. Some of the key initiatives include increasing the use of renewable energy, developing more sophisticated techniques for saving energy and water, and using more sustainable materials in manufacturing. JMKI promotes employee awareness on sustainability through continuous education and awareness programs. This brings sustainability in the organizational culture and frees the scope for innovations and improvements in the work process. Ongoing organization sustains environmental load reduction within the parameters of the best practices followed in other nations. As a premier manufacturer and a socially responsible one, based on the principles of a sustainable approach, Jendamark India always practices the best practices in this domain and acts as an excellent model for good environmental custodianship and an exemplary model for its industry.

11.7.2 Periodic Audits and Performance Evaluation

Jendamark India continually audits in terms of sustaining and upgrading its sustainability practices. These assessments ensure adherence to environmental regulations, audit the effectiveness of sustainable programs, and pinpoint areas to be developed. Strategic assessments of energy consumption, waste generation, and resource consumption ensure that JMKI operates in an efficient and sustainable manner. This process monitoring addresses essential indicators of sustainability performance and thus sets up the specific, measurable targets. It includes some of the items, for instance carbon reduction emissions and achieving zero landfill. This kind of process empowers the organization to clearly spot trends and make choices to take suitable corrective action. Continuous checks and reviews ensure that the organization not only follows best practices but is also reflective of transparency in JMKI. Because of such detailed procedures, Jendamark India continuously enhances its sustainability performance and, as an aftereffect, commits itself to efficient and innovative manufacturing as mentioned in Fig. 11.4.

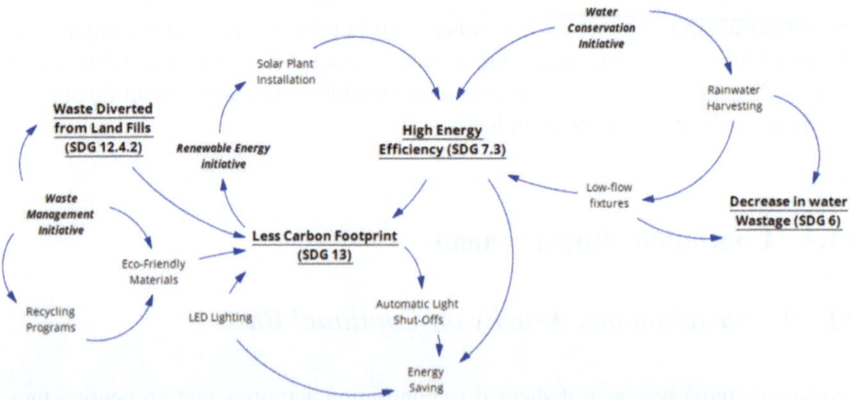

Fig. 11.4 Causal loop diagram of the environmental effect and contribution towards SDGs. *Source* Authors' construction using VENSIM

Figure 11.4 depicts the causal loop diagram of Sustainability Initiatives by Jendamark India. Such systems, other than installing rainwater harvesting facilities and low-flow fixtures, are essential in the achievement of greater standards of energy efficiency SDG 7.3 and reduce the carbon footprint SDG 13. Besides this, the efforts of Jendamark toward the installation of rainwater harvesting and low-flow fixtures decrease the wastage that happens through unnecessary water consumption. The effort is in tune with Sustainable Development Goal 6. Another positive response mechanism that takes place due to automatic light turn-offs and LED lighting installation leads towards energy efficiency and thus reduces carbon emission. The company embraced a circular economy in its manufacturing by recycling and the use of green materials. This leads to a decrease in carbon emission and waste being diverted away from landfills as part of achieving Sustainable Development Goal 12.4.2.

11.8 Artificial Intelligence and Internet of Things in Achieving Energy Efficiency SDG (7.3)

Jendamark India has implemented advanced techniques such as artificial intelligence and Internet of Things (IoT) to enhance their product design and streamline their production process. Generative AI is employed to optimize the weight of their products, as heavier products require a significant amount of energy to move throughout work in progress as mentioned in Fig. 11.5.

Traditional design approaches have not been empirically demonstrated to be efficient in removing unneeded mass from items. However, the use of topology optimization, which employs generative AI, is the most suitable approach for achieving the optimal weight of products. Similarly, the process of precise manufacturing necessitates a significant amount of energy and material for production. IoT-enabled 3D

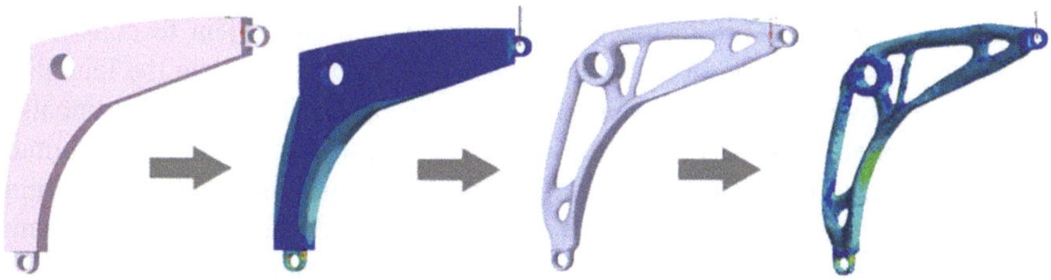

Fig. 11.5 Topology method for weight optimization

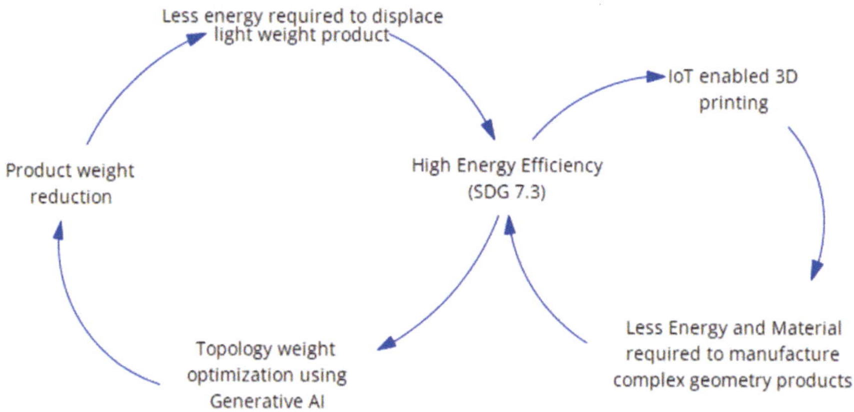

Fig. 11.6 Weight optimization through topology methodology

printing is advancing the production of intricate manufactured goods through the use of additive manufacturing, resulting in reduced energy and material use. Both of these scenarios are decreasing energy consumption improving the overall energy efficiency of Jendamark India and contributing towards SDG goal 7.3 Energy efficiency as mentioned in Fig. 11.6.

11.8.1 Industry 4.0 Technologies (IoT and Machine Learning) Contribution to Optimizing Resource Usage or Reducing Emissions

Jendamark India has mastered the concept of Industry 4.0, particularly the Internet of Things (IoT) and machine learning in the form of its sustainability efforts by enabling real-time monitoring of production and operations which further leads to optimized resource usage, and it eventually reduces carbon emissions across their manufacturing facilities. For instance, sensor-based IoT monitoring system provides cutting-edge solutions for energy consumption in terms of reliability and cost-effectiveness to businesses. Sensors are installed on machinery and equipment to consistently track energy consumption in real-time data. This data will facilitate the identification of ineffective or inefficient operations, ensuring and enabling the adjustments to machines to be set in such a manner that the production schedules will have minimum energy wastage. Another application is predictive analytics using machine learning which predicts the upcoming breakdown of machines and reduces the carbon emissions to avoid further and unnecessary energy consumption. Furthermore, Water tracking systems via sensors can monitor water consumption in cooling systems and usage in the other industrial processes, allowing for the identification of the quantum of water usage at respective units and mitigating the water loss via leaks or excessive water usage can be curbed or negated.

11.8.1.1 Smart Manufacturing Cell

Smart Manufacturing Cell is a dedicated production cell installed with IoT sensors that monitor energy usage, material flow, and machine performance on a real-time basis. Machine learning algorithms analyze the captured data from the sensors and this data is used to identify inefficiencies and suggest adjustments to optimize production processes, leading to significant reductions in energy consumption and waste generation.

Overall, by integrating and implementing automation, IoT, and machine learning technologies into their operations. Jendamark India can make data-driven decisions to minimize resource usage, reduce emissions, achieve sustainable manufacturing practices, and contribute towards sustainable business.

11.9 Conclusion

11.9.1 Summary of Achievements

Jendamark India has been very successful in incorporating sustainability into the organizational vision. It displays the company's commitment towards being environmentally responsible and its commitment to operational excellence. This comprises the installation of a 200 KW rooftop solar energy system, significantly reducing electricity and maintenance costs, and it is able to reduce its carbon footprint by approximately 245 tons per year. This example shows the commitment of JMKI to renewable energy and sustainable practices. From a resource efficiency point of view, JMKI has managed to conserve water using rainwater harvesting systems and installing low-flow fixtures, thus resulting in a considerable reduction in the consumption of water. The automatic light turn-offs coupled with energy-efficient lighting further enhance energy savings and hence show a system based holistic approach toward resource management. The volume of waste after the production process has been saved by significant amounts as well. JMKI has adhered to the concept of zero waste to landfill through high-scale recycling programs and green packaging, which has led to successful diversion of massive volumes from landfills. The generation of hazardous waste has been minimized through environment-friendly materials, such as biodegradable cutting oils. Lean manufacturing approaches have streamlined operations, reduced waste, and enhanced productivity. Continuous improvement and Just-in-Time production technologies have contributed to optimal utilization of resources and reduced excessive inventory accumulation that has resulted in further development in the field of cost efficiency. All sustainability efforts at JMKI are underpinned by comprehensive certifications, which include ISO 14001:2015 certification pertaining to its Environmental Management System while pursuing ISO 50001 related to Energy Management. These would affirm the alignment of the certifications received by JMKI to the leading international standards and regulatory frameworks. Thus, the organization would place it not only as responsible but as a contemporary manufacturer. Jendamark India's sustainability is highly comprehensive and worthy in embracing renewable energy, resource conservation, waste reduction, and application of lean manufacturing principles. Jendamark India has classified energy emissions based on direct and indirect emissions and further controlled and non-controlled emission sources as Scope 1 and Scope 2. Their current CO_2 emission is 498339.10 kg and its further reduction target in 2025 is 25% reduction. Even though data related to a reduction in energy consumption was not readily available but energy consumption resulting from scope 1 and scope 2 is 26543 Kwh which is lower than other organizations in the same industry size.

All these enhance operational efficiencies but also position JMKI as an industry leader for sustainable manufacturing and consequently raise the industry's benchmark.

11.9.2 Outlook and Commitment to Sustainability

Sustainability focus, and commitment to continuous improvement will characterize the future of Jendamark India. In the near term as well as in the future term, the rising numbers of renewable energy projects that it undertakes-from expanding solar capacity to wind or biomass-based sources-of power-should further reduce its reliance on nonrenewable energy sources while also making its supply of energy more robust.

Resource efficiency will be more aggressive. JMKI feels that highly advanced water recycling techniques and better energy usage because of smart technologies and IoT monitoring solutions can be perceived. The data analytics that help them optimize their operations can further minimize the impacts on the environment and maximize efficiency. For that matter, diverted waste to landfills is another concern for them. JMKI aims to take its recycling and composting programs to the next level through research into new methods of waste management that take it forward. Sustainable packaging solutions created from recyclable material that uses minimum material will also form part of the focus. Equal importance is given toward involving stakeholders in the sustainability activities of Jendamark India. This includes new proposals on sustaining employee training programs that tailor sustainability into the culture of stewardship over the environment as mentioned in Fig. 11.7.

JMKI will also add supplier and customer engagement with the goal of promoting more sustainable practices at every value chain level. Certification as well as compliance will be part and parcel of the sustainability policy of JMKI. The organization is looking forward to more certifications in addition to better adherence to growing environmental regulations to ensure that its activities are found to be well aligned with the best international standards. Innovation shall propel JMKI to work hard about its sustainability efforts. The organization is putting much effort into the research and development of increasingly sustainable production methodologies and products. The integration of artificial intelligence and machine learning will allow it to improve its predictive maintenance functionalities, which means minimizing downtime and allowing for efficient task completion. Jendamark India is characterized by some noteworthy perspectives: sustainability, driven by improvements and the

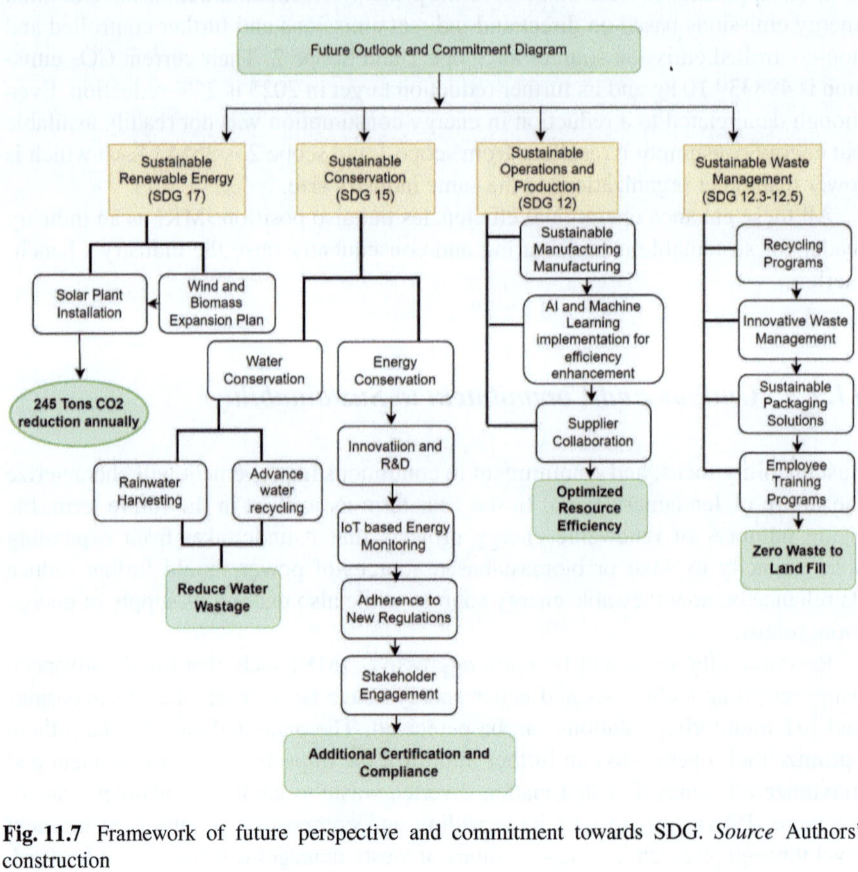

Fig. 11.7 Framework of future perspective and commitment towards SDG. *Source* Authors' construction

involvement of stakeholders in technological progress. Such a better sustainability practice would place JMKI on top as an industry leader by raising the bar on such best practices.

References

Ahmad S, Abdullah A, Talib F (2021) Lean-green performance management in Indian SMEs: a novel perspective using the best-worst method approach. Benchmarking: Int J 28:737–765

Althabatah A, Yaqot M, Menezes B, Kerbache L (2023) Transformative procurement trends: integrating industry 4.0 technologies for enhanced procurement processes. Logistics 7:63

Flint IP, Allwood JM, Serrenho AC (2019) Scrap, carbon and cost savings from the adoption of flexible nested blanking. Int J Adv Manuf Technol 104:1171–1181

Gandhi J, Thanki S, Thakkar JJ (2021) An investigation and implementation framework of Lean Green and Six Sigma (LG&SS) strategies for the manufacturing industry in India. The TQM J 33:1705–1734

Gupta V, Narayanamurthy G, Acharya P (2018) Can lean lead to green? Assessment of radial tyre manufacturing processes using system dynamics modelling. Comput Oper Res 89:284–306

Jyotish NK, Singh LK, Kumar C, Singh P (2023a) Reliability and performance evaluation of safety-critical instrumentation and control systems of nuclear power plant. IEEE Trans Reliab 73:422–437

Jyotish NK, Singh LK, Kumar C, Singh P (2023b) Reliability and performance measurement of safety-critical systems based on petri nets: a case study of nuclear power plant. IEEE Trans Reliab 72:1523–1539

Karedla Y, Mishra R, Patel N (2021) The impact of economic growth, trade openness and manufacturing on CO_2 emissions in India: an autoregressive distributive lag (ARDL) bounds test approach. J Econ, Financ Adm Sci 26:376–389

Il KS, Hur U, Ko YD (2021) Sustainable electric personal mobility: the design of a wireless charging infrastructure for urban tourism. Sustainability 13:1270

Martín-Gómez AM, Agote-Garrido A, Lama-Ruiz JR (2024) A framework for sustainable manufacturing: integrating industry 4.0 technologies with industry 5.0 Values. Sustainability 16:1364

Mattsson J (1992) A service quality model based on an ideal value standard. Int J Serv Ind Manag 3:0

Pansare R, Yadav G, Nagare MR (2024) Integrating operational excellence strategies with Industry 4.0 technologies through reconfigurable manufacturing system practices. The TQM J 36:3–23

Piron M, Wu J, Fedele A, Manzardo A (2024) Industry 4.0 and life cycle assessment: Evaluation of the technology applications as an asset for the life cycle inventory. Sci Total Environ 916:170263

Pulicherla KK, Adapa V, Ghosh M, Ingle P (2022) Current efforts on sustainable green growth in the manufacturing sector to complement "make in India" for making "self-reliant India." Environ Res 206:112263

Srhir S, Jaegler A, Montoya-Torres JR (2023) Introducing a framework toward sustainability goals in a supply chain 4.0 ecosystem. J Clean Prod 138111

Turoń K, Kubik A, Chen F (2021) When, what and how to teach about electric mobility? an innovative teaching concept for all stages of education: lessons from Poland. Energies (Basel) 14:6440

Vanus J, Bilik P (2022) Research on micro-mobility with a focus on electric scooters within smart cities. World Electr Veh J 13:176

Chapter 12
Assessing the Economic Viability of AI in Environmental Management

Mohammad Haseeb, Md. Mominur Rahman, Mustafa Kamal, Sachin Ghai, and Neeru Sidana

Abstract This chapter analyzes the economic implications of incorporating AI into environmental management within India's industrial sector. Regulatory obstacles, insufficient technical expertise, elevated implementation expenses, and inadequate infrastructure impede the extensive adoption of AI in manufacturing, notwithstanding the technology's significant potential for enhancing resource efficiency, mitigating pollution, and promoting sustainability. This study aims to address the gap by analyzing the essential success factors for AI in this context from an economic viewpoint. A total of 254 participants from diverse industrial enterprises in India were surveyed. The research analyzed the interconnections among significant variables utilizing PLS-SEM, or Partial Least Squares Structural Equation Modeling. Economic incentives, legal backing, organizational preparedness, data accessibility and integrity, technical competencies, implementation expenses, and other pertinent elements are highlighted as essential determinants for AI adoption. The report emphasizes the necessity of tackling challenges associated with cost, infrastructure, data accessibility, and organizational readiness to fully harness the transformative potential of AI in environmental management. It moreover necessitates a legislative framework that is more accommodating and targeted funds to promote the utilization

---

M. Haseeb (✉)
School of Economics and Management, and Center for Industrial Economics, Wuhan University, Wuhan, China
e-mail: haseeb.ecb@gmail.com

M. Haseeb · S. Ghai
Department of Management Studies, Graphic Era Deemed to be University, Dehradun, Uttarakhand, India

Md. M. Rahman
Bangladesh Institute of Governance and Management, Dhaka, Bangladesh
e-mail: mominurcou@gmail.com

M. Kamal
Department of Basic Sciences, College of Science and Theoretical Studies, Saudi Electronic University, Dammam, Saudi Arabia

N. Sidana
Amity School of Economics, Amity University, Noida, India
e-mail: nsidana@amity.edu

S. K. Tiwari et al. (eds.), *AI-Driven Environmental Pollution Management*, Climate Risks and Solutions, https://doi.org/10.1007/978-3-031-96243-1_12

of AI. This paper fills a critical gap in the literature by presenting a comprehensive technique for quantifying the potential financial advantages of AI technologies in green efforts. Individuals engaged in the field, legislators, and scholars can all derive significant advantages from these results. The findings indicate a favorable potential for employing AI in sustainable environmental management, contingent upon the resolution of existing issues and the establishment of appropriate conditions for its implementation.

Keywords Environmental management · Artificial Intelligence · Economic viability · Environmental awareness

12.1 Introduction

AI has revolutionized various domains, including environmental management, by offering innovative solutions to complex ecological challenges (Di et al. 2019). Conservation initiatives, pollution reduction, and resource management can all gain from AI's superior monitoring, forecasting, and informed decision-making capabilities facilitated by data-driven insights (Cowls et al. 2021; Vagin et al. 2022). Technology's capacity to deliver precise, rapid, and adaptable solutions underscores its significant role in addressing contemporary and forthcoming environmental challenges (Ayoubi and Tabaa 2023; Cowls et al. 2021). To fully actualize AI's potential, its economic feasibility must be meticulously examined, particularly in resource-limited sectors (Nishant et al. 2020).

The integration of AI in environmental management necessitates substantial investment in infrastructure, expertise, and technological resources; therefore, evaluating the economic viability of this technology is essential (Ghani et al. 2022). Despite AI's undeniable potential to enhance environmental results, enterprises may find it challenging to justify its implementation without a comprehensive cost–benefit analysis (Shalu 2023; Al-Bakri and Kisswani 2024). The potential for efficiency and precision provided by AI is hindered by significant obstacles, such as substantial implementation costs, extensive data requirements, and considerable energy usage. Consequently, the economic analysis of AI is more vital (Rizvi et al. 2021; Cowls et al. 2021). The significant contributions of AI, including its strategic value in addressing critical environmental issues, optimizing resources, and promoting long-term sustainability, ought to be incorporated into economic evaluations alongside direct financial expenditures (Nishant et al. 2020; Somayaji et al. 2022).

Industrial operations in India's manufacturing sector frequently generate significant ecological issues, necessitating the adoption of more sustainable methods; hence, AI-driven environmental management is particularly pertinent in this domain (Ghani et al. 2022; Rizvi et al. 2021). India continues to face persistent issues of overexploitation of natural resources, pollution, and industrial waste, despite its position as a significant global industrial hub (Nayal et al. 2021; Rizvi et al. 2021). Artificial intelligence offers a transformative opportunity to address these challenges by enhancing resource efficiency, minimizing waste, and ensuring adherence to environmental regulations (Ghosh 2022; Somayaji et al. 2022). Nonetheless, the complete potential of AI remains untapped due to skepticism regarding the technology's financial feasibility and the lack of a systematic method to assess its ROI (Nishant et al. 2020). Enhanced adoption decisions may foster sustained industrial advancement if the economic dynamics of AI in this context are comprehensively understood (Akintuyi 2024; Di-Vaio et al. 2020).

Although there is broad consensus on the necessity of economic evaluation, a considerable knowledge gap exists about precise criteria for assessing the financial viability of artificial intelligence in environmental management (Di et al. 2019; Shalu 2023). Organizations encounter difficulties in evaluating the operational and financial impacts of AI implementation in this domain due to the absence of established metrics and supplementary measurements (Jejeniwa et al. 2024; Cowls et al., 2021). This research aims to address the knowledge gap by identifying the elements influencing the financial viability of employing artificial intelligence in environmental management and subsequently evaluating those factors. The proposed technique will not only aid in cost-effectiveness studies but will also illuminate the broader operational and financial implications of AI adoption for businesses (Vagin et al. 2022; Ayoubi and Tabaa 2023).

The primary aim of this study is to establish a foundation for a cost-benefit analysis of AI in environmental management, focusing on the industrial sector. Specifically, we aim to achieve the following objectives: (1) to catalog and delineate the principal factors influencing the economic viability of artificial intelligence in environmental applications; (2) to prioritize the significance and impact of each factor; and (3) to develop a practical evaluation framework for stakeholders to assess the potential success of AI-driven green initiatives. This study aims to address a knowledge gap by offering evidence-based metrics for assessing the long-term sustainability of AI inside the economy, concentrating on the previously specified objectives.

This study significantly enhances the existing knowledge base by establishing a complete framework for the economic evaluation of AI in environmental management, with possible practical applications in the industrial sector. It provides scholars, policymakers, and field professionals with pragmatic guidance by identifying and analyzing important influencing factors. The framework facilitates better informed and economically viable AI applications in environmental management, serving as a foundation for future research to enhance and broaden its scope.

12.2 Literature Review

12.2.1 AI Applications in Environmental Management

Artificial intelligence technologies are revolutionizing data use for the monitoring, forecasting, and optimization of ecological systems in environmental management. AI is augmenting key competencies via advanced algorithms, machine learning techniques, and real-time data processing. Artificial intelligence is essential in managing climate change by predicting weather patterns, simulating climatic impacts, and developing effective mitigation strategies. AI-powered sensors monitor air and water quality in real-time, facilitating prompt responses to environmental risks in pollution management. AI has enabled precise water utilization, forest surveillance, and biodiversity assessment, contributing to resource conservation. Moreover, AI-driven energy efficiency solutions optimize use in both urban and industrial settings. Collectively, these AI-driven innovations enhance decision-making, mitigate environmental impacts, and facilitate more sustainable resource management.

The advantages of integrating AI and ML into environmental management are increasingly evident, as they will enhance data analysis, monitoring, and resource optimization. When used to environmental monitoring, machine learning algorithms can analyze extensive datasets to identify trends and outliers that may have been overlooked by traditional methods. For instance, machine learning can precisely assess both natural and artificial surroundings, offering critical insights into potential hazards and methods for improving these contexts (Shalu 2023). Similarly, Di et al. (2019) demonstrate that machine learning may be employed to evaluate water quality, identify pollution sources, and assist in watershed management through data-driven methodologies. These examples illustrate how artificial intelligence facilitates environmental monitoring to support decisions with data.

Artificial intelligence has achieved considerable advancements in water conservation and several aspects of resource management. To ensure sustainable water utilization, the development of AI-enabled smart water management systems that integrate IoT technologies. These systems provide proactive water conservation through real-time data collection and analysis. Furthermore, Vagin et al. (2022) assert that environmental business practices can be optimized through the application of big data and artificial intelligence to address challenges and improve resource allocation while safeguarding the environment. To mitigate the impacts of climate change and resource scarcity, such technologies are essential.

Integrating investment strategies with environmental sustainability objectives is another domain where AI is influencing the finance sector. According to Olanrewaju et al. (2024), data-driven investment methods can identify possibilities that support net-zero objectives and alleviate associated risks. This approach emphasizes the broader economic advantages of artificial intelligence in environmental management while promoting sustainable practices by directing resources to projects that positively affect the environment. These applications demonstrate the transformative potential of AI across various industries, driving substantial progress toward sustainability.

12.2.2 Economic Assessment Frameworks for AI

Assessing the strategic value, long-term benefits, and financial viability of AI investments necessitates economic evaluation frameworks for AI applications. The predominant method for assessing the advantages and disadvantages of AI is the cost-benefit analysis (CBA), which juxtaposes expenditures (including setup, maintenance, and operational costs) against prospective benefits (such as resource savings, enhanced productivity, and increased efficiency). Likewise, enterprises contemplating substantial first investments in AI can significantly benefit from return on investment (ROI), a fundamental statistic that juxtaposes financial rewards against expenditures. Supplementary metrics, such as the payback period and net present value (NPV), illuminate the overall value generated over time and the rate at which expenditures related to AI can be recovered. These financial methods enable organizations to effectively align financial and environmental objectives, thereby establishing a robust framework for assessing the economic viability of AI.

As artificial intelligence becomes increasingly prevalent in sectors such as healthcare, finance, education, and environmental management, it is essential to evaluate its economic implications. This evaluation necessitates a comprehensive approach that considers both financial and social-ethical implications. Determining the cost-effectiveness of AI in healthcare is challenging due to the complexity of healthcare outcomes and indirect benefits. To comprehensively understand the indirect benefits and enduring consequences of AI-based medical devices, Farah et al. (2024) emphasizes the importance of revising health technology assessments. Despite substantial data supporting AI's effectiveness in healthcare, economic evaluations often lack rigor and methodological soundness, as noted by Schwendicke et al. (2022). Wolff et al. (2020) underscore the necessity for comprehensive, rigorous cost-effectiveness research to assess the total economic impact of AI on healthcare, thereby addressing this knowledge gap.

Xiao (2023) explores economic frameworks to elucidate how AI is transforming economic paradigms and enhancing decision-making and forecasting capacities. For instance, in the banking sector, AI's predictive analytics enhance credit risk assessment, resulting in reduced financial losses from erroneous evaluations of borrower creditworthiness (Brown 2024). Al-Ahmad et al. (2023) advocates for the integration of AI with traditional economic forecasting techniques to enhance the reliability and precision of macroeconomic predictions. Ningsih (2023) illustrates that the economic ramifications of AI extend beyond healthcare and finance to encompass educational technology. Artificial intelligence enhances the efficacy of evaluations and the caliber of training. Moreover, Jejeniwa et al. (2024) examines the influence of AI on developmental economics, particularly how AI-driven credit scoring systems have facilitated access to financial services and mitigated poverty. The numerous applications of AI highlight the necessity for comprehensive economic evaluation frameworks that can assess both direct and indirect impacts across several sectors. By addressing these concerns, organizations and policymakers can attain a deeper understanding of AI's societal, ethical, and economic implications, resulting in more strategic and sustainable utilization of the technology.

12.2.3 Factors Influencing Economic Viability of AI in Environmental Management

AI possesses the capacity to transform environmental management; nevertheless, its feasibility and cost-efficiency are significantly reliant on several essential factors. The initial setup, software acquisition, and ongoing maintenance expenses related to AI adoption provide a significant barrier. These charges are particularly onerous for financially constrained small and medium-sized enterprises (SMEs). Besides financial constraints, the effectiveness of AI systems is significantly reliant on the accessibility and quality of data. Artificial intelligence models depend on substantial quantities of high-quality, accurate data, which can be difficult or expensive to acquire. AI solutions may lack reliability and accuracy, hence constraining their practical utility for enterprises managing incomplete or insufficient data.

The technological complexity of AI systems is another significant element. According to Zhen and Rahman (2024), reporting transparency is positively associated with sustainable performance. High-performance computers and advanced algorithms are necessary to manage the complex data associated with environmental management. While these talents enable precise predictions and pragmatic recommendations, they can be costly and necessitate specialized knowledge for their utilization. This may impose an additional financial strain on companies that lack technical preparedness by necessitating increased expenditures on people training and technical assistance.

The potential of AI for economical environmental management is significantly shaped by public perception and existing regulatory frameworks. Government policies, incentives, and legislation can mitigate financial obstacles and promote AI adoption. Conversely, AI applications may have limitations in scalability due to outdated or too restrictive regulations. Raising public awareness of AI's ability to address environmental issues is equally essential. As awareness of sustainability and climate change increases, firms may discover new market opportunities and be more inclined to adopt AI-driven solutions. Another set of variables influencing AI adoption is the readiness of the infrastructure and the organization. Effective deployment of AI systems necessitates adequate infrastructure, encompassing robust computing resources and reliable connection. Operational inefficiencies can elevate implementation costs and diminish overall effectiveness in areas with constrained technological infrastructure. To enhance the accessibility of AI deployment for a broader spectrum of organizations, economic incentives such as government grants, subsidies, and favorable financing options can mitigate the substantial initial costs.

Due to the substantial energy consumption of numerous AI technologies, particularly those managing extensive data, such as deep learning, energy efficiency is a critical consideration. Operational cost reduction and alignment with sustainability objectives are accomplished by optimizing AI systems for energy efficiency. This reduces their ecological footprint. The implementation of AI in environmental management is more economically feasible when stakeholders collaborate. Governments, technology firms, environmental organizations, and other stakeholders can collaborate to share resources, reduce expenses, and effectively relate sustainable resource management (Rahman and Hossain 2024). Organizations can facilitate the economical use of AI in environmental initiatives by addressing the following factors: data accessibility, technological preparedness, regulatory endorsement, public perception, infrastructure, economic incentives, and energy efficiency. To ensure AI fulfills its potential in enhancing environmental management practices and promoting sustainable solutions, we must consider these issues. Various factors, including technology advancements, organizational readiness, and favorable government legislation, affect the economic sustainability of AI in environmental management. The degree to which AI may be integrated into environmental management strategies is determined by these elements, which subsequently impact both economic outcomes and sustainability objectives.

Technological expertise is essential for evaluating the financial sustainability of artificial intelligence in environmental management. Enhancing operational efficiency can be accomplished through the implementation of adaptive AI technology, enabling real-time monitoring and adjustment of environmental circumstances. This is particularly applicable in precision agriculture. Consequently, agricultural enterprises obtain significant economic advantages from diminished environmental effects, enhanced crop production, and optimal resource management (Akintuyi 2024). Moreover, data-driven decision-making is crucial for sustainable practices across various sectors, such as resource management and agriculture, and the integration of AI with IoT devices facilitates this (Di-Vaio et al. 2020). AI can promote economic sustainability by reducing operational expenses and boosting productivity through automation and advanced analytical insights (Ayoubi and Tabaa 2023).

The implementation of artificial intelligence in environmental management is contingent upon an organization's level of preparedness. Research by Ghani et al. (2022) indicates that for successful AI technology integration, firms must have the requisite skills, cultivate an acceptable culture, and secure leadership support. Research indicates that a culture fostering technological innovation flourishes with robust support from upper management, which is crucial for facilitating the adoption of AI (Wong et al. 2024). Udayanan et al. (2024) assert that firms must possess robust data management systems and a culture that embraces change to enable its infrastructure to accommodate AI applications. To ensure the successful implementation of AI efforts, it is essential to integrate them with the firm's strategic objectives (Jokonya 2024).

Legislative measures and regulatory structures, with organizational factors, are essential in assessing the economic viability of AI implementation in environmental management. Good governance may facilitate AI adoption by providing resources for research and development and establishing regulations that promote sustainable practices (Cowls et al. 2021). Additionally, to address issues related to data privacy, security, and ethical considerations, the government must intervene and create a conducive atmosphere for the use of AI (Nishant et al. 2020). AI possesses the capacity to promote environmental sustainability and yield economic advantages; however, this potential is significantly amplified by the interplay of governmental support and organizational competencies (Al-Bakri and Kisswani 2024).

12.2.4 Challenges and Barriers to AI Implementation

AI has several significant challenges that may hinder its widespread application in environmental management. The substantial cost of initial investment in AI systems constitutes a significant barrier. A significant percentage of this budget is allocated for the acquisition of computers, software, and the skilled personnel required to operate and maintain these systems. These preliminary expenditures may be excessive for some organizations, especially those facing financial constraints. Alongside the financial investment, substantial volumes of high-quality data are typically necessary for AI applications; acquiring this data can be particularly difficult in developing regions or industries with insufficient data infrastructure. Numerous firms may lack the requisite resources or specialized technical expertise to create, maintain, and operate AI systems, hence complicating implementation. Moreover, infrastructural challenges, such sluggish internet, insufficient sensors, or inadequate processing capacity, may significantly hinder the effective implementation of AI systems, particularly in resource-limited areas.

In addition to these technical and practical hurdles, there are environmental and ethical factors to address when using AI. The energy consumption of AI models is a significant concern, particularly for resource-demanding applications such as deep learning and transnational education (Islam et al. 2024). The ecological advantages of AI may be negated by the substantial energy consumption and rising carbon emissions associated with these models, which necessitate significant computational capacity. AI impacts the environment not only during its operation but also during the production and disposal of electronic components, raising concerns around electronic waste and the extraction of raw materials. Addressing these ethical and environmental concerns is increasingly imperative as the use of AI continues to escalate. To guarantee the enduring sustainability of AI technologies in environmental management while reducing their ecological impact, it is essential to create energy-efficient algorithms, advocate for sustainable AI hardware, and implement responsible recycling practices. Numerous challenges impede the integration of AI across several businesses. These issues encompass monetary, technological, ethical, and sociological dimensions. In the healthcare sector and among small and medium-sized enterprises

(SMEs), constrained financial resources provide a significant barrier to the extensive use of AI. Many firms are reluctant to use AI due to the substantial costs associated with acquiring and implementing these technologies. American hospitals have indicated that the expenditures associated with AI systems and inadequate reimbursement procedures have impeded the technology's integration into healthcare environments (Van-Leeuwen et al. 2023). Furthermore, small and medium-sized firms (SMEs) often lack the financial resources to invest in AI systems, despite the critical role these systems play in their growth and competitiveness (Böhme et al. 2024; Oldemeyer et al. 2024). Targeted financial assistance to promote the adoption of AI is essential, as the scarcity of resources exacerbates the issue.

The slow uptake of AI technologies is primarily attributable to technological challenges. Insufficient infrastructure, substandard data, and a lack of technological expertise are prevalent issues across numerous businesses. In healthcare, the efficient implementation of AI applications is obstructed by limitations in clinical data and the necessity for strong data-sharing networks (Borkowski et al. 2022; Oduoye et al. 2024). A further impediment to effective AI implementation in industries such as construction is the sector's fragmented nature, which complicates data collection and integration (Regona et al. 2022). A significant challenge, particularly in regions with limited access to higher education, is the deficiency of trained personnel to develop, construct, and manage AI systems (Barsha and Munshi 2023).

Discourse regarding the utilization of AI focuses on societal and ethical concerns. To foster confidence in AI systems, it is essential to address significant issues such as algorithmic bias and the ethical utilization of data (Benjamens et al. 2020; Jussupow et al. 2022). Zhai et al. (2021) and Bitkina et al. (2020) discovered that healthcare professionals and patients exhibit resistance to the implementation of AI in clinical practice, exacerbated by insufficient understanding of the technology and concerns around job displacement. One method to mitigate these resistance factors is to cultivate trust through transparent communication and the inclusion of stakeholders in the development process (Ciecierski-Holmes et al. 2022). The regulatory framework further complicates the implementation of AI. Prospective adopters exhibit reluctance and uncertainty owing to the absence of defined standards and regulations for AI applications, particularly in sensitive sectors such as healthcare (Yoshida and Kiyuna 2021). To foster innovation and ensure the safe and effective application of AI technologies, regulatory bodies must establish frameworks.

12.2.5 AI in the Indian Manufacturing Sector: Current Trends and Limitations

The manufacturing sector in India is undergoing a profound revolution due to the integration of AI. This change enhances productivity, optimizes operational procedures, and resolves numerous challenges historically associated with conventional manufacturing practices. The industrial sector, a crucial component of India's economy, is increasingly employing AI to enhance production, reduce costs, and improve product quality. This aids the nation in maintaining international competitiveness (Ghosh 2022). Predictive maintenance is prominent among the several AI applications utilized in India's industrial industry. This approach employs machine learning and deep learning algorithms to predict mechanical failures, hence reducing expensive repairs and operational interruptions. Despite being in its nascent stages in India, Ghosh (2022) emphasizes that AI-driven predictive maintenance possesses significant potential to enhance operational efficiency. Artificial intelligence-driven robots and automation have revolutionized manufacturing processes, yielding significant cost reductions and enhanced efficiency (Ghosh 2023). Manufacturers must undergo this shift to be competitive in today's rapid, global economy.

The integration of AI in India's manufacturing sector faces several challenges, notwithstanding the promising advancements. Rizvi et al. (2021) identify challenges in integrating AI into industry, including inadequate infrastructure, a shortage of skilled personnel, and the need for international cooperation. An extensive framework that promotes infrastructure development and talent enhancement is essential for the effective adoption of AI to address these difficulties (Ghosh 2023; Rizvi et al., 2021). Research underscores the significance of organizational adaptability in the integration of AI and IoT technologies, which are essential for optimizing manufacturing supply chain performance (Nayal et al. 2021).

Besides improving operations, the implementation of AI will significantly impact the workforce. The employment landscape in the manufacturing sector is anticipated to evolve with the implementation of AI technology. AI may reduce the demand for unskilled labor while simultaneously creating new opportunities for skilled employment; thus, investment in workforce development and training is essential. This dual effect underscores the necessity for comprehensive employee training to address the problems of an AI-driven industrial environment. The utilization of AI is escalating in India's manufacturing sector, especially in environmental management, where it enhances efficiency and reduces waste. AI assists firms in complying with increasingly stringent environmental regulations by optimizing energy consumption, monitoring emissions, and enhancing waste management processes. Machine learning algorithms are being utilized in supply chain sustainability, production line optimization to minimize resource consumption, and environmental impact forecasting. Furthermore, AI-driven solutions provide real-time monitoring of water and air quality, providing enterprises with a competitive advantage in mitigating pollution and hazardous emissions. The integration of AI into numerous sectors in India could lead to more sustainable production processes, significantly enhancing environmental outcomes.

Despite the significant potential of AI, numerous challenges impede its widespread adoption in India's manufacturing industry. Smaller and medium-sized enterprises (SMEs) often experience challenges related to inadequate infrastructure concerning current computing resources, high-speed internet, and reliable data collection systems. Furthermore, manufacturers encounter difficulties in managing environmental regulations, which can be convoluted and outdated, perhaps failing to fully support or promote the adoption of AI. The high costs of deploying AI technology and the requirement for specialized knowledge further constrain AI's scalability in the business. To surmount these challenges and render AI a practical instrument for advancing sustainable industrial practices, collaboration is essential to enhance infrastructure, modernize regulatory frameworks, and provide financial support.

12.2.6 Research Gap and Need for Economic Viability Assessment

There is a significant deficiency of literature regarding established methodologies for evaluating the economic viability of AI in environmental management, despite its increasing recognition. Despite several research investigating the advantages and disadvantages of AI in this domain, no comprehensive and measurable criteria have been established to assess the financial and operational impacts of AI implementation. Organizations cannot accurately evaluate the ROI and broader economic effects of AI technologies because of this gap. Enterprises find it challenging to assess and compare AI solutions across industries due to the absence of universally recognized measurement standards. The objective of this research is to enumerate and delineate the essential parameters influencing the financial viability of artificial intelligence in environmental management. The stated objective is to assist organizations in assessing the financial feasibility of AI applications by offering a systematic framework for cost-effectiveness analysis and valuable insights.

12.3 Methodology

12.3.1 Data and Sample

This study examines the economic viability of artificial intelligence in environmental management, specifically inside manufacturing businesses in India. The structured survey aimed to collect data from professionals and decision-makers involved in AI implementation and environmental initiatives within these organizations. The survey questionnaire aimed to capture 23 variables, including costs, technical capabilities, data availability, regulatory considerations, and other factors influencing economic feasibility. A structured approach for assessing perceptions and experiences was utilized by documenting responses on a 5-point Likert scale, so ensuring consistency and reliability.

Deb et al. (2022) employed the ten-times rule to determine the minimum sample size required for Partial Least Squares Structural Equation Modeling (PLS-SEM). The poll comprised 23 items, necessitating a minimum of 230 responses for the sample. The study dataset is robust and offers a representative sample of the Indian

Table 12.1 Demographic attributes of the respondents

Demographic attribute	Category	Frequency	Percentage (%)
Industry sector	Automotive	64	25
	Textiles	51	20
	Chemicals and pharmaceuticals	46	18
	Electronics	38	15
	Other manufacturing sectors	55	22
Organizational size	Small enterprises (<50 employees)	76	30
	Medium enterprises (50–249)	115	45
	Large enterprises (250+)	63	25
Job roles	Senior management	89	35
	Middle management	102	40
	Technical/engineering roles	63	25
Experience in industry	Less than 5 years	51	20
	5–10 years	89	35
	More than 10 years	114	45

manufacturing industry, based on the 254 valid responses collected. Armed with this information, we can commence an investigation into the elements that influence the feasibility and cost-efficiency of employing AI in environmental management.

Table 12.1 presents the demographic data of the 254 respondents, encompassing their industry, firm size, job function, and years of experience. A significant majority of respondents (25%) are employed in the automobile industry, while 20% are engaged in textiles, 18% in chemicals and pharmaceuticals, 15% in electronics, and 22% in other manufacturing industries. According to employee count, 30% originate from small enterprises (50 or fewer employees), 45% from medium enterprises (50 to 249 people), and 25% from large enterprises (250 or more employees). Concerning jobs, 35% are employed in upper management, 40% in middle management, and 25% in engineering or technical fields. The subsequent analysis of respondents' years of industry experience is as follows: 20% possess less than 5 years of experience, 35% have 5–10 years, and 45% exceed 10 years. This diverse sample ensures that all relevant perspectives on AI application in environmental management within India's manufacturing sector are sufficiently represented.

12.3.2 Measurement of Variables

This paper provides a comprehensive operationalization of essential criteria to assess the economic viability of AI in environmental management. The objective of the CAI study is to assess if the costs related to the implementation and maintenance of AI systems are justifiable and manageable. The TCAI metric evaluates the efficacy of AI systems in processing complex environmental data and adapting to varying circumstances, emphasizing their ability to provide valuable, actionable insights for decision-making. To ensure precise and reliable predictions for effective environmental management, DAQ evaluates the volume and integrity of data required by AI models.

The variable RSP evaluates the role of governmental regulations and policies in facilitating AI adoption within environmental initiatives. Understanding the importance of public knowledge and involvement in fostering the adoption of AI technologies, EAPS evaluates the degree to which the general populace and diverse stakeholders accept AI's capacity to address environmental issues. When evaluating a company's AI readiness, ORE examines factors such as the availability of qualified personnel and the technological proficiency required to effectively manage AI systems.

Grants, subsidies, and investment opportunities exemplify financial resources that can support AI activities, and EIFA serves as a mechanism to monitor them all. This facilitates a more economical adoption of AI. The efficiency of AI solution deployment is assessed by the IAIC variable, which evaluates the robustness of technological infrastructure, encompassing computer resources and internet connectivity. The EEAI study examines the energy demands of AI systems, their alignment with sustainability goals, and their potential impact on operational costs and conservation initiatives.

Ultimately, the research considers SC, a metric that evaluates the efficacy of cooperation across environmental groups, governments, and organizations in promoting information exchange, expertise transfer, and the effective application of AI in environmental management. EM aims to enhance resource efficiency, reduce environmental footprints, and increase resilience to environmental challenges, with the deployment of artificial intelligence as its dependent variable. This study establishes a systematic approach to evaluate the economic viability and potential impacts of artificial intelligence in environmental management by operationalizing these factors through explicitly defined assessment items (Table 12.2).

12.4 Results and Discussions

12.4.1 Measurement Modelling

Table 12.3 presents the factor loadings for the examined variables, elucidating the degree of association between each assessment item and its respective construct. Hair et al. (2019) and Deb et al. (2022) assert that all factor loadings exceed the recommended threshold of 0.70, signifying that the measurement items exert a significant and dependable influence on their respective constructs. The items evaluating the CAI have loadings ranging from 0.78 to 0.82, indicating a robust association with the construct. Correspondingly, with loadings of 0.84 and 0.87, the TCAI elements

Table 12.2 Measurement items of the variables

Constructs	Descriptions	Measurement items
Cost of AI Implementation (CAI)	CAI captures the financial burden associated with initiating and maintaining AI systems, including setup, operational, and maintenance expenses. High costs may hinder adoption, especially if benefits do not quickly justify the investment	CAI1: "The initial cost of implementing AI systems is manageable within our organization's budget" CAI2: "The operational costs associated with AI applications are justifiable for our environmental goals"
Technical Capabilities of AI (TCAI)	TCAI refers to the ability of AI systems to process complex environmental data accurately and adapt to dynamic environmental conditions. Strong technical capabilities enhance AI's effectiveness in providing actionable insights for environmental management	TCAI1: "The AI systems we use have the capacity to accurately process complex environmental data" TCAI2: "Our AI tools are adaptable to changing environmental conditions"
Data Availability and Quality (DAQ)	DAQ considers the volume and accuracy of data necessary for training and running AI models. Quality data is crucial for AI systems to deliver reliable environmental predictions and inform sustainable management decisions	DAQ1: "We have sufficient data available to effectively train AI models for environmental management" DAQ2: "The data quality we possess is adequate to ensure accurate AI predictions and insights"

(continued)

Table 12.2 (continued)

Constructs	Descriptions	Measurement items
Regulatory Support and Policies (RSP)	RSP assesses the extent to which government regulations and policies facilitate the use of AI in environmental initiatives. Supportive policies, including incentives, can promote AI adoption, while restrictive regulations may pose barriers	RSP1: "Current regulations support the use of AI for environmental management effectively" RSP2: "Government policies provide adequate incentives for adopting AI-driven environmental solutions"
Environmental Awareness and Public Support (EAPS)	EAPS measures the level of public and stakeholder support for using AI in addressing environmental challenges. Positive awareness can drive engagement and funding, enhancing AI's impact on environmental outcomes	EAPS1: "There is strong public support for the use of AI in addressing environmental challenges" EAPS2: "Our stakeholders are well aware of the benefits AI can bring to environmental management"
Organizational Readiness and Expertise (ORE)	ORE indicates an organization's capacity and skill level in implementing and managing AI technology. Higher readiness includes having trained personnel and resources to maximize AI's benefits for environmental management	ORE1: "Our organization has the technical expertise required to implement AI in environmental management" ORE2: "We are well-prepared to handle AI-related projects effectively

(continued)

Table 12.2 (continued)

Constructs	Descriptions	Measurement items
Economic Incentives and Funding Availability (EIFA)	EIFA addresses the accessibility of financial resources, such as grants or subsidies, to support AI projects. Adequate funding can ease implementation costs, making AI more economically viable in environmental contexts	EIFA1: "Our organization has access to sufficient funding for AI-based environmental initiatives" EIFA2: "There are adequate economic incentives available to encourage the use of AI in environmental management"
Infrastructure for AI and Connectivity (IAIC)	IAIC reflects the presence of robust infrastructure, including internet and computing resources, necessary to support AI technologies. Reliable infrastructure is essential for real-time data processing and effective AI deployment in environmental management	IAIC1: "Our current infrastructure can support the AI technologies needed for environmental projects" IAIC2: "We have the necessary connectivity to enable real-time data processing for environmental management"
Energy Efficiency of AI Systems (EEAI)	EEAI assesses the energy demands of AI technologies and their alignment with environmental sustainability goals. Energy-efficient AI systems reduce operational costs and environmental impact, aligning with conservation efforts	EEAI1: "The AI systems we use are energy-efficient and align with our environmental objectives" EEAI2: "Energy consumption by AI tools is a significant consideration in our organization"

(continued)

Table 12.2 (continued)

Constructs	Descriptions	Measurement items
Stakeholder Collaboration (SC)	SC measures the degree of partnership and cooperation among organizations, governments, and environmental groups in supporting AI initiatives. Collaboration can enhance resource sharing, knowledge transfer, and overall, AI project success	SC1: "There is strong collaboration between our organization and stakeholders for AI-driven environmental solutions" SC2: "Our organization has effective partnerships with environmental and technology groups to support AI use"
Environmental Management (EM)	EM represents the effectiveness of strategies and technologies, including AI, in preserving natural resources, mitigating pollution, and promoting sustainability. Improved environmental management indicates positive outcomes such as increased resource efficiency, reduced ecological footprint, and enhanced resilience against environmental challenges	EM1: "The implementation of AI has enhanced the efficiency of our environmental management practices" EM2: "AI-driven tools have significantly improved our ability to monitor and manage environmental resources" EM3: "Using AI in environmental management has led to measurable positive outcomes in sustainability and resource conservation"

proficiently encapsulate the technical attributes of AI. DAQ and RSP exhibit factor loadings between 0.77 and 0.83, signifying their importance and dependability within the model. The factor loadings for the EM construct, the dependent variable, range from 0.85 to 0.88, indicating a high level of explanatory power for the associated measurement items. These findings affirm the measurement model's robustness; all factor loadings above the necessary threshold, supporting the constructs' validity and reliability within the study's framework.

Table 12.3 Factor loadings of the investigated variables

Constructs	Item	Factor loading
CAI	CAI1	0.78
	CAI2	0.82
TCAI	TCAI1	0.84
	TCAI2	0.87
DAQ	DAQ1	0.80
	DAQ2	0.83
RSP	RSP1	0.77
	RSP2	0.81
EAPS	EAPS1	0.79
	EAPS2	0.84
RE	ORE1	0.85
	ORE2	0.86
EIFA	EIFA1	0.81
	EIFA2	0.83
IAIC	IAIC1	0.78
	IAIC2	0.82
EEAI	EEAI1	0.84
	EEAI2	0.80
SC	SC1	0.79
	SC2	0.82
EM	EM1	0.86
	EM2	0.88
	EM3	0.85

This study assessed the components for reliability and validity by Cronbach's Alpha, Composite Reliability (CR), and Average Variance Extracted (AVE), as presented in Table 12.4. The results for Cronbach's Alpha, Composite Reliability, and AVE above the recommended thresholds, demonstrating elevated internal consistency and validity across all dimensions (Hair et al. 2019; Wang et al. 2024; Karim et al. 2023). The findings validate the reliability of the measuring model. The CAI assessment items effectively measure the concept and explain a significant portion of its variance, evidenced by Cronbach's Alpha, Composite Reliability, and AVE values of 0.76, 0.85, and 0.74, respectively. The TCAI have strong reliability and convergent validity, evidenced by a Cronbach's Alpha of 0.81, Composite Reliability of 0.89, and an Average Variance Extracted (AVE) of 0.79. The EM construct, as the dependent variable, demonstrates optimal ratings with a Cronbach's Alpha of 0.85, Composite Reliability of 0.91, and AVE of 0.82, signifying that its measurement items effectively represent the construct. These results illustrate the reliability and validity of the study's constructs.

Table 12.4 Reliability and validity analysis

Construct	Cronbach's Alpha	CR	AVE
CAI	0.76	0.85	0.74
TCAI	0.81	0.89	0.79
DAQ	0.78	0.86	0.75
RSP	0.74	0.83	0.71
EAPS	0.79	0.87	0.77
ORE	0.82	0.89	0.79
EIFA	0.80	0.88	0.76
IAIC	0.75	0.85	0.72
EEAI	0.77	0.85	0.73
SC	0.78	0.86	0.74
EM	0.85	0.91	0.82

The Heterotrait-Monotrait Ratio (HTMT) is a metric utilized to assess the uniqueness of structures inside the model. Table 12.5 presents the investigation of discriminant validity based on this metric. The constructs exhibit adequate discriminant validity, since all HTMT values fall below the recommended threshold of 0.90 (Hair et al. 2022). This outcome supports the conceptual differentiation of the notions by confirming their adequate distinctiveness and minimal association. An instance is the somewhat acceptable HTMT value of 0.65 between TCAI and CAI. Comparative analysis of the dependent variable, EM, against other constructs, including DAQ (0.66) and ORE (0.72), indicates HTMT values between 0.59 and 0.75. These figures further substantiate that, despite the interrelations among concepts, they remain distinct and should be regarded accordingly. The findings support the structural model and reinforce the notion that each component contributes uniquely to the research. The measurement model is more reliable as it complies with the set HTMT threshold, hence enhancing the credibility of the path analysis results and the inferences drawn from the correlations among the variables.

12.4.2 Structuring Modelling

Table 12.6 presents the results of the path analysis, highlighting the direct relationships between EM, the dependent variable, and the independent components. All independent variables—RSP, DAQ, Technical Capabilities of AI (TCAI), CAI, ORE, EIFA, IAIC, EEAI, and SC—exhibit a positive and statistically significant impact on EM, with p-values below the 0.05 threshold. In the realm of Environmental Management, the Technical Capabilities of AI (TCAI) exert the greatest influence. This finding indicates a significant association between improved environmental management methods and advancements in AI's technical capabilities. The coefficients of 0.30 and 0.28 demonstrate a substantial positive correlation between ORE

Table 12.5 Discriminant validity using HTMT

Construct	CAI	TCAI	DAQ	RSP	EAPS	ORE	EIFA	IAIC	EEAI	SC	EM
CAI											
TCAI	0.65										
DAQ	0.58	0.69									
RSP	0.53	0.61	0.63								
EAPS	0.47	0.55	0.50	0.60							
ORE	0.62	0.71	0.64	0.59	0.58						
EIFA	0.49	0.58	0.56	0.57	0.60	0.66					
IAIC	0.55	0.62	0.59	0.64	0.63	0.67	0.69				
EEAI	0.50	0.57	0.55	0.60	0.58	0.61	0.63	0.70			
SC	0.48	0.56	0.54	0.59	0.62	0.60	0.58	0.63	0.64		
EM	0.59	0.68	0.66	0.64	0.65	0.72	0.70	0.75	0.69	0.63	

and CAI. Our findings underscore the essential requirement of organizational readiness for implementing AI and financial viability in achieving effective environmental management outcomes.

Conversely, EEAI's coefficient is the lowest at 0.19, although still statistically significant, signifying the energy efficiency of AI systems. Financial, technical, regulatory, and collaborative elements collectively influence the economic feasibility and efficacy of AI in environmental management. The persistent statistical significance across all variables underscores the model's comprehensive nature. Although energy efficiency remains a significant consideration, its direct influence on Environmental Management is somewhat less prominent than other variables. To effectively utilize AI technology for sustainable environmental results, it is essential to consider various

Table 12.6 Path analysis

Independent variable	Relationship	Coefficient	t-value	p-value
Cost of AI Implementation	CAI → EM	0.28	3.42	0.000
Technical Capabilities of AI	TCAI → EM	0.32	4.03	0.000
Data Availability and Quality	DAQ → EM	0.25	2.87	0.004
Regulatory Support and Policies	RSP → EM	0.22	2.65	0.008
Environmental Awareness and Public Support	EAPS → EM	0.21	2.47	0.014
Organizational Readiness and Expertise	ORE → EM	0.30	3.56	0.000
Economic Incentives and Funding Availability	EIFA → EM	0.27	3.11	0.002
Infrastructure for AI and Connectivity	IAIC → EM	0.24	3.01	0.003
EEAI	EEAI → EM	0.19	2.18	0.029
Stakeholder Collaboration	SC → EM	0.23	2.71	0.007

Table 12.7 Evaluating the model

Statistic	Value
R-square (R^2)	0.75
Adjusted R-square (Adjusted R^2)	0.73
Q-square (Q^2)	0.68

factors, including financial viability, technical proficiency, organizational preparedness, regulatory backing, and teamwork. This comprehensive strategy is justified by these findings.

Table 12.7 presents the model's explanatory and predictive capabilities as quantified by R^2, Adjusted R^2, and Q^2 statistics. The model's independent variables account for 75% of the variation in EM, as indicated by an R-squared value of 0.75, reflecting strong explanatory power. The model's robustness is further substantiated by an Adjusted R-squared value of 0.73, after considering the ratio of predictors to sample size. This outcome indicates a little reduction in explanatory power, even when considering potential overfitting. Furthermore, the model effectively forecasts outcomes for the dependent variable and accounts for a substantial proportion of the variation (Q-squared = 0.68), demonstrating significant predictive validity. These findings substantiate the proposed framework's efficacy in evaluating the financial viability of artificial intelligence in environmental management, affirming its overall reliability and robustness. The model effectively elucidated the complexities of the correlation between AI adoption and environmental management results, as demonstrated by its strong explanatory and predictive capabilities.

To evaluate the model's fit, we can examine Table 12.8, which presents the outcomes utilizing the three principal fit indices: SRMR, NFI, and RMSE. The SRMR value, being below the recommended threshold of 0.08, suggests a high congruence between the predicted and observed correlation matrices, indicating a favorable model fit. The NFI score of 0.92 indicates a superior fit relative to the baseline null model, exceeding the threshold of 0.90. The model's capacity to accurately represent the relationships among variables is further corroborated by the RMSE value of 0.08, which falls within the acceptable range. Collectively, these fit indices demonstrate that the proposed model is robust and dependable in elucidating AI's economic viability in environmental management, as it is well-defined.

Table 12.8 Model fit statistics

Fit statistic	Value
Standardized Root Mean Square Residual (SRMR)	0.06
Normed Fit Index (NFI)	0.92
Root Mean Square Error of Approximation (RMSE)	0.08

12.4.3 Discussions

This study emphasizes the economic feasibility of AI in environmental management by underscoring the significance of many factors in this domain. The TCAI exhibited the highest effect coefficient of 0.32 among all evaluated categories. Consequently, enterprises must emphasize the enhancement of their AI competencies to attain improved environmental outcomes; AI systems capable of analyzing complex environmental data and adapting to evolving conditions are essential. ORE emphasizes the significance of internal factors such as skilled personnel and technology resources in environmental management, with a coefficient of 0.30. The importance of financial factors was further emphasized by other variables with considerable impacts, such as EIFA (0.27) and CAI (0.28). Although incentives and funding exist to facilitate AI deployment, firms must consider the long-term sustainability of AI adoption alongside initial costs. Successful implementation of AI in environmental applications necessitates a robust technical foundation, as evidenced by the role of IAIC (0.24). This foundation must encompass reliable computing resources and internet access.

Furthermore, SC (0.23) and EEAI (0.19) signify that these components remain important, but to a diminished degree. Energy-efficient artificial intelligence systems offer numerous advantages, including environmental protection and reduced operational costs. Increased accessibility to AI-driven environmental solutions will occur if stakeholders collaborate efficiently to optimize resources and disseminate information. Organizations, politicians, and stakeholders aiming to leverage AI for effective environmental management can significantly benefit from the study's pragmatic findings. Initially, organizations must invest in cutting-edge AI systems capable of processing complex environmental data due to the significant influence of Technical Capabilities of AI (TCAI). To improve decision-making and advance sustainability policies, enterprises must invest in high-performance AI technologies capable of adapting to evolving environmental conditions.

Enhancing capacity is essential, and the significance of ORE underscores this necessity. Organizations can enhance their internal capabilities and leverage AI's potential by investing in training programs and hiring skilled experts that can effectively integrate AI into environmental management systems. Personnel in the environmental sector will be more equipped to facilitate the integration of AI if governmental bodies and industry organizations collaborate to develop specialized training programs. The paper highlights that EIFA and CAI are pivotal in the adoption of AI. Financial incentives such as tax exemptions, grants, and subsidies might mitigate the financial obstacles to AI implementation; hence, authorities should contemplate their introduction or expansion. To make informed decisions that maximize long-term value, companies must meticulously assess the cost-benefit trade-offs of AI implementation and explore potential funding sources.

The significance of robust digital and energy infrastructures is paramount, as indicated by the findings of the IAIC and EEAI studies, which urge governments and technology firms to prioritize this matter. Governments should emphasize the establishment of reliable communication networks while promoting the adoption of AI technologies that decrease energy consumption. This approach will enhance the sustainability and scalability of AI programs, augmenting their use in environmental management. Finally, cross-sector coalitions are crucial, as evidenced by the impacts of SC. Corporations, non-governmental organizations (NGOs), and governmental entities should collaborate to address environmental issues through the utilization of AI. Partnerships may accelerate innovation and create synergies, leading to more effective and efficient solutions for important environmental issues.

12.5 Conclusions

This paper examines the financial viability of utilizing AI for environmental management, using the industrial sector in India as a case study. This study enhances existing knowledge regarding sustainability and technological innovation by identifying the key characteristics influencing AI adoption and its efficacy in addressing environmental issues. The findings indicate that optimizing AI's potential to enhance environmental outcomes relies significantly on technical capabilities, organizational preparedness, and stakeholder engagement. The paper emphasizes the necessity of robust infrastructure, the advancement of energy-efficient technologies, and the establishment of supportive legislative laws to eliminate existing barriers to AI adoption. To assist companies in reconciling financial constraints with sustainability objectives, economic factors such as implementation expenses and funding availability are incorporated. This paper addresses a gap in the literature by offering quantitative metrics and a pragmatic technique for assessing the impact of artificial intelligence on sustainable development. It provides enterprises with the resources necessary to leverage AI beneficially, assisting them in achieving their financial and environmental objectives.

This research offers several important contributions to the fields of sustainable technological innovation and environmental management. This study establishes a comprehensive approach for evaluating the financial viability of AI in environmental management to address a significant gap in existing research. The study assesses the financial viability and environmental objectives of AI by delineating and implementing key variables, hence providing quantifiable structures. The study elucidates the significance of context in AI adoption within India's manufacturing sector, offering valuable insights into how AI environmental initiatives are influenced by sector-specific challenges such as infrastructural limitations and regulatory constraints. The findings provide a pragmatic perspective for those aiming to integrate AI into sustainability efforts, and their focus on context renders them particularly relevant to developing economies facing similar economic and environmental challenges. Finally, the study offers pragmatic recommendations for organizations, legislators, and stakeholders, so contributing to both practice and policy. It emphasizes the necessity of fostering stakeholder participation, enhancing technical competencies, and providing financial incentives to promote the adoption of AI technology for sustainable development. The study enhances scholarly discourse by integrating theoretical concepts with actual evidence. It also equips practitioners with pragmatic strategies to enhance AI investments for efficient environmental management.

References

Akintuyi OB (2024) Adaptive AI in precision agriculture: a review: investigating the use of self-learning algorithms in optimizing farm operations based on real-time data. Res J Multidisc Stud 7(02):016–030. https://doi.org/10.53022/oarjms.2024.7.2.0023

Ayoubi H, Tabaa Y (2023) Artificial intelligence in green management and the rise of digital lean for sustainable efficiency. In: E3S web of conferences, vol 412, EDP Sciences. p 01053. https://doi.org/10.1051/e3sconf/202341201053

Al-Bakri A, Kisswani M (2024) The role of artificial intelligence in supporting green economy in GCC: its economic, managerial and necessary legal framework impact. Artif Intell Legal Econ Prospects Challenges 66(3):759–778. https://doi.org/10.21608/jelc.2024.342122

Al-Ahmad ASM, Judijanto L, Tooy D, Putra P, Hermansyah M, Kumalasanti M, Agit A (2023) Integration of artificial intelligence and macro-economic analysis: a novel approach with distributed information systems. EAI Endorsed Trans Scalable Inf Syst. https://doi.org/10.4108/eetsis.4452

Brown M (2024) Influence of artificial intelligence on credit risk assessment in banking sector. Int J Modern Risk Manag 2(1):24–33. https://doi.org/10.47604/ijmrm.2641

Barsha S, Munshi SA (2023) Implementing artificial intelligence in library services: a review of current prospects and challenges of developing countries. Library Hi Tech News 41(1):7–10. https://doi.org/10.1108/lhtn-07-2023-0126

Benjamens S, Dhunnoo P, Meskó B (2020) The state of artificial intelligence-based FDA-approved medical devices and algorithms: an online database. NPJ Digital Med 3(1):118. https://doi.org/10.1038/s41746-020-00324-0

Bitkina OV, Jeong H, Lee BC, Park J, Park J, Kim HK (2020) Perceived trust in artificial intelligence technologies: a preliminary study. Human Factors Ergon Manuf Service Ind 30(4):282–290. https://doi.org/10.1002/hfm.20839

Borkowski AA, Jakey CE, Thomas LB, Viswanadhan N, Mastorides SM (2022) Establishing a hospital artificial intelligence committee to improve patient care. Federal Practitioner 39(8):334. https://doi.org/10.12788/fp.0299

Böhme C, Graf-Pfohl C, Meusinger K (2024) Human centered implementation process of AI in SMEs–conditions for success. In: First working conference on artificial intelligence development for a resilient and sustainable tomorrow, p 83. https://doi.org/10.1007/978-3-658-43705-3_7

Cowls J, Tsamados A, Taddeo M, Floridi L (2023) The AI gambit: leveraging artificial intelligence to combat climate change—opportunities, challenges, and recommendations. In: Ai & society, pp 1–25. https://doi.org/10.1007/s00146-021-01294-x

Ciecierski-Holmes T, Singh R, Axt M, Brenner S, Barteit S (2022) Artificial intelligence for strengthening healthcare systems in low-and middle-income countries: a systematic scoping review. NPJ Digital Med 5(1):162. https://doi.org/10.1038/s41746-022-00700-y

Di-Vaio A, Boccia F, Landriani L, Palladino R (2020) Artificial intelligence in the agri-food system: rethinking sustainable business models in the COVID-19 scenario. Sustainability 12(12):4851. https://doi.org/10.3390/su12124851

Di Z, Chang M, Guo P (2019) Water quality evaluation of the Yangtze River in China using machine learning techniques and data monitoring on different time scales. Water 11(2):339. https://doi.org/10.3390/w11020339

Deb BC, Rahman MM, Rahman MS (2022) The impact of environmental management accounting on environmental and financial performance: empirical evidence from Bangladesh. J Account Organ Chang 19(3):420–446

Farah L, Borget I, Martelli N, Vallee A (2024) Suitability of the current health technology assessment of innovative artificial intelligence-based medical devices: scoping literature review. J Med Internet Res 26:e51514. https://doi.org/10.2196/51514
Ghani EK, Ariffin N, Sukmadilaga C (2022) Factors influencing artificial intelligence adoption in publicly listed manufacturing companies: a technology, organisation, and environment approach. Int J Appl Econ Finance Acc 14(2):108–117. https://doi.org/10.33094/ijaefa.v14i2.667
Ghosh M (2022) AI-driven predictive maintenance in Indian manufacturing: enhancing operational efficiency. Innov Res Thoughts 8(4). https://doi.org/10.36676/irt.v8.i4.1512
Hair JF, Risher JJ, Sarstedt M, Ringle CM (2019) When to use and how to report the results of PLS-SEM. Eur Bus Rev 31(1):2–24
Islam ME, Polas MRH, Rahman MM (2025) Sustainability and AI in transnational higher education. In: Bridging global divides for transnational higher education in the AI era. IGI Global Scientific Publishing, pp 379–400. https://doi.org/10.4018/979-8-3693-7016-2.ch018
Jussupow E, Spohrer K, Heinzl A (2022) Identity threats as a reason for resistance to artificial intelligence: survey study with medical students and professionals. JMIR Formative Res 6(3):e28750. https://doi.org/10.2196/28750
Jejeniwa TO, Mhlongo NZ, Jejeniwa TO (2024) AI solutions for developmental economics: opportunities and challenges in financial inclusion and poverty alleviation. Int J Adv Econ 6(4):108–123. https://doi.org/10.51594/ijae.v6i4.1073
Karim KS, Islam ME, Ibrahim AM, Pan SH, Rahman MM (2023) Online marketing trends and purchasing intent: advances in customer satisfaction through PLS-SEM and ANN approach. Adv Decis Sci 27(4):1–31. https://doi.org/10.47654/v27y2023i4p24-54
Nishant R, Kennedy M, Corbett J (2020) Artificial intelligence for sustainability: challenges, opportunities, and a research agenda. Int J Inf Manage 53:102104. https://doi.org/10.1016/j.ijinfomgt.2020.102104
Ningsih F (2023) Classtime.com as an AI-based testing platform: analysing ESP students' performances and feedback. J Languages Language Teach 11(3):390–404. https://doi.org/10.33394/jollt.v11i3.8286
Nayal K, Kumar S, Raut RD, Queiroz MM, Priyadarshinee P, Narkhede BE (2022) Supply chain firm performance in circular economy and digital era to achieve sustainable development goals. Bus Strateg Environ 31(3):1058–1073. https://doi.org/10.1002/bse.2935
Oduoye MO, Fatima E, Muzammil MA, Dave T, Irfan H, Fariha FNU, Marbell A, Ubechu SC, Scott GY, Elebesunu EE (2024) Impacts of the advancement in artificial intelligence on laboratory medicine in low-and middle-income countries: challenges and recommendations—a literature review. Health Sci Rep 7(1):e1794. https://doi.org/10.1002/hsr2.1794
Olanrewaju OIK, Oduro P, Babayeju OA (2024) Exploring capital market innovations for net zero goals: a data-driven investment approach. Finance Accounting Res J 6(6):1091–1104. https://doi.org/10.51594/farj.v6i6.1244
Oldemeyer L, Jede A, Teuteberg F (2024) Investigation of artificial intelligence in SMEs: a systematic review of the state of the art and the main implementation challenges. Manag Rev Q 1–43. https://doi.org/10.1007/s11301-024-00405-4
Rahman MM, Hossain ME (2025) Synergy of governance, finance, and technology for sustainable natural resource management. J Open Innov Technol Market Complexity 11(1):100468. https://doi.org/10.1016/j.joitmc.2025.100468
Regona M, Yigitcanlar T, Xia B, Li RYM (2022) Opportunities and adoption challenges of AI in the construction industry: a PRISMA review. J Open Innov Technol Market Complexity 8(1):45. https://doi.org/10.3390/joitmc8010045
Rizvi AT, Haleem A, Bahl S, Javaid M (2021) Artificial intelligence (AI) and its applications in Indian manufacturing: a review. Current Adv Mech Eng Select Proc ICRAMERD 2020:825–835. https://doi.org/10.1007/978-981-33-4795-3_76
Shalu GS (2023) Environmental monitoring with machine learning. EPRA Int J Multidisc Res 208–212. https://doi.org/10.36713/epra13330

Schwendicke F, Grano C, de Oro J, Garcia Cantu A, Meyer-Lückel H, Chaurasia A, Krois J (2022) Artificial intelligence for caries detection: value of data and information. J Dent Res 101(11):1350–1356. https://doi.org/10.1177/00220345221113756

Udayanan AR, Bargavi N, Awasthi S, Deshmukh SV, Jadhav DY (2024) Determinants influencing the adoption of artificial intelligence in driving effective human resource management. J Inf Educ Res 4(2). https://doi.org/10.52783/jier.v4i2.828

Van-Leeuwen KG, de Rooij M, Schalekamp S, van Ginneken B, Rutten MJ (2024) Clinical use of artificial intelligence products for radiology in the Netherlands between 2020 and 2022. Eur Radiol 34(1):348–354. https://doi.org/10.1007/s00330-023-09991-5

Vagin SG, Klimenko VA, Telegina ZA, Aleksashina TV (2022) Improving environmental decision-making in environmental business-management using big data and AI. Front Environ Sci 10:951306. https://doi.org/10.3389/fenvs.2022.951306

Wang C, Rahman MM, Siddik AB, Wen ZG, Sobhani FA (2024) Exploring the synergy of logistics, finance, and technology on innovation. Sci Rep 14(1):21918. https://doi.org/10.1038/s41598-024-72409-9

Wong JW, Yap KHA (2024) Factors influencing the adoption of artificial intelligence in accounting among micro, small medium enterprises (MSMES). Quantum J Soc Sci Humanities 5(1):16–28. https://doi.org/10.55197/qjssh.v5i1.323

Wolff J, Pauling J, Keck A, Baumbach J (2020) Systematic review of economic impact studies of artificial intelligence in health care. J Med Internet Res 22(2):e16866. https://doi.org/10.2196/16866

Xiao Y (2023) The multifaceted relationship between AI and economics: impacts, challenges, and insights. J Econ Manag Sci 6(3):p1–p1. https://doi.org/10.30560/jems.v6n3p1

Yoshida H, Kiyuna T (2021) Requirements for implementation of artificial intelligence in the practice of gastrointestinal pathology. World J Gastroenterol 27(21):2818. https://doi.org/10.3748/wjg.v27.i21.2818

Zhen T, Rahman MM (2024) Greening emerging economies: enhancing environmental, social, and governance performance through environmental management accounting and green financing. Sustainability 16(11):4753

Zhai H, Yang X, Xue J, Lavender C, Ye T, Li JB, Xu L, Lin L, Cao W, Sun Y (2021) Radiation oncologists' perceptions of adopting an artificial intelligence–assisted contouring technology: Model development and questionnaire study. J Med Internet Res 23(9):e27122. https://doi.org/10.2196/27122

Chapter 13
Quantifying the Economic Benefits of Improved Environmental Quality in BRICS Economies

Mohd Shuaib, Shoaib Ansari, and Ashulekha Gupta

Abstract Although enhanced environmental quality postures challenges for traditional models of economic growth, also provides various opportunities for sustainable growth through numerous factors. Using a comprehensive analysis of environmental indicators and economic indicators, this chapter investigates the relationship between environmental quality and economic performance, including trade openness, government expenditure, and foreign investment in BRICS economies from 1990 to 2022. By applying numerous econometric techniques such as cross-sectional dependence, stationarity tests and cointegration test, we found the existence of cross dependency, stationarity, and cointegration between the variables across these economies. Furthermore, the findings of Common Correlated Effects Mean Group (CCEMG) estimator reveal that an improvement in environmental quality would also benefit economic growth BRICS economies. Furthermore, trade openness, government expenditure, and FDI positively impact the economic growth, whereas natural resources negatively affect economic growth BRICS economies. Based on these findings, the chapter recommends that BRICS policymakers focus on environmentally sustainable economic policies to reduce emissions and encourage investments in green technology and renewable energy. Given the negative impact of natural resource reliance, an emphasis on research and development for sustainable resource management is advised. These strategies aim to balance economic growth with environmental sustainability in the region.

M. Shuaib (✉)
School of Economics and Management, and Center for Industrial Economics, Wuhan University, Wuhan, China
e-mail: shuaib6593@gmail.com

M. Shuaib · A. Gupta
Department of Management Studies, Graphic Era Deemed to be University, Dehradun, Uttarakhand, India
e-mail: ashulekha1806@gmail.com

S. Ansari
Department of Economics, AKI's Poona College of Arts, Science and Commerce, SPPU, Pune, Maharashtra, India
e-mail: sansari83@myamu.ac.in

S. K. Tiwari et al. (eds.), *AI-Driven Environmental Pollution Management*, Climate Risks and Solutions, https://doi.org/10.1007/978-3-031-96243-1_13

Keywords Environmental quality · Economic growth · BRICS economies · CCEMG approach

13.1 Introduction

Emerging or growing economies signify to developing countries that possess rapid potential for growth and significant role in the world's economy. The economic expansion of these emerging economies has significantly affected the global economic structure and is assuming a gradually significant role in international relations. Developing countries frequently have greater populations, ample natural resources, and inferior costs of labor, although making notable development in industrialization, technological innovation and urbanization. Because of the rapid advancement of economic development, developing countries are challenged with the environmental pollution problem and consumption of resources. A global agenda adopted by the United Nations in 2015, defining the sustainable development goals (SDGs) for 2030 (Assembly 2015). Therefore, environmental problems are a huge challenge around the world, as well as developing countries. It has captivated the attention of academia, corporate bodies, governments, and international organizations. This is due to a continuing drop in global environmental quality. Akadiri et al. (2019) ascribed deterioration of environment to carbon dioxide emissions. It suggests that a rise in CO_2 emissions by energy consumption diminishes environmental quality. Conversely, Mesagan and Vo (2023) postulate that raised carbon emissions are frequently related to economic activities. Because of this, the association amid economic growth and environmental quality has concerned the interest of scholars across various disciplines, however the empirical evidence is inconsistent, likely due to differences in data selection, methodology, and geographical focus.

For example, Kahia et al. (2019) and Ahmad and Du (2017) have claimed that there is a straight connection amid environmental quality and economic activities. It represents that economic development measures are escorted by a consistent rise in CO_2 emissions which accordingly disturb the environmental quality. In order to boost growth, economies use energy sources such as coal, fossil fuel and gas, which are perceived to be cost-effective yet have significant environmental repercussions. The purpose is to optimize the benefits derived from economic activity (Olunkwa et al. 2021). This argument thus presented the environmental Kuznet curve's first-order condition, indicating that at the primary stage of economic growth, CO_2 emissions increases, affecting the environmental quality negatively due to the insufficient wealth to implement less carbon-emitting technologies to push their needed economic growth (Olaoye and Dauda 2022).

Raising other lines of argument, according to studies investigated by Salahuddin and Gow (2019), do not support the evidence for a direct or indirect association of environmental quality with economic growth upon which environmental policies aimed at preventing environmental pollution and improving environmental quality could be based. The influence of environmental quality and economic growth on each other still needs to be established. In this regard, Abdouli and Hammami (2017) showed a uni-direction causality from economic growth to the environment. For the above evidence, economic activity that rises on production, distribution, and completion scales acts toward degrading the quality of the environment by destroying the very delicate balance of biodiversity, deforestation for the construction of industries and production plants, and carbon emissions arising from the thick energy consumption. Again, while Saud et al. (2019) and Akadiri et al. (2019) provide evidence of such causal relationships, they differ from their predecessors in finding a two-way influence between environmental quality and economic performance. It indicates that economic performance threatens natural environments on the one hand while quality of the environment might limit the performance of economies on the other hand. This indicates that stricter environmental policies aimed at improving environmental quality might hinder the unfettered operation of economic activity and thus impede growth.

Building upon the profound insights extracted from our extensive study, we chart a course for future research objectives that not only address critical gaps in policy implementation, impact assessment, and global transferability but also resonate with the sustainable development goals (SDGs) set forth by the United Nations. Our first stride involves a detailed exploration of the challenges encountered related to environmental quality during the rapid economic growth within BRICS economies. This chapter focusses on identifying effective strategies to tackle these problems, easing effective policy operation. Concurrently, we initiate an extensive assessment of the enduring impact of environmental quality on economic growth. This multi-faceted analysis significantly contributes to unraveling the enduring viability of the positive outcomes generated by our proposed measures. Furthermore, this study utilized sophisticated and current techniques for data validation and regression analyses. A variety of diagnostic procedures were utilized in this study, encompassing the stationarity tests such as cross-sectional augmented Dickey-Fuller (CADF) and cross-sectional Im Pesaran (CIPS)," "cross-sectional dependence tests," and "Kao, Pedroni, and Westerlund cointegration." In addressing the issues identified through these tests, we applied the "Common Correlated Effects Mean Group (CCEMG) Regression to quantifying the economic benefits of improved environmental quality in BRICS economies."

The structure of this chapter is framed in five sections as follows: Section 13.1 is an introduction, which explains background, objectives and contribution of the study. Section 13.2 of this chapter reviews all relevant existing literature. Moreover, the Sect. 13.3 deals with research design, giving an overview of the sources of data and analytical methods employed. Furthermore, the Sect. 13.4 covers the results and findings of the study, summarizing the main outcomes of the analysis. Finally, the Sect. 13.5 of this chapter provides a conclusion, summing up the insights and implications emerging from the study, and suggestions for further research.

13.2 Review Literature

The literature shows poor evidence that empirically researched impacts of environmental degradation on economic growth in higher countries in CO_2 emissions are scanty. The study in option revealed one-way causality resulting from CO_2 emissions to economic development from 1965 to 2006 in South Africa. The analysis suggests a revolting policy on evolving energy replacements for coal, the basic source of carbon emissions. Insofar as it directed not to ascertain long-term equilibrium relation and causality among CO_2 emissions and economic growth, Ghosh (2010) indicated a two-way causal relation between CO_2 emissions and economic growth during 1971–2006. Findings imply that reductions in CO_2 emissions in the short run may result in lower economic growth in India. Furthermore, Menyah and Wolde-Rufael (2010) found quite a bit of evidence of two-way short-run causality between emissions and growth in South Africa. Pao and Tsai (2010) analyzed causal linkages among pollution emissions and economic growth for the BRIC countries (Brazil, Russia, India, and China) during the period 1971–2005 and, for Russia, used the period 1990–2005. In general, the results suggest the implementation of investment plans in energy supply and in energy efficiency along with policies for the conservation of energy in BRIC economies to reduce emission without crippling their output.

A wide array of literature is available that analyzes trade openness and growth in specific country contexts or on a wider international basis. At the macro level, it has been pointed out that trade openness positively affects economic growth for two different two-panel datasets: 56 countries from 1951 to 1998 and 105 economies across the world. The results reveal that trade openness has a significant influence on economic growth and is eternal and sizably important (Karras 2003). Zahonogo (2016) discussed the nexus of trade openness and growth in Sub-Saharan countries and found that trade openness affects economic growth positively in a non-linear manner. The import and export impact of trade liberalization on growth has been studied in 7 African economies. The results showed a strongly positive role of trade on growth in these economies considered, with similar outcomes in country-specific studies (Dava 2012). Furthermore, studies regarding trade and economic growth have also investigated in Ghana as well as Nigeria and findings indicated that trade openness has a significant and positive effect on Ghana credibly and a significantly negative effect on Nigeria. An effective income generation by free non-restricted trade was expressed by Yeboah et al. (2012), across 38 African economies. Others on this subject, particularly in country-specific research, include Keho (2017), which presents findings that trade openness has a positive long- and short-term economic growth effect in Ivory Coast. Another study is also investigated for Nigeria and indicated a unidirectional causal linkage running to economic growth from trade openness (Olufemi 2004). According to Polat et al. (2015), the relationship between trade and economic growth in South Africa and suggested that trade negatively affects economic growth. Furthermore, Musila and Yiheyis (2015) examined the relationship between economic growth and trade openness in Kenya and noted that export and imports enhance growth.

Furthermore, government expenditure is a dual-edged sword for economic growth as it has positive and negative effects depending on the countries and sectors involved. Econometric findings indicate that government expenditure enhances growth through infrastructure, education, and health spending in addition to moderating income disparity. However, it is noteworthy that the impact of such expenditures highly depends on proper distribution and application of the money. For instance, government spending in India has shown positive correlation with both short- and long-term GDP growth mainly because of income redistribution that boosts the purchasing power of poor people (Dash 2024). On the other hand, Nigeria's recurrent expenditures have detrimental effects on the short run growth even if it can have a favorable effect on the GDP in the long run as noted by Jibir et al. (2023). Sector-wise impacts also differ; the level of government expenditure on agriculture and health sectors has a positive relationship with the GDP in Tunisia, but the similar relationship for military and education expenses is negative where the coefficient of determination (R-square) = 0.93 (Abdelli et al. 2024). The Tanzanian government should consequently invest more in infrastructure and education since these have a higher return on investment as compared to expenditure which has less impact as stated by Magai and Masele (2024). As suggested in the literature, it is wise to pursue productive expenditure with high growth density in areas like public investment in infrastructure, education and health (Jibir et al. 2023; Bangura 2024). For instance, the government of Sierra Leone identifies investment in capital and education as critical factors for economic growth (Bangura 2024). In conclusion, it can be noted that government expenditure plays a role in enhancing economic growth; nevertheless, its effectiveness depends on the proper planning, execution, and application to realize the maximum benefits.

In addition, rents from natural resources are one of the largest components for growth and are directly participating in the economy. This must be achieved in sustainable and effective usage of natural resources while excessive usage of it degrades this situation. The gross domestic product growths of both Iran and Pakistan have suffered because of the wastage investment and rents unproductively in their respective countries (Yasmeen et al. 2021). Furthermore, Hussain et al. (2021) asserted that productivity of natural resource depends with technological advancement in order to champion for economic welfare and is also resilient to policy for future activities. Consequently, the structure of the firm defines how effectively natural resources are utilized. Another published work conducted by Wu et al. (2018) concerned firm-oriented and resource-concerned with contribution on economic growth. Firm-oriented indication perceives natural resources as beneficial to long-run economic growth while the firm-oriented body of evidence is cynical of and abrasive towards such assertions. In the view of Xu and Tan (2020) despite the enhancement of the industrial structure in China, the overemphasis on the industrial sector including heavy industries has been a drawback to the optimal utilization of the country's natural resource endowment. However, high growth rates and natural resource efficiency has been realized through exploiting productive base in natural resource endowment. In the past, Li et al. (2021) noted that there has been a decrease in the secondary industry and increase in resource utilization in China due to the improved industrial structure. As noted earlier, this resulted in a slowdown of economic development. However, in some other fields, green economic development has stagnated. Thanks to the proliferation of numerous low-quality industries that employ inaccessible resource in efficient ways, a resource curse is in existence today (Cheng et al. 2020). Another study by Yang et al. (2021) showed that both an increase and a decrease in oil rents improved economic growth.

Moreover, new economy countries are constant in their desire for FDI. The FDI also offers help to the industry to open a new high level of technological advancement, cheap and trained labor force for better and sharper growth as well as productivity in the sector. In general, FDI presents the beneficiary economy with skilled management, physical wealth and sophisticated technology thus enhancing the recipient nation's economic growth; Keller and Yeaple (2009). The existing literature review has also shown that there is a large geographical variation in China. In addition, shore regions received five times more foreign investment than inland regions as well as seven times of foreign investment per capita as studied by Ouyang and Fu (2012). Moreover, according to Zhang (2014), industrial efficiency referred to a country's ability to manufacture and export products effectively to other nations. FDI as part of capital has a significant positive effect on industrial performance of China. Its impact is particularly noticeable in low-technology industries but is also observed in medium and high technology industries. However, economic growth may be attainable by industry keenness against that solidarity is probable with effective foreign investment as well as industrial attentiveness. From a regional perspective, industrialization with foreign investment is proportional to the degree of value creation required for micro firms to contribute to regional gross domestic product than the small firms in the country (Canes et al. 2021).

13.3 Model, Data and Methods

13.3.1 Model

In this chapter, we quantify the economic benefits of environmental quality in BRICS economies. Based on this, this study utilized GDP as an indicator of economic growth, carbon dioxide emissions as proxy of environmental quality, and others control indicators such as trade openness, government expenditure, natural resources, and foreign investment in the model. To estimate this relationship, we constructed a functional form of the model outlined below.

$$GDP_{it} = f(EQ_{it}, TO_{it}, GE_{it}, NR_{it}, FDI_{it}) \tag{13.1}$$

where, $GDP_{it}, EQ_{it}, TO_{it}, GE_{it}, NR_{it},$ and FDI_{it} denote economic growth, environmental quality, trade openness, government expenditure, natural resources, and foreign investment, respectively for country i at the time t. We have transformed data of all variables into natural log to overcome from the problem of variability and heteroskedasticity. Boutabba (2014). Therefore, Eq. (13.1) is transformed into Eq. (13.2) as below:

$$\ln GDP_{it} = \alpha_0 + \beta_1 \ln EQ_{it} + \beta_2 \ln TO_{it} + \beta_3 \ln GE_{it} + \beta_4 \ln NR_{it} + \beta_5 \ln FDI_{it} + \mu_{it} \tag{13.2}$$

where α_0 shows intercept, μ_{it} represents the error term, and β_1, β_2, β_3, β_4, and β_5 stand for undermined coefficients.

13.3.2 Data Description

This research assesses BRICS national data from 1990 to 2021. The World Development Indicators (WDI) is the source of the data for this study. Variable descriptions and a consolidated data summary can be found in Table 13.1. The period of the study was chosen according to the availability of data, yielding 165 observations per variable of economic and environmental indicators. To quantify the economic benefits of improved environmental quality in BRICS economies, this study utilized economic growth (GDP; constant 2015 US$) as the dependent variable for economic benefits and CO_2 emissions (total, Mt CO_2e) included as an independent variables for environmental quality. Furthermore, this study included controlling factors such as trade openness, which is measured through exports and imports of goods and services (constant 2015 US$), government expenditure specified by government final consumption expenditure, natural resources shown by total natural resources rents (% of GDP) and foreign investment represented by net inflows (BoP, current US$). This study obtained data from the World Bank database from 1990 to 2020 for BRICS economies. Table 13.1 provided the descriptions and source of the studied variables.

Table 13.1 Variables description

Symbol	Variable	Unit measurement	Source
LnGDP	Economic growth	GDP (constant 2015 US$)	WDI
LnEQ	Environmental quality	Carbon dioxide (CO_2) emissions (total, Mt CO_2e)	WDI
LnTO	Trade openness	Exports and Imports of goods and services (constant 2015 US$)	WDI
LnGE	Government expenditure	General government final consumption expenditure	WDI
LnNR	Natural resources	Total natural resources rents (% of GDP)	WDI
LnFDI	Foreign direct investment	Foreign direct investment, net inflows (BoP, current US$)	WDI

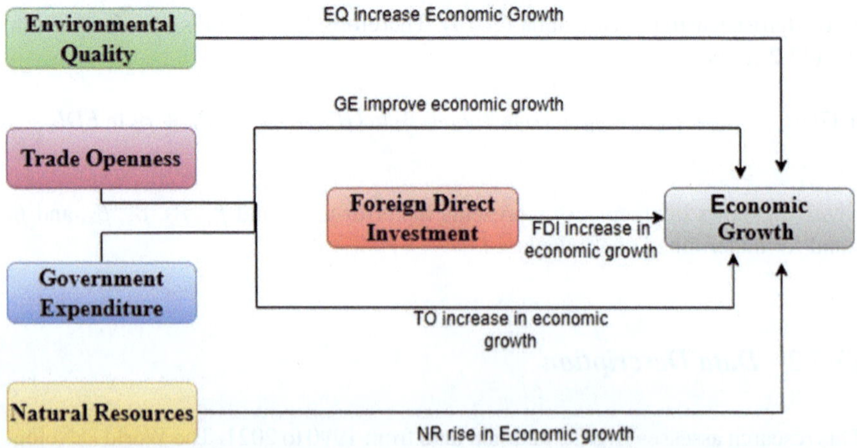

Fig. 13.1 The conceptual framework

The mainstream literature asserts that environmental quality, trade openness, government expenditure, natural resources, and foreign investment enhance economic growth. As depicted in Fig. 13.1, these factors collectively subsidize economic progress. However, the positive influence of these variables on economic growth is a central theme in the existing research. Environmental quality, by fostering a healthier ecosystem, trade openness by enabling market expansion, government expenditure by stimulating demand, natural resources by providing essential inputs, and foreign direct investment by bringing capital and technology, all play pivotal roles in driving economic growth.

13.3.3 Descriptive Statistics and Correlation Matrix

Table 13.2 display the descriptive statistics of the estimated variables. The findings of the descriptive statistics show that the natural logarithm of GDP has a mean of 27.791, a standard deviation of 1.049, and a range from 25.913 to 30.424, indicative of moderate variability. Environmental quality (LnEQ), measured via CO_2 emissions, has a mean of 7.062 and ranges from 5.392 to 9.451. Trade openness (LnTO) shows a mean of 26.291, and government expenditure (LnGE) a mean of 25.567, with both displaying limited range. Natural resource rents (LnNR) show more variability with a mean of 1.428, while foreign direct investment (LnFDI) ranges significantly with a mean of 23.078.

Table 13.3 presents the correlation matrix, illustrating the relationships among different variables. This analysis is used to assess the associations, as well as the strength and direction, between variables. For instance, LnGDP exhibits a strong positive correlation with LnEQ (0.789) and LnFDI (0.633). Trade openness (LnTO)

Table 13.2 Descriptive statistics

Variable	Obs	Mean	Std. Dev.	Min	Max
LnGDP	165	27.791	1.049	25.913	30.424
LnEQ	165	7.062	1.105	5.392	9.451
LnTO	165	26.291	0.829	23.231	27.969
LnGE	165	25.567	0.745	24.336	26.622
LnNR	165	1.428	0.732	−0.146	3.068
LnFDI	165	23.078	2.636	1.866	26.564

Table 13.3 Matrix of correlations

Variable	LnGDP	LnEQ	LnTO	LnGE	LnNR	LnFDI
LnGDP	1.000					
LnEQ	0.789	1.000				
LnTO	0.376	0.211	1.000			
LnGE	0.365	−0.007	0.735	1.000		
LnNR	−0.261	0.030	0.236	0.146	1.000	
LnFDI	0.633	0.417	0.340	0.267	−0.147	1.000

also shows a positive correlation with LnGE (0.735) and LnFDI (0.340). Interestingly, LnNR demonstrates a negative correlation with LnGDP (−0.261).

13.3.4 Methodology

The systematic approach to empirical analysis was utilized by the study as displayed in Fig. 13.2. A cross-sectional dependence test is an important component of empirical research. To this end, Pesaran CD, Pesaran scaled LM, and Breusch-Pagan LM tests were used to test for cross-sectional dependence. The second step was carrying out the stationarity test using Pesaran CADF and CIPS tests. With unit root results confirmed, we then performed our cointegration test, using the Kao, Pedroni, and Westerlund tests for this purpose. The CCEMG estimator was used for long-term analysis.

13.3.4.1 Cross-Sectional Dependence

Most econometric methods generally operate on the independence of the cross-sectional units of panel data from each other at a particular time. However, this forms an unrealistic assumption in aspects of economic interaction. The BRICS trade agreements, among many other interlinkages, play an active role in this. For

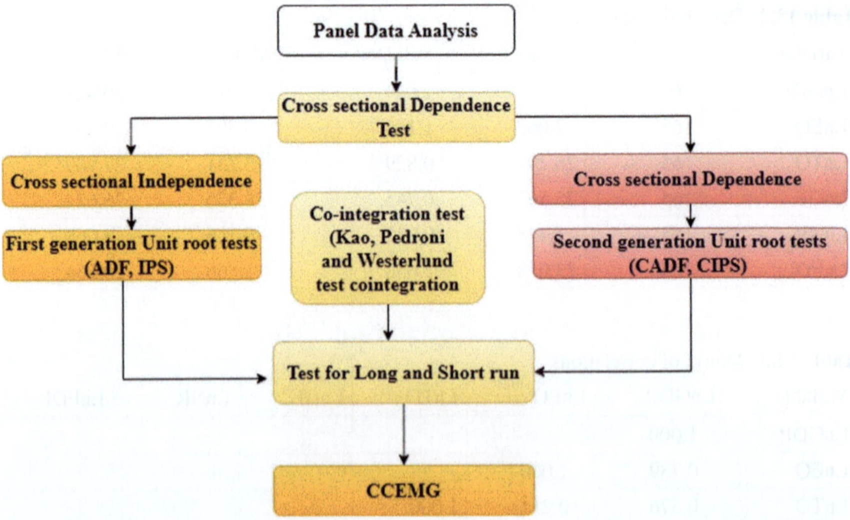

Fig. 13.2 Estimation flow-chart

example, trade relations are established through the Southern African Customs Union (SACU), Southern Common Market (MERCOSUR), and the Latin American Trade bloc, which have indeed gone forward to become the India-MERCOSUR and SACU trilateral agreements (WTO 2013). Thus, such tests are important in conducting empirical analyses to ensure accurate results. In other words, for empirical studies to be free from inaccuracies and assured of reliability in the estimates obtained, it is now opined that allowance should be made for correlation bias while performing the regression analysis (Pesaran 2004, 2007; Hussain et al. 2020). Here, we adopted both the Pesaran scaled LM test and Breusch-Pagan LM test to inspect cross-sectional dependence based on the suggestions of Breusch and Pagan (1980). For large T-size and small N-size cases, Pesaran (2007) developed a scaled version of the Lagrange multiplier (LM) test to give proper efficiency, especially among large N size and small T-sized cases. Nevertheless, Pesaran and Pedroni (2004) developed the Pesaran CD test to address size distortion problems that are often encountered with the Breusch-Pagan LM test, the scaled LM test, and the Godfrey-Breusch LM (B-G LM) test. It can, without difficulty, handle size distortions with datasets that are large for N with fixed T.

13.3.4.2 Unit Root Tests

Afterward, confirming the model's cross-sectional dependency, we proceeded to take the unit root tests of stationarity of the estimated series. For testing the stationarity of the estimated series, the study applied two types of unit root tests namely Cross-Sectional Augmented Dickey-Fuller (CADF) and Pesaran Cross-Sectionally

Augmented Im, Pesaran, and Shin (CIPS). These tests allow cross-sectional dependence by averaging over lags and differences for each unit. The CIPS stationarity was devised by Pesaran (2007). It enhances the "Dickey-Fuller (ADF)" regression by introducing cross-section mean of lagged values of the explained indicators and the first differences for every unit in the cross-section. The resulting augmented regressions are known as "Cross-Sectional ADF (CADF). Thus, the equation of the CADF test is demonstrated as:

$$\Delta Y = \alpha_{it} + \beta_i Y_{t-1} + \gamma_i \overline{Y}_{t-1} + \delta_i \Delta \overline{Y}_t + \mu_{it} \quad (13.3)$$

where t shows period of time (t = 1,2, 3, … T), i stands for number of cross-sections (i = 1,2,3, … N), the term $\overline{Y}_t$ now signifies the cross-sectional mean, which is calculated as $\overline{Y}_t = N^{-1}\sum_{i=1}^{N} Y_{it}$. The introduction of this mean should moderate contemporaneous correlation among cross-sectional units. Therefore, the null hypothesis of Eq. (13.3) is $\beta_i = 0$ for all i while the alternative hypothesis argues that $\beta_i < 0$ for some i.

Finally, the CIPS test, as explicated by Pesaran (2007), is given as follows: $CIPS(N, T) = N^{-1}\sum_{i=1}^{N} t_i(N, T)$. . In this context, (N, T) signifies the β t-statistics.

13.3.4.3 Panel Cointegration Test

After testing the stationarity of the series, further investigation, namely cointegration analysis is conducted. Heterogeneous panel co-integration analysis proposed by Pedroni (1997, 1999) will be used to test for the co-integration of the data. The approach introduced by Pedroni is a residual-based test that allows for heterogeneous cross-sectional effects, slope coefficients, and individual linear trends across countries. Two types of residual-based tests were developed by Pedroni (1999). In the first category, there are four tests: panel v-statistic, panel rho-statistic, panel PP-statistic, and panel ADF-statistic—which are asymptotically distributed as standard normal and are based on pooling residuals within the within-dimension framework. The second category consists of three tests: group rho-statistic, group PP-statistic, and group ADF-statistic, which also are under the asymptotic standard normal but are based on pooling residuals within the between-dimension framework. These statistics compute averages of the cross-sectional units' individually estimated coefficients, contingent upon their short-run dynamics, fixed effects, deterministic trends, and slope coefficients (Pedroni 2004). Pedroni (1999, 2004) has shown, through Monte Carlo simulations, that the panel ADF-statistic and group ADF-statistic are the most reliable tests because they have better small-sample properties when compared with others. The paper also uses the residual-based co-integration test proposed by Kao (1999). The null hypothesis allows for no co-integration in the dynamic panel while the alternative hypothesis enables for co-integration in the dynamic panel. So, we applied the Kao (1999), Pedroni et al. (2004) and Westerlund (2007) panel cointegration tests in this chapter.

13.3.4.4 Panel Long-Run Parameter Estimation

This chapter applies the Common Correlated Effects Mean Group (CCEMG) method for long-run estimation to address cross-sectional dependence (CSD) and slope heterogeneity. The CCEMG estimator provides consistent estimates via unobservable core variable factors that allow time-varying and cross-sectional effects to differ across members of the panel, and it was introduced by Pesaran (2006). The method is handled practically when a large number of panel members share common component (Chudik and Pesaran 2015). This method admits a finite number of "strong" unobserved traits as well as an unrestricted number of "weak" factors (El Anshasy and Katsaiti 2014). It has been acknowledged in much of the literature as an effective estimation method for nonstationary panel variables in business and organizational contexts. Through the incorporation of the cross-sectional averages of the dependent and independent variables into the regression equation, the CCEMG estimator controls for unobservable common factors (Pesaran 2006). This method is capable of maintaining strong robustness properties by allowing for the possibility that only a few strong factors unite those panel members along with a freely large count of weak ones Chudik et al. (2011). Furthermore, they allow for the correction of both slope heterogeneity and unit roots in at least some aggregate unobservable shocks (Kapetanios et al. 2011).

The formal structure of the CCEMG method is presented in the form of the general Eq. (13.4) below.

$$Y_{it} = \alpha_i + \beta_i X_{it} + \gamma_i g_t + \alpha_i \overline{Y}_{it} + \delta_i \overline{X}_{it} + \varepsilon_{it} \tag{13.4}$$

where Y_{it} and X_{it} are variables or observables; β_i is the estimate of each coefficient country by country; g_t is the shared fixed effect that is the same for different countries but escapes observation; α_i is the constant term; and ε_{it} is the stochastic term or error term.

13.4 Results and Discussion

13.4.1 CSD Tests Findings

Cross-sectional dependency is a serious and an early step in the preliminary findings of empirical analysis. This study applied several tests such as the Breusch-Pagan LM test, Bias-Corrected Scaled LM test, and Scaled LM test & CD test of Pesaran to check for cross-sectional dependence of the series. All these tests' findings for the BRICS panel are offered in Table 13.4. Results for each test prescribe stringent rejection of the null hypothesis of no cross-sectional dependency at significance level 1%, implying very strong cross-section dependence amid the estimated variables.

Table 13.4 Cross sectional analysis

Series	Breusch–Pagan LM	Pesaran scaled LM	Bias-corrected scaled LM	Pesaran CD
	Statistic	Statistic	Statistic	Statistic
LnGDP	281.626*	60.737*	60.659*	16.716*
LnEQ	184.474*	39.013*	38.935*	8.125*
LnTO	211.589*	45.076*	44.998*	13.996*
LnGE	215.994*	46.061*	45.983*	14.104*
LnNR	111.127*	22.612*	22.534*	9.698*
LnFDI	203.069*	43.171*	43.093*	14.077*

Note *Denotes the 1% significance level

13.4.2 Findings of Unit Root Tests

After determining the CSD amid the indicators, we examined the variables' stationarity using the Pesaran CADF and CIPS unit root tests. The outcomes from these methods are shown in Table 13.5. From the results, the null hypothesis was rejected at both intercept and intercept trend levels. Stationarity was established at 1%, 5%, and 10% levels of significance at both intercept and intercept-with-trend conditions at level and first difference. This posture signifies that all variables are integrated at mixed order.

13.4.3 Cointegration Tests Findings

We applied the Kao, Pedroni, and Westerlund tests to examine cointegration amid the indicators. The tests' outcomes are detailed in Table 13.6. The outcomes of Kao test show the rejection of null hypothesis for no cointegration amid the variables, indicating the existence of cointegration. Furthermore, the outcomes of Pedroni test also indicate cointegration among the variables. Similarly, the outcomes of Westerlund test further support the findings of Kao and Pedroni tests for existence of cointegration amid the variables. Therefore, the findings of the Kao, Pedroni and Westerlund techniques permit an analysis of long-run examination for model. Accordingly, we employed CCEMG technique to establish the long-run relationships amid the indicators.

Table 13.5 Stationarity analysis

Variables	Intercept				Intercept and Trend			
	CADF		CIPS		CADF		CIPS	
LnGDP	−1.464		−1.464		−2.678**		−2.678***	
Δ LnGDP	−3.629*		−3.629*		−3.580*		−3.580*	
LnEQ	−1.908		−1.908		−2.842*		−2.842**	
Δ LnEQ	−4.177*		−4.177*		−4.162*		−4.162*	
LnTO	−2.244**		−2.244**		−2.517		−2.517	
Δ LnTO	−3.846*		−3.846*		−3.944*		−3.944*	
LnGE	−3.265*		−3.265*		−3.193*		−3.193*	
Δ LnGE	−3.499*		−3.499*		−2.917*		−2.917*	
LnNR	−2.364*		−2.364*		−2.834*		−2.834*	
Δ LnNR	−4.050*		−4.033*		−4.050*		−4.050*	
LnFDI	−2.502*		−2.502*		−2.858*		−2.858**	
Δ LnFDI	−3.283*		−3.283*		−3.050*		−3.050*	
Critical values	1%	5%		10%	1%	5%		10%
	−2.24	−2.11		−2.05	−2.79	−2.59		−2.46

Note *, **, and *** represent 1%, 5%, and 10% significance level, respectively

Table 13.6 Cointegration analysis

Kao test for cointegration	
Test statistic	Stats
Mod. Dickey–Fuller	2.478
Dickey–Fuller	0.208**
Aug. Dickey–Fuller	10.601*
Unadj. Mod. Dickey–Fuller	−7.844*
Unadj. D-F	−7.957*
Pedroni test for cointegration	
Mod. variance ratio	−6.679*
Mod. Phillips–Perron	4.405*
Phillips–Perron	−4.440*
Aug. Dickey–Fuller	−3.618*
Westerlund test for cointegration	
Variance ratio	1.352**

Note *Shows 1% and **signify 5% significance level

13.4.4 CCEMG Technique Findings

In this chapter, we analyzed how environmental quality, trade openness, government spending, natural resources, and foreign investment influence economic growth across BRICS countries. Our findings demonstrate that all these factors, with the exception of natural resources, have a significant and positive impact on economic growth in the BRICS nations. Specifically, a one-percent improvement in environmental quality is linked with a 0.26% rise in GDP for BRICS economies, highlighting the positive role of environmental improvements in driving economic growth. These findings align with those from prior studies, such as those by Ali et al. (2022) and Nguyen and Khominich (2023), who also observed similar positive relationships between environmental quality and economic performance. Additionally, the data indicate that a 1% increase in trade openness enhance to a 0.15% GDP growth in the BRICS region. This suggests that more open trade policies can contribute significantly to economic performance by enhancing market access and competitive dynamics. These findings are consistent with previous research by Burange et al. (2019) and Banday et al. (2021), which originated a positive relation of trade openness and economic growth. Government expenditure also plays a dynamic part in fostering economic growth, as evidenced by our findings. A one-percent rise in government spending is estimated to boost GDP by 0.43%, emphasizing the importance of fiscal policy and government investment in driving economic activity. Our findings are also similar with prior studies, including Shah et al. (2016) and Hashmi et al. (2021), which also noted a positive influence of government expenditure on economic growth. FDI emerged as another significant driver of economic growth within the BRICS nations. Our analysis shows that a 1% rise in FDI correlates with a 0.86% rise in GDP, underscoring FDI's role in supporting domestic production, technology transfer, and overall economic growth. This result is consistent with other researches, such as those by Bose and Kohli (2018) and Nistor (2015), who reported similar positive impacts of FDI on economic performance. Conversely, our analysis revealed an inverse relationship between natural resources and GDP growth in BRICS. Specifically, a one-percent increase in natural resource use appears to decrease GDP by approximately 0.037%. This suggests that an over-reliance on natural resources may have a detrimental effect on economic performance, possibly due to factors such as resource depletion or adverse environmental impacts. Our findings in this regard align with those of Baloch et al. (2019) and Sicen et al. (2022), who also identified a negative relationship between natural resource exploitation and economic growth (Tables 13.7).

The findings of the regression analysis using the CCEMG estimator can be expressed in the form of an equation as follows:

$$\begin{aligned} GDP = {} & 0.264 * LnEQ + 0.153 * LnTO + 0.438 * LnGE \\ & - 0.037 * LnNR + 0.868 * LnFDI \\ & - 4.543 * AdjustedTerm + \varepsilon \end{aligned}$$

Table 13.7 Regression analysis (CCEMG)

Variable	Coefficient	Std. errs	z-value	P > |z|
LnEQ	0.264*	0.083	3.190	0.000
LnTO	0.153*	0.043	3.540	0.000
LnGE	0.438**	0.179	2.450	0.014
LnNR	−0.037**	0.016	−2.320	0.021
LnFDI	0.868**	0.372	2.330	0.020
Adjusted Term	−4.543*	1.280	−3.550	0.000
Wald chi^2	48.390 (0.000)			

*$p < 0.01$, **$p < 0.05$

where LnGDP is the dependent variable, while LnEQ, LnTO, LnGE, LnNR, and LnFDI are the independent variables. The coefficients represent the estimated impact of each variable on the dependent variable, the adjusted term is included to account for any common correlated effects and ϵ represents the error term.

13.5 Conclusions and Policy Implications

This chapter quantifies the economic benefits in BRICS economies of increased environmental quality, with economic growth chosen as the dependent variable and CO_2 emissions as the key environmental indicator. Other controls variables considered included trade openness, government expenditure, natural resources, and foreign investment. Various econometric tests, which adduced the raw data from 1990 to 2022, were used to test for cross-section dependence, stationarity, and cointegration. The findings were that dependence in the cross-section is statistically significant, while the variables were of mixed-order integration and cointegration existed among the series. Furthermore, the results of CCEMG regression analysis indicate that improved environmental quality, trade openness, government spending, and FDI are positively associated with economic growth while dependence on natural resources is negatively correlated with growth. It emphasizes the need for sustainable environmental policies to combine economic growth in BRICS economies.

The chapter thus provides crucial policy insights for BRICS economies and executives in developing countries. It refers to the development of strategies where environmental protection should be at the forefront of the developing countries/mind of policymakers to promote eco-friendly technologies and lower dependency on non-renewable energies. Likewise, it stresses the important role investment in research and development would have in the economic growth of the region. Hence governments should encourage sustainable business practices through the provision of incentives and support. Overall, this chapter provides valuable insights in the crafting of policies that foster balance in the field of economics and environmental protection in the BRICS economies.

This chapter has some limitations, such as time span of the data used and the estimation method. Future variables that have been suggested for this study are those that will look to investigate environmental policies and growth mechanisms through which positive effects of environmental quality and trade openness can be maximized through government spending and FDI on economic growth in all BRICS countries. This might involve study into the influence of government regulations with incentives to promote eco-friendly investments and policies that promote sustainability. Generalizability could also be improved through extending the time series analysis or through extension over more countries. A further comparison could therefore be drawn between BRICS economies or other regions with one another to yield the best practices instrumental in supporting sustainable growth.

References

Abdelli H, Gheraia Z, Sekrafi H, Abid M (2024) The asymmetric impact of government expenditure on economic growth: evidence from a NARDL model. Int J Adv Appl Sci 11(3):92–101
Abdouli M, Hammami S (2017) Investigating the causality links between environmental quality, foreign direct investment and economic growth in MENA countries. Int Bus Rev 26(2):264–278
Ahmad N, Du L (2017) Effects of energy production and CO_2 emissions on economic growth in Iran: ARDL approach. Energy 123:521–537
Akadiri SS, Bekun FV, Taheri E, Akadiri AC (2019) Carbon emissions, energy consumption and economic growth: a causality evidence. Int J Energy Technol Policy 15(2–3):320–336
Ali N, Phoungthong K, Techato K, Ali W, Abbas S, Dhanraj JA, Khan A (2022) FDI, green innovation and environmental quality nexus: new insights from BRICS economies. Sustainability 14(4):2181
Assembly G (2015) Sustainable development goals. SDGs transform our world 2030(10.1186)
Baloch MA, Mahmood N, Zhang JW (2019) Effect of natural resources, renewable energy and economic development on CO_2 emissions in BRICS countries. Sci Total Environ 678:632–638
Banday UJ, Murugan S, Maryam J (2021) Foreign direct investment, trade openness and economic growth in BRICS countries: evidences from panel data. Transnational Corporations Rev 13(2):211–221
Bangura FS (2024) The impact of government fiscal expenditure on Sierra Leoneâ€™ s economic growth. J Sci Rep 7(1):1–14
Bose S, Kohli B (2018) Study of FDI trends and patterns in BRICS economies during the period 1990–2015. Emerging Econ Stud 4(1):78–101
Boutabba MA (2014) The impact of financial development, income, energy and trade on carbon emissions: evidence from the Indian economy. Econ Model 40:33–41
Breusch TS, Pagan AR (1980) The Lagrange multiplier test and its applications to model specification in econometrics. Rev Econ Stud 47(1):239–253
Burange LG, Ranadive RR, Karnik NN (2019) Trade openness and economic growth nexus: a case study of BRICS. Foreign Trade Rev 54(1):1–15
Canes C, Aurelia V, Tanesia JPY, Hasudungan A, Lukas E (2021) Analysing the influence of industrial value-added of small and micro-sized firms on regional FDI in Indonesia. Int J Bus Stud 5(3):157–165
Cheng Z, Li L, Liu J (2020) Natural resource abundance, resource industry dependence and economic green growth in China. Resour Policy 68:101734
Dash KK (2024) Analyzing the effect of governments spending on India’s economic growth. Int J Multidisc Res 6(2)

Dava E (2012) Trade liberalization and economic growth in the SADC: a difference in difference analysis. In: IESE conference paper

Ghosh S (2010) Examining carbon emissions economic growth nexus for India: a multivariate cointegration approach. Energy Policy 38(6):3008–3014

Hashmi SM, Khushik AG, Gilal MA, Yongliang Z (2021) The impact of GDP and its expenditure components on unemployment within BRICS countries: evidence of Okun's law from aggregate and disaggregated approaches. SAGE Open 11(2):21582440211023424

Hussain J, Khan A, Zhou K (2020) The impact of natural resource depletion on energy use and CO_2 emission in belt & road initiative countries: a cross-country analysis. Energy 199:117409

Hussain M, Ye Z, Bashir A, Chaudhry NI, Zhao Y (2021) A nexus of natural resource rents, institutional quality, human capital, and financial development in resource-rich high-income economies. Resour Policy 74:102259

Jibir A, Abdu M, Bappayaya B, Isah A (2023) Disaggregated impact of government expenditure on economic growth: evidence from Nigeria. Lapai J Econ 7(1):15–32

Kahia M, Jebli MB, Belloumi M (2019) Analysis of the impact of renewable energy consumption and economic growth on carbon dioxide emissions in 12 MENA countries. Clean

Kao C (1999) Spurious regression and residual-based tests for cointegration in panel data. J Econometrics 90(1):1–44

Kapetanios G, Pesaran MH, Yamagata T (2011) Panels with non-stationary multifactor error structures. J Econometrics 160(2):326–348

Karras G (2003) Trade openness and economic growth can we estimate the precise effect? Appl Econom Int Dev 3

Keho Y (2017) The impact of trade openness on economic growth: the case of Cote d'Ivoire. Cogent Econ Financ 5:1332820

Keller W, Yeaple SR (2009) Multinational enterprises, international trade, and productivity growth: firm-level evidence from the United States. Rev Econ Stat 91(4):821–831

Li Q, Liu S, Yang M, Xu F (2021) The effects of China's sustainable development policy for resource-based cities on local industrial transformation. Resour Policy 71:101940

Magai PS, Masele RE (2024) Government expenditure and economic growth in Tanzania. Tanzanian Econ Rev 13(2):60–76

Menyah K, Wolde-Rufael Y (2010) Energy consumption, pollutant emissions and economic growth in South Africa. Energy Econ 32:1374–1382

Mesagan EP, Vo XV (2023) Does natural resource rent and consumption interplay worsen Africa's pollution? Heterogeneous panel approach with cross-sectional dependence. Resour Policy 82:103562

Musila JW, Yiheyis Z (2015) The impact of trade openness on growth: the case of Kenya. J Policy Model 37:342–354

Nguyen DH, Khominich IP (2023) The measurement of green economic quality in the BRICS countries: should they prioritize financing for environmental protection, economic growth, or social goals? Russian J Econ 9(2):183–200

Nistor P (2015) FDI implications on BRICS economy growth. Proc Econ Finance 32:981–985

Olaoye O, Dauda RO (2022) Energy use, financial development and pollution in selected African countries. J Econ Impact 4(3):188–195

Olufemi SM (2004) Trade openness and economic growth in Nigeria: further evidence on the causality issue. South Afr J Econ Manag Sci 7(2):299–315

Olunkwa CN, Adenuga JI, Salaudeen MB, Mesagan EP (2021) The demographic effects of Covid-19: any hope for working populations? BizEcons Q 15(1):3.12

Ouyang P, Fu S (2012) Economic growth, local industrial development and inter-regional spillovers from foreign direct investment: evidence from China. China Econ Rev 23(2):445–460

Pao HT, Tsai CM (2010) CO_2 emissions, energy consumption and economic growth in BRIC countries. Energy Policy 38(12):7850–7860

Pedroni P (1997) Panel cointegration: asymptotic and finite sample properties of pooled time series tests, with an application to the PPP hypothesis: new results. Working paper, Indiana University

Pedroni P (1999) Purchasing power parity tests in cointegrated panels. Working paper, Indiana University

Pedroni P (2004) Panel cointegration: asymptotic and finite sample properties of pooled time series tests with an application to the PPP hypothesis. Economet Theor 20(3):597–625

Pesaran MH (2004) General diagnostic tests for cross section dependence in panels. Economics 1240(1):1 (Cambridge Working Papers)

Pesaran MH (2006) Estimation and inference in large heterogeneous panels with a multifactor error structure. Econometrica 74(4):967–1012

Pesaran MH (2007) A simple panel unit root test in the presence of cross-section dependence. J Appl Economet 22(2):265–312

Polat A, Shahbaz M, Rehman IU, Satti SL (2015) Revisiting linkages between financial development, trade openness and economic growth in South Africa: fresh evidence from combined cointegration test. Qual Quant 49:785–803

Salahuddin M, Gow J (2019) Effects of energy consumption and economic growth on environmental quality: evidence from Qatar. Environ Sci Pollut Res 26(18):18124–18142

Saud S, Chen S, Haseeb A (2019) Impact of financial development and economic growth on environmental quality: an empirical analysis from Belt and Road Initiative (BRI) countries. Environ Sci Pollut Res 26(3):2253–2269

Shah SA, He C, Yu M, Xiaoqin W (2016) Government expenditure, defense expenditure and economic growth: a causality analysis for BRICS. Eur J Econ Stud 4:447–458

Sicen L, Khan A, Kakar A (2022) The role of disaggregated level natural resources rents in economic growth and environmental degradation of BRICS economies. Biophys Econ Sustain 7(3):7 (Technol Environ Policy 21(4):871–885)

Westerlund J (2007) Panel cointegration tests of the Fisher effect. J Appl Economet 23(2):193–233

World Trade Organisation Centre (2013) BRICS trade policies, institutions and areas for deepening cooperation

Wu S, Li L, Li S (2018) Natural resource abundance, natural resource-oriented industry dependence, and economic growth: evidence from the provincial level in China. Resour Conserv Recycl 139:163–171

Xu L, Tan J (2020) Financial development, industrial structure and natural resource utilization efficiency in China. Resour Policy 66:101642

Yang J, Rizvi SKA, Tan Z, Umar M, Koondhar MA (2021) The competing role of natural gas and oil as fossil fuel and the non-linear dynamics of resource curse in Russia. Resour Policy 72:102100

Yasmeen H, Tan Q, Zameer H, Vo XV, Shahbaz M (2021) Discovering the relationship between natural resources, energy consumption, gross capital formation with economic growth: can lower financial openness change the curse into blessing. Resour Policy 71:102013

Yeboah O-A, Naanwaab CB, Saleem S, Akuffo AS (2012) Effects of trade openness on economic growth: the case of African countries

Zahonogo P (2016) Trade and economic growth in developing countries: evidence from sub-Saharan Africa. J Afr Trade 3(1–2):41–56

Zhang KH (2014) How does foreign direct investment affect industrial competitiveness? Evidence from China. China Econ Rev 30:530–539

Chapter 14
Impact of Artificial Intelligence on Sustainable Financial Services: An Empirical Analysis

Ahsanuddin Haider

Abstract This paper analyses the environmental impact of Artificial Intelligence (AI) in financial services and emphasizes opportunities and challenges for sustainability. In past few decades, AI has come up as a prominent advancement in numerous sectors such as agriculture, finance, transport, energy, healthcare media and education. AI emerges as a powerful mechanism to reshape global sustainability also. The main objective of the paper is to discuss ethical AI and environment sustainability, and AI's applicability in financial services for social betterment. In order to maintain harmonization between AI and Sustainability, it becomes important to have an progressive insight to the specific contexts such as environmental, social and ethical implications. Regardless of AI seamlessly developing, the concern arises for its potential negative impacts on environment and sustainability, including ethical challenges. This research helps to investigate the environmental impact of AI in financial services, mutual benefits, challenges and opportunities for AI and Sustainability. AI plays a crucial role in maintaining sustainability, on the other hand sustainable practices are important for the development of Artificial Intelligence. This paper highlights the strategic utilization of AI in financial services to advance the critical sustainability objectives.

Keywords Artificial intelligence · Sustainability · Environment · Financial services

14.1 Introduction

The digital revolution has by this time altered our lives, how we labor and get in touch with one another. And this is just the beginning. However, the very technologies that promise to make billions of people happier and healthier also create public policy

A. Haider (✉)
College of Administrative and Financial Sciences, Saudi Electronic University, Riyadh, Saudi Arabia
e-mail: a.haider@seu.edu.sa

S. K. Tiwari et al. (eds.), *AI-Driven Environmental Pollution Management*, Climate Risks and Solutions, https://doi.org/10.1007/978-3-031-96243-1_14

challenges for global citizens and governments. Recent events have made security and privacy a primary focus for everyone from the DOD to Facebook—whether about concerns of election meddling, data breaches, or cyber-attacks as technology changes how we think not just national security but maybe even democracy itself (Ojokoh et al. 2020). In today's time Artificial Intelligence (AI) has become integral part and leading exponent that could change how we live our lives or even do jobs. AI technologies are being leveraged more and more in different sectors of industry to automate processes, streamline operations or better decision-making (Deepalakshmi and Sc n.d.). The most important benefit that AI tools provide is the capacity to automate tasks that are routine and take up a lot of workers' time and attention so they can concentrate on more strategic, creative work.

The use of AI tools for enhanced risk management and detection of fraud is becoming common in the finance industry. There are several benefits that can be obtained from the application of AI within the sector of financial services. It enables organizations to quickly and accurately analyze large amounts of data which in turn increases productivity and efficiency. Cost reduction is another advantage as tasks can be done more affordably with automated systems than with manual labor (Habib et al. 2021). Moreover, AI also helps to enhance the accuracy of financial operations which decreases the margin of error and improves reliability. However, in addition to these benefits, there are several financial costs associated with the AI-powered automation (Khan 2021). One of the main concerns is career destruction. AI systems are becoming ever more advanced which means that jobs that could be done by people are now being done by systems thus decreasing the number of jobs available in some regions (Lu 2018).

Environmental sustainability priorities the preservation of life support systems, such as the atmosphere, essential for economic production and human existence. It is crucial to preserve and maintain these regions for human survival on Earth. The foremost global framework for international collaboration, "the 2030 Agenda for Sustainable Development and its Sustainable Development Goals (SDGs)," is founded on the principle of sustainability. A multitude of governments, enterprises, and organizations have pledged to sustainable objectives (Khan et al. 2023). They are achieving objectives such as minimizing environmental footprints, making green investments, and preserving resources. Sustainable policies emphasis the future impacts of business practices or specific policies on ecosystems, humanity, and the wide economy. The presumption often lines up with the concept that, absent significant alterations to global governance, the world will encounter unalterable destruction. The world community has transitioned to sustainable behaviors and strategies to combat climate change, biodiversity decline, and pollution. This paper explores the impact of AI in financial services, highlighting opportunities and challenges for sustainable growth (Gandomi et al. 2023).

14.1.1 Role of AI in the Financial Services

The Financial services are enduring a remarkable transformation by virtue of generative AI, which is stimulating transformation and optimizing processes. Artificial intelligence is upgrading risk management, and customer experience, revolutionizing capital markets, by its extensive assortment of applications (Arshad Khan and Alhumoudi 2022). Dealing with hurdles and potential presented by AI, the banking industry is strategically moving ahead in order to cope up with an AI-enabled future.

Artificial intelligence (AI), especially generative AI (GenAI), has emerged as the key to revolutionary transformation in the ever-evolving financial services industry, changing the banking industry's operational and strategic horizons. The ability of GenAI to produce fresh, original content is not just a small step forward; rather, it is a fundamental shift that is driving banking towards a more inventive and efficient future (Minhaj and Altaf Khan 2022).

This entails setting aside funds for new projects that use AI and technology to improve decision-making and reduce expenses. Purchasing software that enables a single login to arrange internal financial transfers and monitor all of your bank accounts is a real-world example. This can lessen the amount of money that sits in your accounts overnight. Given the numerous technological upheavals which have transpired over the decade, including digital payments, the introduction of UPI, Aadhaar-based KYC, fintech, buy-now-pay-later platforms, and more, it is especially important to embrace innovation at this time. AI-driven personal finance, broader block chain use cases, and the Reserve Bank of India's Central Bank Digital Currency (CBDC) are all highly anticipated in the future. The organization may suffer if the finance department does not adjust swiftly (Khan et al. 2024).

14.1.2 Environmental Impact of AI

Although artificial intelligence (AI) is being heralded as a game-changer, there is a serious worry about its impact on the environment beneath its revolutionary potential. AI technology creation, upkeep, and disposal all have significant carbon footprints. We look at how society may effectively address these issues and work towards a sustainable future where environmental preservation and artificial intelligence coexist through proactive measures and moral behavior.

14.1.3 Positive Impact of AI

AI may prove to be extremely useful for environmental monitoring and assisting governments, corporations, and individuals in making more environmentally friendly decisions. Additionally, it can increase efficiency. For instance, UNEP employs AI to identify methane emissions from petrol and oil facilities, which are a major contributor to climate change. These kinds of developments are giving rise to optimism that AI may be able to assist the world in addressing at least some of the triple global crisis—the loss of nature and biodiversity, pollution and waste, and climate change. As AI-powered self-driving cars become more common, more individuals may choose to drive rather than ride their bikes or utilize public transportation, which would increase greenhouse gas emissions. Experts refer to these impacts as higher-order effects (Zamani et al. 2023).

Cost and time savings are the main drivers of AI's growth in the financial services industry. In order to increase productivity and save operating expenses, numerous financial institutions are using AI-powered technology to automate and streamline time-consuming, manual procedures like fraud detection, risk management, and customer support. As a result, in order to remain competitive and increase their profitability, numerous financial firms have resorted to artificial intelligence (Luan et al. 2020). Even though artificial intelligence (AI) is known to the financial industry, recent developments and the increased interest in cutting-edge technologies like ChatGPT, Bard, and general GPT encourage progressive financial companies to investigate new applications for the technology. AI is a natural match for growth in the financial industry because of the enormous volume of data produced by financial transactions, which creates the ideal setting for its development.

14.1.4 Negative Impact of AI

Data centers, especially those operated by cloud service providers, host the majority of extensive AI deployments. The environment may significantly deteriorate due to these data centers. Their electronics require an astonishing quantity of raw materials: 800 kg are necessary to produce a 2 kg computer. As stated in Navigating New Horizons, rare earth elements, often extracted using environmentally detrimental methods, are essential for the microchips that power artificial intelligence. The second issue is that data centers produce electronic waste, often containing hazardous substances such as lead and mercury.

Moreover, water is utilized to cool electrical components in data centers (Rahmani et al. 2021), both during construction and after operation. One estimate suggests that AI-related infrastructure may soon consume six times more water globally than Denmark, a country with a population of six million. Considering that 25% of individuals currently lack access to sanitation facilities and potable water, this constitutes a significant concern (Data and Analytics 2000).

Finally, data centers that accommodate AI technology necessitate substantial energy to operate their advanced electronics, which is predominantly sourced from fossil fuels, resulting in greenhouse gas emissions that contribute to global warming. The International Energy Agency reports that a query conducted through ChatGPT, an AI-driven virtual assistant, consumes 10 times more electricity than a Google search. Despite little global statistics, the agency forecasts that by 2026, data centers in Ireland, a technological hub, could represent almost 35% of the country's energy usage as a result of AI advancements.

The number of data centers has increased from 500,000 in 2012 to 8 million, partly due to the expansion of AI, and experts foresee that the demands of this technology will rise more.

14.1.5 Sustainable Environmental and Sustainable Financing

Environmental sustainability can be understood as the rational use of natural resources to fulfill present needs with protecting the capacity of environment for future generations. Although it has nothing to do with internal finance, environmental preservation practices are known to convey a positive signal to investors and lower the costs of external funding. The first crucial step in addressing the environmental effects of AI's exponential growth is for ML researchers and practitioners to cultivate and embrace a sustainability mentality. (SAE) Overspending on environmental projects would deplete internal resources and negate the advantages of internal funding. To achieve sustainable growth, businesses should employ sensible leverage and make appropriate finance decisions. Lenders use environmental performance as a key metric to assess corporate risks and provide the right interest rate. In the field of environmental protection, technology is essential. Through capital accumulation and greater R&D expenditure, environmental protection companies can reallocate their resources to high-return industries like design and technological innovation. To raise the standard of environmental performance and boost production efficiency through technological innovation, businesses should increase their budget for environmental investments. Their investments in technology will be compensated for by this competitive edge (Najafabadi et al. 2015).

In terms of artificial intelligence and robotics, there is little question that mankind has made tremendous strides forward. After WWII, in the 1950s, researchers began to examine the feasibility of artificial intelligence (AI) by seeing how well computers could mimic human thought processes. Projects involving robotics and chess games were examples of the ten years of intense study on artificial intelligence that began in the 1960s. Latest research have developed neural networks and sophisticated algorithms that replicate human behavior in areas including learning, reasoning, problem-solving, computational intelligence, and the use of mathematical tools to simulate the natural environment. Over the next two decades, AI, block chain, machine learning, and big data will have a more profound impact on financial reporting as it pertains to companies (Inci 2023). To be more specific, it is now obvious that automation will soon supplement human accountants' job thanks to the convergence of AI and block chain technology. Even while AI can't replace human brainpower, it can produce results that are just as accurate. For artificial intelligence (AI) to be beneficial in accounting and business problem-solving, we need to assess its capabilities and shortcomings. We must also identify the education and experience levels necessary for accountants to effectively operate AI.

As accountants begin to utilize associated AI technology, they will be empowered to refine their expertise and provide investors with superior financial analysis. The amalgamation of computer programs, imperfect models, forecasting models, and information analytics under artificial intelligence (AI) presents significant potential for enhancing sustainable finance. Investor organizations are demanding more comprehensive financial disclosures from companies and accountants. A fundamental concept at the intersection of sustainable finance and the Sustainable Development Goals (SDGs) is sustainable finance. This definition is rather limited as it solely takes into account ESG criteria. Thus, this SI posits that AI systems, applications, and models ought to be included in the concept of sustainable finance, alongside all other activities and elements that contribute to financial sustainability. Additionally, businesses should create patented, high-end, high-tech products. It is advised that a policy mix of financial supporting policies, environmental protection mechanisms, and sustainable financial development be established from a governance perspective. Voluntary corporate environmental protection behavior can be sustained and encouraged by the right policies (Goodell et al. 2021). As soon as feasible, not just in the small pilot areas but across the entire region, they ought to raise the bar for the protection of intellectual property rights and strengthens their funding environment. Both short- and long-term economic success are significantly impacted by debt financing. Long-term debt is preferred by businesses over short-term debt in order to increase their financial sustainability (EP).

14.2 Literature Review

Making investments in sustainable development and artificial intelligence are becoming more widely recognized, according to a comprehensive literature assessment. As a result, many concerns have been raised regarding the impact of AI on companies and the part it should play in assisting stakeholders and investors in making the best decisions and in bolstering organizational efforts to safeguard users' and customers' privacy by avoiding attacks via the internet. Researching how AI will change financial reporting is the first and most important step. When it comes to accounting for hazards as well as possibilities connected to environmental, social, and governance (ESG), AI helps investors gather, analyze, and understand more data than ever before. This data is crucial for both company and portfolio analysis. In addition, it paves the way for sustainable investors to sift through big data, which contains information vital to ESG investment in vast quantities (Cao 2021).

Organizations may make data-driven decisions to advance sustainability programs, discover areas for development, and obtain meaningful insights into their environmental footprint by utilizing AI. Second, AI's impact is increased when combined with other cutting-edge technologies like block chain and the Internet of Things (IoT), which allow for secure verification of sustainability claims, transparent supply chain traceability, and real-time environmental parameter monitoring. Additionally, by automating environmental reporting procedures, improving accuracy, and lessening the workload on organizations, AI makes it easier to comply with changing regulatory standards. (LAI) AI and ML have the potential to significantly improve financial stability in the financial services sector. These benefits mostly show up as increased operational effectiveness in the provision of financial services, along with better capacity to track and reduce systemic and regulatory risks. (PSA) The connection between financial choices and economic performance is impacted by environmental performance. Debt financing has a stronger detrimental impact on short-term performance and a less pronounced effect on long-term performance when EP initiatives are high. Additionally, high EP lessens the effect of internal funding on output (EP) Just acknowledging the significance of AI is insufficient; companies also need to make investments in developing the skills, partnerships, and infrastructure required to fully utilize its disruptive potential. Furthermore, the research and application of AI must continue to be guided by ethical issues. It is crucial to make sure that AI technologies respect the values of justice, openness, and accountability as they are incorporated more and more into decision-making processes, especially in delicate domains like environmental reporting and assessment. AI has a bright future in environmental impact assessment and reporting, but to reach its full potential, stakeholders from all areas of the ecosystem must work together. We can use AI to create a more resilient and sustainable future for future generations if we embrace innovation, encourage teamwork, and maintain ethical standards. AI generates both enthusiasm and worry (Soni 2021). While AI and machine learning offer significant benefits, they also pose unique risks. The success of addressing the aforementioned difficulties is crucial for the future of AI in the banking industry.

Banks and insurance businesses can leverage their traditions and values to strengthen their market position in the digital era. (AIF) To address biases, some solutions such as inclusive data practices, algorithmic audits and transparency measures, ethical AI design principles, regulatory frameworks, and stakeholder alliance to promote consensus and accountability. (DEC) Foresight is essential for creating sustainable company plans that prioritize financial rewards and environmental responsibility. AI, IoT, and predictive modeling will accelerate innovation and sustainability in the financial services business. Financial institutions must adopt sustainable finance technology to meet the demands of environmentally concerned investors and consumers. Transparency and accountability are key. (IFS) AI automation has the ability to alter the financial services industry by increasing efficiency, improving client experiences, and promoting sustainable growth. Financial institutions can innovate, distinguish, and lead in a digital-first world by using AI responsibly, proactively solving problems, and leveraging collaborative relationships. Financial institutions can design an inclusive, resilient, and technologically advanced future by strategically investing in AI talent, ethical governance, and new applications. (EIA) AI can enhance the alignment of financial practices with environmental, social, and governance (ESG) criteria, thereby contributing to the broader objectives of sustainable development. The most common dimension was environmental, which includes research on how AI interacts with the environment, including energy usage optimization, carbon emission assessment, and the energy costs associated with operating big machine learning models. Closely behind comes the social dimension, which examines moral issues, educational ramifications and equality concerns pertaining to AI (OECD 2021). (AIS) By optimizing the effectiveness and speed of trading procedures, artificial intelligence (AI) and machine learning (ML) could improve financial market accessibility. High-risk transactions and intricate financial procedures may benefit from the application of AI and ML. It is determined that certain financial institutions would profit from increased risk by adopting management techniques. (PSA) Going forward, we must take a comprehensive approach to the AI ecosystem to guarantee its environmentally sustainable growth. From data collection to model exploration and experimentation, training, optimization, and run-time inference, we need to examine the entire machine learning pipeline. (SAE).

Objectives

- To investigate the environmental effect of AI in financial services
- To explore the challenges in adoption of AI in the sector of financial services
- To find out the opportunities in adoption of AI in the sector of financial services
- To emphasize the importance of AI and sustainability.

14.3 Hypothesis's Development

14.3.1 AI and Sustainable Finance

Making investments that take ESG (environmental, social, and governance) factors into account are known as sustainable finance, and they help to ensure long-term economic growth. Finding a middle ground between monetary expansion, social progress, and ecological preservation is central to sustainable finance. Green bonds, social bonds, environmental investing, including sustainability-linked loans are all parts of sustainable finance (Khan and Alhumoudi 2024).

As we work towards a future based on sustainable finance, industry plays an essential role in mitigating the effects of the world's extraordinary rate of change. In any case, are sustainability and finance complementary? A growing number of institutions and other financial organizations are looking to AI to facilitate greener and more effective operations in an effort to tackle the myriad environmental problems we face today.

H01: AI has a positive effect on Sustainable finance.

H05: Sustainable finance is influenced by the interconnectedness between AI and financial service.

H06: Sustainable finance is influenced by the interconnectedness between AI and Financial Institution.

H07: Sustainable finance is influenced by the interconnectedness between AI and Investors.

14.3.2 AI and Financial Services and Financial Institutions

Financial organizations and banks have been quick adopters of AI in recent years, with the goal of better serving customers, cutting costs, and enhancing risk management. At some point, this will serve as a primer on sustainable finance. Accenture predicts that by 2025, the financial sector will have increased productivity by $140 billion thanks to AI (Haider et al. 2024).

The capacity of AI to look across massive amounts of data for associations and patterns can be useful to financial institutions, which in turn improves decision-making and the bottom line. Important ethical and regulatory concerns, including data privacy and security, transparency, and prejudice, are brought up by the deployment of AI in banking. Financial institutions must exercise extreme caution when implementing AI systems, and they must also hold themselves accountable.

H02: AI has a positive effect on financial services.
H03: AI has a positive effect on Financial Institution.
H04: AI has a positive effect on Investors.

14.4 Research Methodology

14.4.1 Sampling and Data Collection

Participants in Saudi Arabia who were 20 years old or older and had some background knowledge in sustainable finance, financial markets, and artificial intelligence were the subjects of the study. The decision was based on our target demographic's potential earliest adopters. For this study, the researchers used a convenience sample. The survey was administered online, using a platform that safeguarded the participants' privacy, from May 2024 to September 2024. Out of 485 forms that were gathered, 33 were not included since the individuals listed did not possess the necessary prior knowledge in sustainable finance, financial markets, or artificial intelligence (Minhaj et al. 2024).

14.4.2 Measurement of Variables

A Likert scale was used to assess these attributes; a score of 1 signifies strong disagreement and a score of 5 indicates strong agreement. The factors were modified from earlier research to fit the current study.

The Likert scale is a type of ordinal scale commonly used in surveys and questionnaires to measure attitudes, opinions, perceptions, or behaviors. It typically asks respondents to rate statements based on how much they agree or disagree with them. While Likert scales are ordinal in nature, they allow for quantification, which makes statistical analysis feasible.

14.4.3 Data Analysis

The validation of models using the PLS method allowed us to evaluate the structural model and find out how reliable and valid the measurement scales were. This study employed the software application Smart-PLS 3, which utilized a bootstrapping technique with 5000 samples, to ascertain the parameters' statistical significance. In case you are conducting research with the main goal of making predictions and determining what factors impact investor behavior, then Partial Least Squares (PLS) is the way to go.

14.5 Finding and Analysis

Out of 485 total opinions, 452 were chosen to be included in the study's results in order to facilitate the handling of data. To statistical examination of the information. For this study, researchers used SmartPLS 3. In this section, the findings and conclusions drawn from the investigations are detailed.

14.5.1 Background Information of the Respondents

The entirety of the info provided hereunder originates from authentic sources, and the next section displays the representative group of persons whose contributed throughout the enquiries. Table 14.1 highlights what they said related to the investigation's predefined characteristics.

Table 14.1 summarizes the socioeconomic and cultural traits of the participants. Factors such as gender, age, level of education, employment position, and monthly income are part of this category. The entirety of 63.93% of the collection was male, with 36.07% being female, in accordance to the records.

According to the results, 31.63% of the people who took part in the survey were in the 31–40 age bracket. Following is the breakdown of the participants' ages: 2.5% were in the 21–30 age bracket, while 28% were 40 and beyond. Additionally, respondents beneath the age of 20 made up 14.38% of the total.

Out of all the participants, 10.61% were categorised as Bachelor (UG), 27–87% as Graduation (G), 34.15% as Postgraduate (PG), and 22.34% as PDHs thereafter.

Table 14.1 Respondent initial statistics (N = 452)

Basis	Categories	F	%
Gender	M	289	63.93
	F	163	36.07
Age group	Up to 20 years	65	14.38
	21–30 years	117	25.88
	31–40 years	143	31.63
	41 and above	127	28.09
Educational qualification	UG	48	10.61
	G	126	27.87
	PG	177	39.15
	PDH	101	22.34
Occupational status	Government employees	89	19.69
	Private employees	156	34.51
	Business and self employees	198	43.80
	Students	09	1.99
Monthly income	≤ Rs. 10000	56	12.38
	Rs. 10,000–20,000	145	32.07
	Rs. 20,001–40,000	198	43.80
	> Rs. 40,000	53	11.72

The following groups comprise the respondents: 19.69% government employees, 34.51% private sector workers, 43.81% individuals who are self-employed or have their own business, and 1.99% academics.

According to the data, 12.38% of the participants had monthly earnings below Rs. 10,000, while 32.07% had incomes between Rs. 10,000 and 20,000. In addition, 11.72% of the participants had earnings higher than Rs. 40,000, while 43.80% of the people surveyed indicate earnings between Rs. 20,000 and 40,000.

14.5.2 Measurement Model Evaluation

With the help of "internal consistency, convergent validity, and discriminant authenticity," the measurement framework were validated.

The inquiry's underlying constructs—artificial intelligence, sustainable finance, financial services, financial institutions, and investors—are shown as rings in Fig. 14.1. There are eight assertion symbols determining artificial intelligence, five assessing sustainable finance, three evaluating financial services, three evaluating financial institutions, and three reviewing financiers, all arranged within the

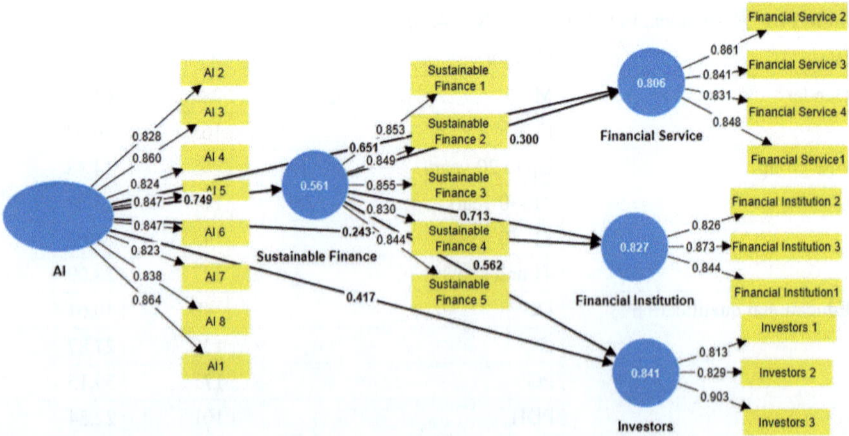

Fig. 14.1 Measurement model from SmartPLS 3.0

arrows pointing to the particular substances or elements. Every item's or construct's transferring factor measurements are also extremely near to their matching indicators.

Whenever every element of a specific AI, sustainable finance, financial service, financial institution, or investor construct have an average score higher than 3, as shown in Table 14.2, then it seems like there will be a good response. The coefficient loadings for some components in each construct are greater than the cutoff value of 0.70. Essentially suggested constructs are discussed thoroughly in every assertion.

Each of the five configurations met the required benchmark restriction, as shown in Table 14.3, with "Composite Reliability" (C.R.) values over 0.7 and "Average Variance Extracted" (AVE) values over 0.5 (Ventre and Kolbe 2020). Results showing internal consistency were somewhat higher than 0.7 on the "Cronbach's Alpha and rho-a" scale (Khan and Minhaj 2022). In response to this requirement, the concept of "convergent validity" was (Khanifar et al. 2012).

14.5.3 Discriminant Validity Result

To check for discriminant validity, researchers used the "Fornell-Larcker and cross-loading criteria," whereby find out how different the statistic is from similar ideas in the nomological cloud.

To calculate the "Fornell-Larcker" standard, as demonstrated in Table 14.4, it is necessary to determine the square roots of the "Average Variance Extracted" numbers corresponding to the relevant ideas. When compared to the associations across the different groups, such as financial services (0.820), financial institutions (0.809), investors (0.831), and AL (0.798), the results reveal that sustainable finance has the

Table 14.2 Mean, SD, and loadings of constructs

Construct	Item	Mean	SD	Loading
Artificial intelligence (A1)	AI 1	4.01	0.65	0.91
	AI 2	3.98	0.77	0.78
	AI 3	3.77	0.81	0.77
	AI 4	3.94	0.74	0.81
	AI 5	3.32	0.71	0.84
	AI 6	4.11	0.79	0.87
	AI 7	3.86	0.74	0.82
	AI 8	3.11	0.67	0.84
Sustainable finance (SF)	SF 1	3.80	0.78	0.70
	SF 2	3.88	0.81	0.81
	SF 3	4.04	0.89	0.84
	SF 4	4.30	0.87	0.79
	SF 5	4.11	0.80	0.76
Financial services (FS)	FS 1	3.34	0.76	0.84
	FS 2	4.0	0.83	0.78
	FS 3	3.02	0.76	0.77
Financial institution (FI)	FI 1	3.67	0.75	0.76
	FI 2	3.89	0.76	0.86
	FI 3	4.01	0.73	0.80
Investors	Investors 1	3.03	0.89	0.78
	Investors 2	4.15	0.81	0.79
	Investors 3	4.02	0.77	0.75

Table 14.3 Convergent validity result

Factor	Cronbach's alpha	Rho-A	C.R.	A.V.E.
Artificial intelligence (AI)	0.752	0.815	0.879	0.682
Sustainable finance (SF)	0.769	0.861	0.897	0.711
Financial services (FS)	0.709	0.789	0.793	0.743
Financial institution (FI)	0.844	0.821	0.810	0.744
Investors	0.822	0.773	0.863	0.771

strongest relationship. It was possible to confirm discriminant validity by applying the "Fornell-Larcker" approach (Fornell and Lacker 1981).

Table 14.5 shows that in every column, the forces applied for all builds were significantly greater than the cross-loadings for the additional constructs (Khan and Minhaj 2022), as shown by the need for cross-loading. The cross-loading requirement was implemented to establish discriminant validity (Henseler et al. 2015).

Table 14.4 Discriminant validity–Fornell-Larcker criterion

Factors	Artificial intelligence	Sustainable finance	Financial services	Financial institution	Investors
Artificial intelligence	**0.798**				
Sustainable finance	0.654	**0.811**			
Financial services	0.631	0.710	**0.820**		
Financial institution	0.744	0.790	0.761	**0.809**	
Investors	0.751	0.702	0.765	0.741	**0.831**

Table 14.5 Discriminant validity–loading and cross-loading criterion

Factor	Artificial intelligence (AI)	Sustainable finance (SF)	Financial services (FS)	Financial institution (FI)	Investors
AI 1	**0.711**	0.541	0.711	0.509	0.555
AI 2	**0.802**	0.654	0.666	0.611	0.609
AI 3	**0.791**	0.576	0.540	0.598	0.711
SF 1	0.509	**0.781**	0.601	0.711	0.560
SF 2	0.568	**0.804**	0.550	0.721	0.519
SF 3	0.616	**0.811**	0.610	0.692	0.721
SF 4	0.670	**0.764**	0.751	0.590	0.654
SF 5	0.509	**0.730**	0.690	0.633	0.690
FS 1	0.432	0.650	**0.831**	0.490	0.589
FS 2	0.439	0.609	**0.809**	0.609	0.498
FS 3	0.555	0.639	**0.819**	0.611	0.609
FI 1	0.701	0.690	0.701	**0.765**	0.544
FI 2	0.654	0.560	0.680	**0.778**	0.699
FI 3	0.444	0.619	0.578	**0.790**	0.701
FI 4	0.509	0.604	0.701	**0.819**	0.723
Investors 1	0.550	0.560	0.511	0.711	**0.779**
Investors 2	0.611	0.563	0.590	0.609	**0.781**
Investors 3	0.654	0.690	0.701	0.676	**0.832**

14.5.4 Structural Equation Model

The structural model investigation has to take multicollinearity into account, even if it can significantly impact the reliability of the conclusions (Khan and Minhaj 2021). Although the "Variance Inflation Factor" (VIF) values varied between 1.303 and

2.267, the researchers concluded that the model did not display multicollinearity. The practicality of the theory was then determined by 5000 bootstrapping recalculations of the structural model.

You can see the PLS-SEM model in Fig. 14.2. Whenever the tested values surpass the stated regression weight criterion of 1.96, it's immediately evident that every path is deemed significant at a 5% or higher significance level. As a result, the path parameter that is anticipated has some significance. The results of the structural equation modelling (SEM) information are shown in Table 14.6.

The results of the investigation show that proposition H01 proved correct (Table 14.6). There is a favorable correlation between Artificial Intelligence therefore Sustainable Finance ($\beta = 0.798$, t-value $= 3.98$, and $p < 0.001$).

Table 14.7 displays the outcome of the H02 test. Artificial intelligence is also positively and specifically linked to financial services, as shown by a $\beta = 0.543$, t-value $= 6.910$, and $p < 0.001$.

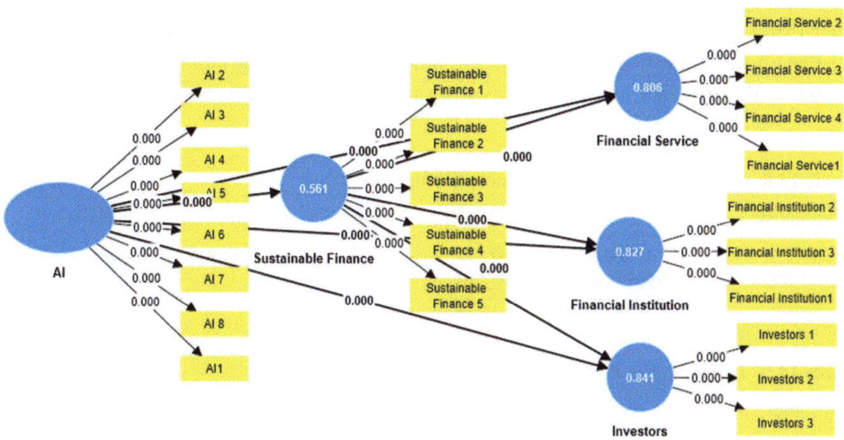

Fig. 14.2 Structural equation model (SEM)

Table 14.6 Direct impact of AI on sustainable finance

Hypothesis	Path	B	t-value	p-value	Result
H01	AI → Sustainable finance	0.798	3.98	$P \leq 0.001$	Supported

Table 14.7 Direct impact of AI on financial services

Hypothesis	Path	B	t-value	p-value	Result
H02	AI → financial services	0.543	6.910	$P \leq 0.001$	Supported

Results showing acceptance of H03 can be seen in Table 14.8. There has been a positive and immediate association between AI while financial institutions (β = 0.511, t-value = 3.761, and $p < 0.001$).

As you can see in Table 14.9, we ran tests on H04. Additionally, AI is favorably associated with investors (beta = 0.401, $t = 2.76$, $p < 0.001$).

Results from testing corroborated assumptions H05, H06, and H07, as shown in Table 14.10. When it comes to artificial intelligence, sustainable finance impacts financial services indirectly or through a mediator (β = 0.287; t-value = 2.479; $p < 0.005$). Sustainable finance has an indirect or intermediary effect on financial institutions in relation to artificial intelligence (β = 0.311, t-value = 2.390, and $p < 0.005$). Similarly, investors are indirectly affected by sustainable finance in the context of artificial intelligence (β = 0.367; t-value = 2.891; $p < 0.005$).

The results may indicate that AI improves efficiency, reduces operational costs, and makes decision- making faster and better. For example, contemplate how AI has recently allowed financial services to incorporate sustainable practices, such as automating assessments of ESG (Environmental, Social and Governance) criteria, measuring carbon footprints, or assessing climate risk. The research states, any new AI-driven financial products advancing sustainability like green bonds, sustainable investment portfolios, or impact investing underscore these benefits.

Table 14.8 Direct impact of AI on financial institution

Hypothesis	Path	B	t-value	p-value	Result
H03	AI → financial institutions	0.511	3.761	$P \leq 0.005$	Supported

Table 14.9 Direct impact of AI on investors

Hypothesis	Path	B	t-value	p-value	Result
H04	AI → investors	0.401	2.76	$P \leq 0.001$	Supported

Table 14.10 Mediating or Indirect impact of sustainable finance and hypothesis testing

Hypothesis	Path	B	t-value	p-value	Result
H05	AI → sustainable finance → financial services	0.287	2.479	$P \leq 0.005$	Supported
H06	ML → sustainable finance → financial institutions	0.311	2.390	$P \leq 0.005$	Supported
H07	AI → sustainable finance → investors	0.367	2.891	$P \leq 0.005$	Supported

14.6 Discussion

Also, new market dynamics might emerge as a result of hedge funds using AI and Sustainable finance methodologies to create their own unique trading methods. Improving the speed and accuracy of data processing could help reduce the buildup of macro-financial pricing imbalances by making it easier to spot and fix price discrepancies sooner.

The use of artificial intelligence and Sustainable finance methods has the ability to increase risks, but only whether investors adhere to legislation without doing risk assessments they can utilize them to reduce capital or margins while increasing anticipated earnings on capital. Increased influence, more stringent stability buffers as well, and accelerated maturity transition could be possible with the help of AI and Sustainable finance in optimization processes as opposed to the alternative.

The integration of AI with Sustainable finance principles has the potential to bring about significant improvements in the financial services sector, which is crucial for maintaining financial stability. Better operational efficiencies in the provision of financial services and greater capacities for the monitoring and mitigation of regulatory and systemic risks are the most obvious manifestations of these benefits. Greater efficiency in the financial system could be achieved through the improvement of credit risk information processing and the decrease of client interaction expenses. Risk management, fraud detection, and regulatory compliance could be enhanced with artificial intelligence (AI) and the Sustainable finance applied to internal (back-office) applications, all while reducing costs. With the integration of Sustainability applications and improved AI-derived data processing, portfolio managers may be able to make financial markets far more efficient and resilient.

A strong potential for AI-driven improvements and increased efficiency through better analysis, better prediction, and automation. This, in turn, can assist businesses more. We can talk about the way AI can help to bring better solutions to ensure risk management frameworks for sustainability. For clear trending predictors in risks related to climate, AI products can provide more accurate analytics to help financial service companies take data-driven decisions on investments that are climate friendly. We could also talk about the ethics of AI with respect to sustainable finance. AI systems that have been poorly designed can pose challenges such as potentially exacerbating biases or the need for data privacy, transparency in decision making by AI, etc.

14.7 Practical Implication

This capability to track investments and measure their environmental price, more sustainable financial products could be developed. It would enhance sustainability in the long term by enabling funding in initiatives that solve global problems, such as climate change. Financial institutions are faced with ever evolving regulations related to sustainability and AI tools can be leveraged to make compliance easier in these changing environments by automating compliance checks, reporting on sustainability goals, and confirming that sustainability practices align with regulatory standards. AI can also serve to build the synergies needed between the public and private sectors to spur the implementation of sustainable finance. The research could focus on discussion on AI-enabled platforms that are driving collaboration across multiple sectors to drive environmental sustainability. AI can analyze large quantities of data, and improve transparency in financial services, enhancing clients and investor's ability to see how their investments contribute to Sustainable Development Goals (SDGs).

The use of AI and the integration of sustainable finance principles into the trade sector could significantly alter the breadth and depth of "directional" trading. Assuming all goes according to plan; the financial system may gain from the diverse trading apps being created among market players. One example is the use of Sustainable finance in robo-advisors, which could lead to more tailored investment recommendations based on user preferences and less reliance on algorithmic trading strategies. These apps could extend the pool of potential investors inside capital markets by removing the barriers that prevent ordinary people from taking part in investment activities. There may be positive and negative outcomes in terms of maturity change, liquidity, and leverage that could result from financial market players like hedge funds and market makers utilising AI and sustainable finance.

Financial market liquidity could be enhanced through the optimization of trading activities' velocity and efficacy made possible by combining AI with sustainable finance. High-risk transactions and complicated financial activities may be better identified with the use of artificial intelligence and Sustainable finance, and certain financial institutions may be able to benefit from enhanced risk management strategies. Innovative types of credit portals that link borrowers to financiers effectively (often referred to as FinTech credit) are made possible by these tools, which could reduce banks' reliance on traditional bank loans, lead to lower bank debt, and encourage a more distributed risk distribution across the financial system.

14.8 Conclusions

Immediate detection and correction of pricing discrepancies and, under ideal circumstances, a reduction in the frequency of congested transactions, can bring about this enhancement. The implementation of use cases by policymakers and examiners holds the promise of more comprehensive systemic risk assessments and increased supervisory efficacy in the financial markets. The future may bring even more third-party dependence reason being network effects and the adaptability of new technologies. This could lead to the emergence of novel entities with far-reaching consequences for the system. A set of non-binding guidelines for the ethical application of AI that addresses the environment has been embraced by more than 190 nations. Additionally, laws have been developed in the US and the EU to lessen the negative effects of AI on the environment. However, according to Radwan, such policies are rare.

Governments are rushing to create national AI policies, yet they seldom ever consider sustainability and the environment. The absence of environmental safeguards is just as harmful as the absence of other AI-related protections.

Some of the most well-known IT businesses are increasingly focusing on providing AI services that also align with sustainability goals. Engaging with third-party providers with a solid reputation, wide reach, and accessibility could be beneficial for financial institutions, much like in other platform-based businesses. It is possible for oligopolies or natural monopolies to exist. Whenever technology companies manage to establish a stranglehold on certain segments of the financial sector, the competition worries—which are serious from an efficiency standpoint—could lead to threats to the stability of the financial system. In the case of a major disruption or bankruptcy faced by a large organisation, the possible systemic effects of interdependencies and reliance on third parties become obvious.

In the present day, it is possible that the majority of companies offering AI and sustainability solutions for financial services are either not knowledgeable enough about the relevant regulations or are acting outside of the purview of such regulations. In situations when financial institutions rely on AI and sustainable finance service providers for critical operations and restrictions about outsourcing are either nonexistent or poorly understood, these suppliers may run their businesses without proper supervision. Similarly, if the makers of these tools start selling financial services to regular people or institutions, it's possible that some of the money changing hands would happen outside of the rules. Upholding the monitoring of AI applications and sustainable finance is crucial. It is possible that, if the underlying technology develops further, it will find applications outside of the ones discussed in this paper. It is critical to keep an eye on these developments and update this evaluation from time to time for future use.

14.9 Future Research and Limitation

This study will be helpful for further research to apply probability sample in result of that we may compare them with those researches that have employed the technique of non-probability sampling. Only Saudis with established AI and sustainable finance systems were included for this study. Additional research with opposing viewpoints and data analysis using the proposed methodology is necessary. For financial and time reasons, we were reluctant to conduct a comprehensive mixed-method study. Further investigations can employ a mixed-methods approach to obtain more precise results.

First of all, the study's small sample size makes broad assumptions about Saudis as a whole questionable. If businesses in Saudi Arabia can't connect with the rural class, which makes up a larger percentage of the population, then the efficacy of AI and ML becomes quite questionable. Thus, the rural class of Saudi Arabia may be included in future research that expands the study's scope. When choosing participants for this study, we also took their practicality into consideration. Without using probability sampling methods, subsequent study may not be able to apply the results to a broader population.

References

Arshad Khan M, Alhumoudi HA (2022) Performance of e-banking and the mediating effect of customer satisfaction: a structural equation model approach. Sustainability 14(12):7224. https://doi.org/10.3390/SU14127224

Cao L (2021) AI in finance: challenges, techniques and opportunities. SSRN Electron J 1(1):1–40. https://doi.org/10.2139/ssrn.3869625

Data SB, Analytics D (2000) Big data, data analytics and artificial intelligence in accounting: an overview, vol 61, no 2, pp 1–34

Deepalakshmi S, Sc M (no date) Impact of artificial intelligence in e-payments

Fornell, Lacker (1981) Discriminant validity assessment: use of Fornell & Larcker criterion versus HTMT criterion. J Phys Conf Ser Paper 890(1):1–6

Gandomi AH, Chen F, Abualigah L (2023) Big data analytics using artificial intelligence. Electronics (Switzerland) 12(4):3–7. https://doi.org/10.3390/electronics12040957

Goodell JW et al (2021) Artificial intelligence and machine learning in finance: identifying foundations, themes, and research clusters from bibliometric analysis. J Behav Exp Finance 32:100577. https://doi.org/10.1016/j.jbef.2021.100577

Habib S, Hamadneh NN, Khan MA (2021) Influence of electronic word of mouth (Ewom) and relationship marketing on brand resonance: a mediation analysis. Sustainability (Switzerland) 13(12). https://doi.org/10.3390/su13126833

Haider A et al (2024) The role of e-banking, mobile-banking, and e-wallet with response to e-payment and customer trust as a mediating factor using a structural equation modelling approach. J Infrastruct Policy Dev 8(9):6644. https://doi.org/10.24294/jipd.v8i9.6644

Henseler J, Ringle CM, Sarstedt M (2015) A new criterion for assessing discriminant validity in variance-based structural equation modeling. J Acad Mark Sci 43(1):115–135. https://doi.org/10.1007/s11747-014-0403-8

Inci AC (2023) Artificial intelligence in finance. In: Contemporary issues in quantitative finance, pp 266–274. https://doi.org/10.4324/9781003213697-19

Khan MA, Alhumoudi H et al (2024) The present scenario of artificial intelligence and machine learning in financial services: an empirical study. https://doi.org/10.24294/jipd.v8i11.8818

Khan MA, Husain S et al (2024) To explore the impact of corporate culture and leadership behaviour on work performance, mental health and job satisfaction of employees: an empirical study. https://doi.org/10.24294/jipd.v8i11.6417

Khan MA, Minhaj SM (2021) Performance of online banking and direct effect of service quality on consumer retention and credibility of consumer and mediation effect of consumer satisfaction. Int J Bus Inf Syst 1(1):1. https://doi.org/10.1504/IJBIS.2021.10043829

Khan MA, Minhaj SM (2022) Dimensions of e-banking and the mediating role of customer satisfaction: a structural equation model approach. Int J Bus Innov Res 1(1):1. https://doi.org/10.1504/IJBIR.2022.10045447

Khan MA (2021) Netizens' perspective towards electronic money and its essence in the virtual economy: an empirical analysis with special reference to Delhi-NCR, India

Khan MA et al (2023) Importance of social networking sites and determining its impact on brand image and online shopping: an empirical study. Sustainability 15(6):5129. https://doi.org/10.3390/su15065129

Khanifar H et al (2012) Factors influencing the intendancy of e-banking: an integration of TAM & TPB with eservice quality. J Appl Sci Res 8(3):1842–1852

Lu L (2018) Decoding Alipay: mobile payments, a cashless society and regulatory challenges. Butterworths J Int Bank Financ Law 33(1):40–43. https://ssrn.com/abstract=3103751

Luan H et al (2020) Challenges and future directions of big data and artificial intelligence in education. Front Psychol 11:1–11. https://doi.org/10.3389/fpsyg.2020.580820

Minhaj SM, Altaf Khan S (2022) The revolutionary impact of micro-finance and role of financial institutions on agriculture income of farmers: an empirical analysis's. Int J Bus Innov Res [Preprint]. https://doi.org/10.1504/IJBIR.2022.10052373

Minhaj SM et al (2024) Investor sentiment and the function of blockchain technology in relation to digital currencies: the here and now and the future investor. https://doi.org/10.53555/kuey.v30i5.3942

Najafabadi MM et al (2015) Deep learning applications and challenges in big data analytics, pp 1–21. https://doi.org/10.1186/s40537-014-0007-7

OECD (2021) Artificial intelligence, machine learning and big data in finance: opportunities, challenges, and implications for policy makers. OECD Business and Finance Outlook 2020: Sustainable and Resilient Finance, pp 1–72. https://www.oecd.org/finance/financial-markets/Artificial-intelligence-machine-learning-big-data-in-finance.pdf

Ojokoh BA et al (2020) Big data, analytics and artificial intelligence for sustainability. Sci Afr 9:e00551. https://doi.org/10.1016/j.sciaf.2020.e00551

Rahmani AM et al (2021) Artificial intelligence approaches and mechanisms for big data analytics: a systematic study, pp 1–28. https://doi.org/10.7717/peerj-cs.488

Soni P (2021) A study on artificial intelligence in finance sector, vol 9, no 5, pp 2320–2882. Available at: www.ijcrt.org

Ventre I, Kolbe D (2020) The impact of perceived usefulness of online reviews, trust and perceived risk on online purchase intention in emerging markets: a Mexican perspective. J Int Consum Mark 32(4):287–299. https://doi.org/10.1080/08961530.2020.1712293

Zamani ED et al (2023) Artificial intelligence and big data analytics for supply chain resilience: a systematic literature review, pp 605–632

Part V
Technological Innovations for Sustainability

Chapter 15
Innovative Energy-Saving Techniques in Construction and Heavy Machinery

Gyan Wrat, Prabhat Ranjan, Joseph T. Jose, and Rakesh Kumar

Abstract The construction and heavy machinery industries are significant contributors to global energy consumption and carbon emissions, necessitating innovative energy-saving techniques. This paper explores advancements in energy-efficient engines, hybrid power systems, alternative fuels, advanced materials, and regenerative energy technologies. Artificial Intelligence (AI) and automation play a pivotal role in optimizing energy usage, predicting energy loads, and reducing emissions. Additionally, integrating renewable energy sources and sustainable construction practices enhances efficiency while minimizing environmental impact. AI-driven analytics, predictive maintenance, and smart grid technologies further contribute to energy conservation. Despite the benefits, challenges such as high implementation costs, data security concerns, and skill shortages hinder widespread adoption. Addressing these barriers through policy support and technological advancements can accelerate the transition toward energy-efficient construction and heavy machinery operations. This study highlights the importance of adopting innovative energy-saving strategies to improve operational efficiency and sustainability.

Keywords Energy efficiency · Hybrid systems · Artificial intelligence · Alternative fuels · Regenerative energy · Predictive maintenance · Sustainable construction

G. Wrat (✉)
Aalborg University, Aalborg, Denmark
e-mail: gyan007wart@gmail.com

P. Ranjan
Padmabhooshan Vasantdada Patil Institute of Technology, Pune, India

J. T. Jose
NLC India Ltd., Neyveli, Tamil Nadu, India

R. Kumar
PP Savani University, Dhamdod, Gujarat, India

S. K. Tiwari et al. (eds.), *AI-Driven Environmental Pollution Management*, Climate Risks and Solutions, https://doi.org/10.1007/978-3-031-96243-1_15

15.1 Introduction

The construction and heavy machinery industries are pivotal to economic development but are also among the highest energy consumers and emitters of greenhouse gases. Innovations in energy-saving methods are crucial for reducing operational costs and minimizing environmental impact. In today's era of globalization, the world is rapidly changing with a growing population, leading to increased pollution and global warming, which adversely affect the environment. The lack of energy and fuel-efficient equipment makes energy conservation a top priority. Numerous studies have attempted to lower fuel use and save costs, but much more has to be done. Equipment manufacturers are now focusing on hybrid vehicles that dual means of power generation and achieve higher power transmission efficiency. Advanced hybrid propulsion technology is a promising alternative for achieving fuel economy. The Hydraulic energy-saving techniques enhance system efficiency by optimizing fluid power usage and reducing energy losses (Wrat et al. 2024) These methods include controller and energy recovery systems, leading to significant cost and environmental benefits.

Energy consumption in the construction and heavy machinery industries is a significant concern due to the substantial amount of energy resources these sectors consume. According to Pretty (2018) buildings alone account for approximately one-third of total primary energy resources, making building energy efficiency a crucial factor in addressing the increasing energy demands of the sector. To achieve energy efficiency, it is essential to consider the impact of building user behavior on energy consumption. Belussi et al. (2019) emphasize the importance of understanding the performance of zero energy buildings and energy efficiency solutions, highlighting the need for strategies that influence building occupant behaviors such as eco-feedback, social interaction, and gamification. By combining the qualitative analysis of habitation patterns alongside the quantitative evaluation of energy savings, these methods aim at boosting energy efficiency.

The review conducted by Ruparathna et al. (2016) provides insights into improving the energy efficiency of existing building stock, particularly commercial and institutional buildings. The analysis highlights the necessity for creative ideas to solve the problem of attaining and sustaining energy-efficient behavior without sacrificing building occupants' comfort while critically examining a variety of energy efficiency methods. Moreover, the study by Eleftheriadis, Mumovic, and Greening. Eleftheriadis et al. (2017) discusses life cycle energy efficiency in building structures, focusing on current developments and future outlooks based on Building Information Modeling (BIM) capabilities. The review emphasizes the need for interdisciplinary approaches in order to obtain fresh perspectives on the dynamic character of occupant energy behavior and the variables affecting human behavior (Fig. 15.1).

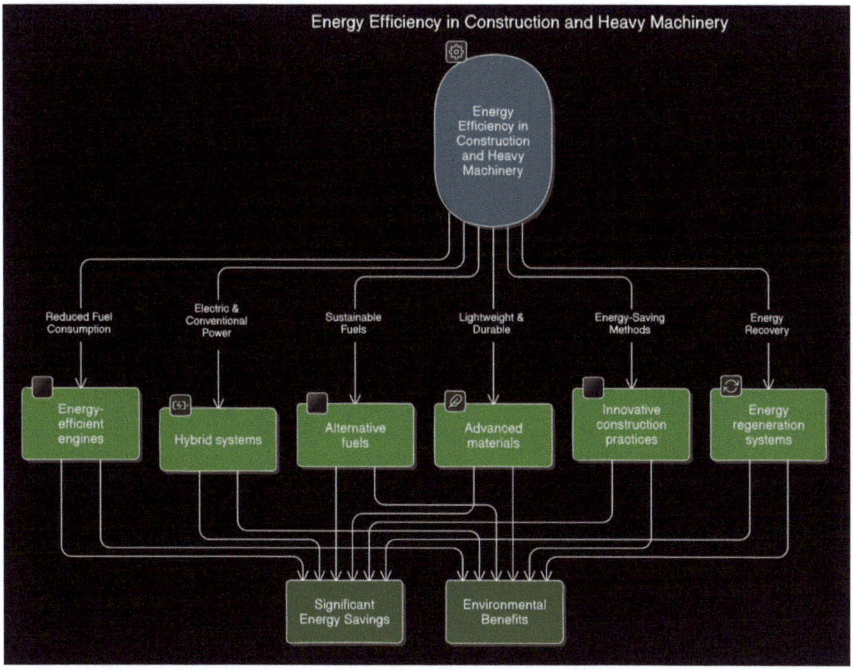

Fig. 15.1 Graphical representation of the energy efficiency of HEMM

15.2 Hybridization in Automobiles and Construction Machineries

Modern diesel engines have undergone significant advancements to improve fuel efficiency and reduce emissions. Technologies such as turbocharging, direct fuel injection, and exhaust gas recirculation (EGR) have been instrumental in these improvements. Turbocharging improves the breathing capacity of the engine, enabling thorough fuel combustion thereby improves engine efficiency. By delivering fuel straight into the combustion chamber, direct fuel injection increases fuel atomization and air mixing, which raises combustion efficiency.

Hybrid engines, which combine internal combustion engines with electric motors, offer substantial energy savings. These systems can recover and store energy during braking and other operations, which can then be used to power the machinery. Hydraulic energy-saving techniques further enhance system efficiency by optimizing fluid power usage and reducing energy losses. These methods include variable displacement pumps and energy recovery systems, leading to significant cost and environmental benefits. In hybrid construction and agricultural machinery (HCAM), energy storage is primarily managed through batteries (Pollet et al. 2012; Somà 2017), super-capacitors, hydraulic accumulators, and flywheels, as illustrated in Fig. 15.2.

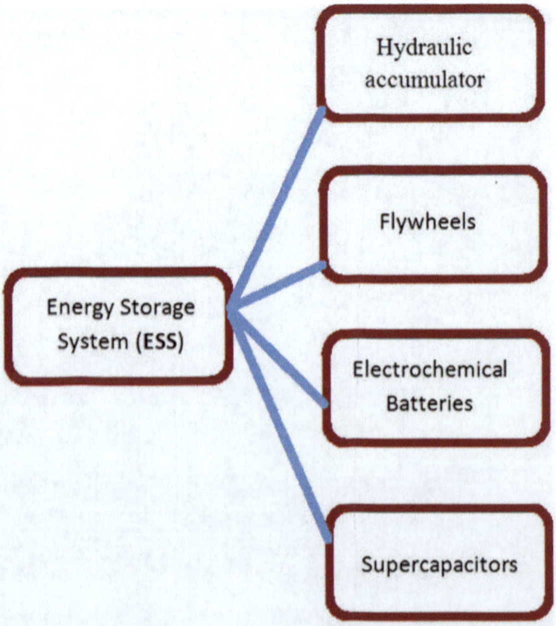

Fig. 15.2 Energy storage for hybrid construction and agricultural machinery

15.3 Hybrid Hydraulic System

Hybrid power systems in cars require a huge amount of electrical energy when starting and accelerating but can then run at a comparatively uniform speed. In contrast, construction equipment often undergoes time-varying and quick speed fluctuations. In hybrid excavators, the charging and discharging of the accumulator occurs in seconds, requiring synchronization with the rapid speed of the charging and discharging phenomena (Fig. 15.3).

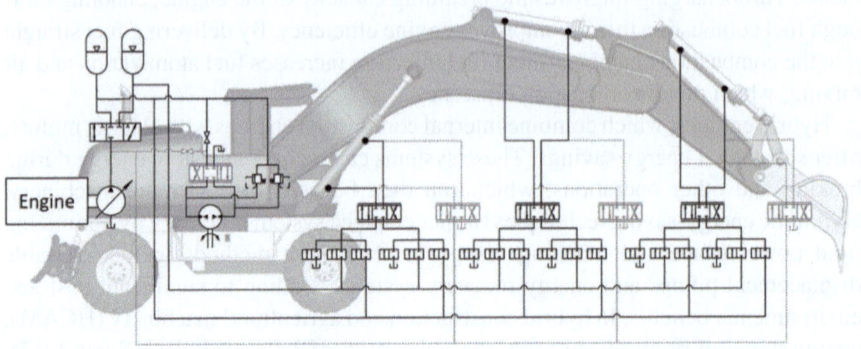

Fig. 15.3 Hybrid Hydraulic Excavator

Table 15.1 The comparison of fuel cost between three configurations and load level (Inoru 2025)

System	Load level		
	Heavy (cost of fuel) (%)	Medium (cost of fuel) (%)	Light (cost of fuel) (%)
Traditional	100	100	100
Parallel	70.5	74.7	79.3
Series	70.5	74.7	79.3

Zhang et al. (2005) conducted a simulation study comparing various hybrid systems—series, parallel, and series–parallel—for a 20-ton excavator. Their results indicated that combining a hybrid system with a separate driving system, motor regeneration, and meter-out control achieved the highest energy savings, up to 58%. However, implementing motor regeneration at ground level remains challenging. Future research should focus on integrating a parallel hybrid system, meter-out control, separate driving, and an electric motor for rotating actuators (Table 15.1).

Wang et al. (2009) analyzed a 5-ton excavator to determine the optimal hybrid powertrain. Their study compared series, parallel, and traditional configurations, finding that the parallel hybrid system offered superior fuel efficiency and economy in both heavy and light modes, while both parallel and series hybrids performed better in medium mode. The parallel hybrid was deemed the best overall in terms of performance and cost. Kanezawa et al. (2001) from the University of Japan investigated a series-type hybrid system that improved exhaust emissions, noise, and fuel economy. Their experiments demonstrated a 35% reduction in fuel consumption compared to traditional systems through energy regeneration during deceleration or descent. Nanjo et al. (2004) developed a simulation model for a hybrid excavator to assess fuel consumption in practical operations. The hybrid system, which includes a closed-loop boom, series hybrid powertrain, electric motor, and hydraulic pump-motor combination, achieved energy savings of over 40%.

15.4 Alternative Fuels

A viable substitute for traditional diesel fuels is biodiesel, which is made from renewable resources like vegetables and animal fats. Its use not only helps in significantly lowering carbon emissions but also enhances engine lubrication. A comparative study by Sheehan et al. (1998) highlighted that biodiesel reduces CO2 emissions by 78% compared to traditional diesel and provides a slight improvement in fuel efficiency. This makes biodiesel a compelling choice for reducing environmental impact while maintaining operational performance.

Another alternative fuel gaining traction is natural gas, available in liquefied (LNG) or compressed (CNG) forms. Natural gas combustion results in fewer pollutants and greenhouse gases than diesel, contributing to cleaner air. The shift towards natural gas in construction equipment has been notably beneficial. According to

Table 15.2 Performance and emission analysis (Kutala et al. 2024)

Fuel type	CO_2 emissions (g/kWh)	Fuel efficiency (L/h)
Diesel	500	25
Biodiesel	110	24.5

Green Energy Group (Yandri et al. 2025) companies adopting natural gas have reported a 30% decrease in operational costs and a 50% reduction in particulate matter emissions. This demonstrates natural gas' effectiveness in both improving environmental outcomes and offering cost savings in the construction sector. Both biodiesel and natural gas represent significant steps towards more sustainable and efficient fuel options in heavy machinery (Table 15.2).

15.5 Advanced Materials

The adoption of lightweight materials, including high-strength steel, aluminum, and composites, can notably reduce the weight of construction machinery, resulting in lower fuel consumption. Switching from traditional steel to high-strength steel and aluminum in equipment frames has demonstrated a 15% improvement in fuel efficiency (Hassan and Saeed 2024). Additionally, the use of wear-resistant materials extends the lifespan of machinery components, decreasing the need for frequent replacements and reducing energy consumption. For instance, employing hardened steel in excavator buckets has led to a 40% increase in component lifespan, which correlates with a reduction in both energy and material usage (Jagadeesh 2023). These advancements in materials not only enhance operational efficiency but also contribute to more sustainable and cost-effective machinery performance.

15.6 Energy Savings Analysis

A study comparing traditional construction methods with prefabrication revealed a 30% reduction in energy use due to decreased transportation and on-site activities (Auti and Patil 2019). Specifically, traditional construction methods exhibited an energy consumption of 100 kWh/m^2, whereas prefabrication methods consumed only 70 kWh/m^2. Building Information Modeling (BIM) facilitates precise planning and optimization of construction processes, thereby reducing material waste and energy consumption. The implementation of BIM in construction projects has led to a 20% reduction in project timelines and a 15% reduction in energy use (Baracho et al. 2025). Furthermore, integrating renewable energy sources such as solar panels and wind turbines on construction sites can significantly offset energy consumption from non-renewable sources. A case study demonstrated that incorporating solar power on a construction site resulted in a 25% reduction in grid electricity consumption

(Ergun et al 2025). Furthermore, excess energy produced by renewable sources can be stored using energy storage devices like batteries, guaranteeing a steady power supply and lowering dependency on fossil fuels. The adoption of energy storage systems in construction machinery has shown a 20% reduction in fuel consumption and operational costs (Zhang et al. 2019). Consequently, the integration of solar power and batteries has achieved a 25% reduction in fuel consumption and 20% cost savings, underscoring the potential benefits of renewable energy integration in construction projects. The Hydraulic energy-saving techniques enhance system efficiency by optimizing fluid power usage and reducing energy losses (Wrat and Das 2023) (Fig. 15.4).

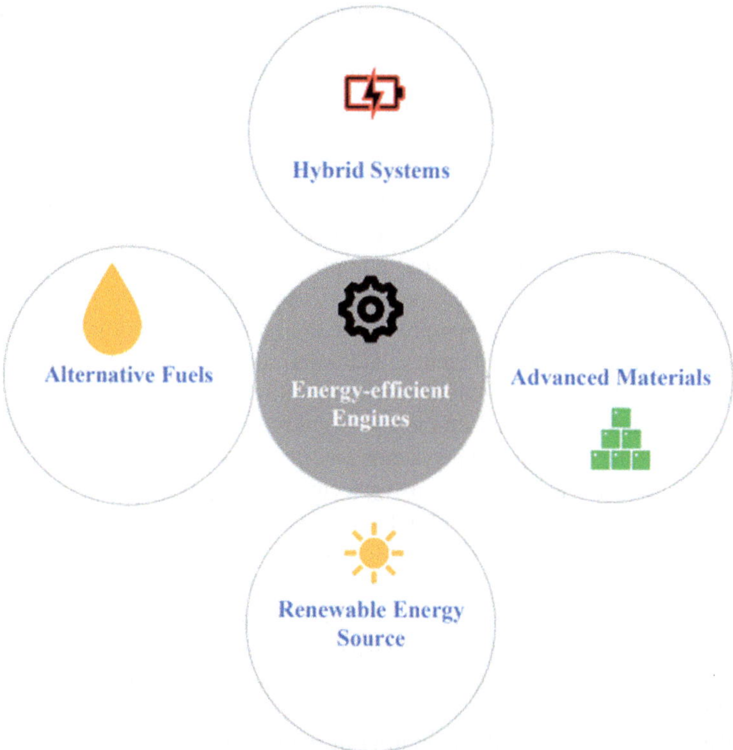

Fig. 15.4 Energy efficient system

15.7 Artificial Intelligence and Automation in Energy Management

AI-driven energy load prediction plays a critical role in minimizing energy waste and reducing emissions. Machine learning algorithms analyzes historical energy consumption patterns, weather data, noise pollution and grid demand to optimize energy distribution (Tiwari et al. 2022, 2023). According to Wang et al. (2024) deep learning models such as long short-term memory (LSTM) networks and convolutional neural networks (CNNs) significantly improve load forecasting accuracy, thereby reducing reliance on fossil fuel-based peaking power plants.

Furthermore, reinforcement learning-based optimization models have demonstrated the ability to manage real-time energy loads effectively. For instance, Q-learning and deep Q-networks (DQN) are being utilized in smart grids to dynamically balance demand and supply (Bousnina and Guerassimoff 2021). AI systems also facilitate demand-side management by integrating renewable energy sources such as solar and wind into power grids, further minimizing carbon footprints.

AI enhances predictive maintenance in energy-intensive sectors by identifying potential failures in machinery before they occur, reducing unplanned downtimes and optimizing energy consumption. Predictive maintenance models, powered by AI, utilize sensor data from industrial equipment and heavy machinery to detect anomalies and assess wear-and-tear patterns (Ucar et al. 2024).

In fleet management, AI-driven predictive analytics enable optimal routing and scheduling, leading to fuel efficiency and lower emissions. Intelligent transport systems, integrated with AI, analyze real-time traffic data and vehicle performance metrics to enhance operational efficiency. A study by Inoru (2025) and highlights that AI-enabled fleet management systems contribute to a 15–20% reduction in fuel consumption, thus lowering greenhouse gas emissions significantly.

The use of AI in monitoring and reducing emissions from heavy machinery has gained significant attention. AI-powered sensors and Internet of Things (IoT) devices continuously track emissions, providing real-time data for regulatory compliance and environmental sustainability (Kutala et al. 2024; Bouakline et al. 2024). Advanced AI algorithms, such as hybrid deep learning models, analyze pollutant levels and predict emission trends to implement corrective actions before exceeding permitted thresholds.

Furthermore, AI-integrated automation in industrial processes enables adaptive control mechanisms, optimizing combustion efficiency and reducing excess fuel consumption. For example, reinforcement learning-based control systems adjust fuel–air mixtures in engines dynamically, leading to a substantial reduction in nitrogen oxides (NOx) and particulate matter emissions (Sanusi et al. 2019).

Governments and regulatory bodies play a crucial role in accelerating the adoption of AI-driven energy-efficient technologies. Policies promoting AI-powered smart grids, intelligent transportation systems, and energy-efficient industrial equipment encourage sustainable development. Incentivizing AI-based solutions through tax credits, research grants, and carbon pricing mechanisms can significantly drive innovation in green construction and manufacturing. Moreover, implementing stringent emission regulations that mandate AI-driven monitoring and optimization tools ensure compliance and promotes responsible energy consumption. AI-enhanced sustainability frameworks, when backed by robust policies, can lead to the large-scale adoption of cleaner technologies in sectors such as construction, transportation, and manufacturing.

15.8 Future Work

The adoption of energy-saving technologies faces several technical and economic barriers, including high initial costs, lack of infrastructure, and integration challenges with existing systems (Shen et al. 2024). Additionally, the integration of AI with renewable energy in machinery remains complex due to data processing limitations, cybersecurity risks, and interoperability issues among various industrial components. However, advancements in AI-driven predictive maintenance and smart energy management can enhance efficiency and sustainability. Future trends in energy-saving technologies focus on decentralized energy systems, smart grids, and AI-optimized industrial automation to reduce energy consumption (Yu and Zhou 2024). The rise of blockchain in energy transactions and the Internet of Things (IoT) for real-time monitoring will further revolutionize the sector. Research must address scalability, affordability, and policy frameworks to accelerate adoption. Overcoming these challenges will drive sustainable industrial transformation and energy efficiency in the coming years.

15.9 Conclusion

Innovative energy-saving techniques in construction and heavy machinery are essential for improving efficiency and reducing environmental impact. This study examined key strategies, including energy-efficient engines, hybrid power systems, alternative fuels, advanced materials, and AI-driven energy optimization. These advancements enhance operational sustainability while minimizing emissions and energy waste. AI applications, such as predictive maintenance and energy load forecasting, enable smarter energy management and extend machinery lifespan. Furthermore, integrating renewable energy and regenerative systems supports sustainable construction practices.

Despite these benefits, challenges such as high implementation costs, data security concerns, and the need for skilled personnel remain. Addressing these issues through industry collaboration, policy support, and continued research can facilitate broader adoption. By embracing these technologies, the construction and heavy machinery sectors can achieve substantial energy savings, lower operational costs, and contribute to global sustainability goals. Future research should focus on refining AI models, improving renewable energy integration, and developing regulatory frameworks to support energy-efficient innovations.

References

Auti SD, Patil JR (2019) Prefabrication technology-a promising alternative in construction industry. Int J Sci Res 8(8):240–244

Baracho RMA, da Silva Santiago LG, Vidigal MJM, Porto MF (2025) Building information modeling (BIM) for planning and construction. Archit Struct Constr 5(1):1–9

Belussi L, Barozzi B, Bellazzi A, Danza L, Devitofrancesco A, Fanciulli C, Ghellere M et al (2019) A review of performance of zero energy buildings and energy efficiency solutions. J Build Eng 25:100772

Bouakline O, Elidrissi A, Khomsi K, Leghrib R (2024) A hybrid deep learning model–based LSTM and modified genetic algorithm for air quality applications. Environ Monit Assess 196(12):1–25

Bousnina D, Guerassimoff G (2021) Deep reinforcement learning for optimal energy management of multi-energy smart grids. In: International conference on machine learning, optimization, and data science. Springer International Publishing, Cham, pp 15–30

Eleftheriadis S, Mumovic D, Greening P (2017) Life cycle energy efficiency in building structures: a review of current developments and future outlooks based on BIM capabilities. Renew Sustain Energy Rev 67:811–825

Ergun S et al (2025) Large-scale renewable energy integration: tackling technical obstacles and exploring energy storage innovations. Sustainability 17(3):1311. https://doi.org/10.3390/su17031311

Hassan HZ, Saeed NM (2024) Advancements and applications of lightweight structures: a comprehensive review. Discov Civ Eng 1(1):47

Inoru (2025) How does AI in fleet management work to optimize routes and fuel consumption? (Online). Available at: https://www.inoru.com/blog/ai-in-fleet-management/. Accessed 3 Feb 2025

Jagadeesh P, Puttegowda M, Girijappa YGT, Sathyanarayana K, Rangappa SM, Siengchin S, Abu Hassan S (2023) Lightweight and sustainable materials for structural applications. In: Lightweight and sustainable composite materials. Woodhead Publishing, pp 197–217

Kanezawa Y, Daisho Y, Kawaguchi T (2001)Increasing efficiency of construction machine by hybrid system. In: JSAE (Society of Automotive Engineers of Japan) annual congress, vol 100, pp 17–20

Kutala S, Awari H, Velu S, Anthonisamy A, Bathula NJ, Inthiyaz S (2024) Hybrid deep learning-based air pollution prediction and index classification using an optimization algorithm. AIMS Environ Sci 11(4)

Nanjo T, Imanishi E, Kagoshima M (2004) 20044744 power simulation for energy saving in hybrid excavator. Trans Soc Automot Eng Jpn 35(4):101–106

Pollet BG, Staffell I, Shang JL (2012) Current status of hybrid, battery and fuel cell electric vehicles: from electrochemistry to market prospects. Electrochim Acta 84:235–249

Pretty J (2018) Intensification for redesigned and sustainable agricultural systems. Science 362(6417):eaav0294

Ruparathna R, Hewage K, Sadiq R (2016) Improving the energy efficiency of the existing building stock: a critical review of commercial and institutional buildings. Renew Sustain Energy Rev 53:1032–1045

Sanusi I, Mills A, Trodden P, Kadirkamanathan V, Dodd T (2019) Reinforcement learning for condition-based control of gas turbine engines. In: 2019 18th European control conference (ECC). IEEE, pp. 3928–3933

Sheehan J, Camobreco V, Duffield J, Graboski M, Shapouri H (1998) An overview of biodiesel and petroleum diesel life cycles
Shen Q, Wen X, Xia S, Zhou S, Zhang H (2024) AI-based analysis and prediction of synergistic development trends in US photovoltaic and energy storage systems. Ann Appl Sci 5(1)
Somà A (2017) Trends and hybridization factor for heavy-duty working vehicles. Hybrid Electr Veh 3–32
Tiwari SK, Kumaraswamidhas LA, Garg N (2022) Comparison of SVM and ARIMA model in time-series forecasting of ambient noise levels. In: Advances in energy technology: select proceedings of EMSME 2020. Springer, Singapore, pp 777–786
Tiwari SK, Kumaraswamidhas LA, Garg N (2023) Assessment of noise pollution and associated subjective health complaints in Jharia Coalfield, India: a structural equation model analysis. Noise Mapping 10(1):20220172
Ucar A, Karakose M, Kırımça N (2024) Artificial intelligence for predictive maintenance applications: key components, trustworthiness, and future trends. Appl Sci 14(2):898
Wang D, Guan C, Pan S, Zhang M, Lin X (2009) Performance analysis of hydraulic excavator powertrain hybridization. Autom Constr 18(3):249–257
Wang X, Wang H, Bhandari B, Cheng L (2024) AI-empowered methods for smart energy consumption: a review of load forecasting, anomaly detection and demand response. Int J Precis Eng Manuf Green Technol 11(3):963–993
Wrat G, Das J (2023) Energy saving in off-road vehicle using leakage compensation technique. FPSI 2023 (Online). Available at: https://www.researchgate.net/publication/374118311_Energy_Saving_in_Off-road_Vehicle_using_leakage_compensation_technique
Wrat G, Ranjan P, Jose JT, Kumar R (2024) Modelling and analysis of the hydraulic system parameters under different loading conditions. In: Dynamics of transportation ecosystem, modeling, and control. Springer Nature Singapore, Singapore, pp 9–21
Yandri E, Uhanto U, Hilmi E, Saiful R, Ariati R (2025) Evaluating effects of biodiesel blends on engine durability and maintenance strategies in heavy equipment applications. Available at SSRN 5117280
Yu X, Zhou Y (2024) Machine learning and artificial intelligence-distributed renewable energy sources: technologies, perspectives, and challenges. In: Advances in digitalization and machine learning for integrated building-transportation energy systems, pp 17–30
Zhang W, Wang J, Du S, Ma H, Zhao W, Li H (2019) Energy management strategies for hybrid construction machinery: Evolution, classification, comparison and future trends. Energies 12(10):2024
Zhang YT, Wang QF, Xiao Q (2005) Simulation research on energy saving of hydraulic system in hybrid construction machinery. In: Proceedings of the Sixth international conference on fluid power transmission and control. Hangzhou, pp 509–513

Chapter 16
Sustainable Dozer Technologies by Evaluating Servo and Hydraulic Pilot Actuated Control Systems

Gyan Wrat, Joseph T. Jose, Prabhat Ranjan, and Rakesh Kumar

Abstract The mining sector is under increasing pressure to adopt sustainable practices that minimize environmental impacts while maintaining operational efficiency. This chapter examines the environmental impact of sustainable dozer technologies in mining, specifically comparing servo and hydraulic pilot actuated control systems. Crawler-mounted dozers are crucial in coal mining operations for tasks like pushing heavy overburden and constructing haul roads. Additionally, the integration of machine learning (ML) and deep learning (DL) models in dozer optimization is emerging as a transformative approach to enhancing operational efficiency. These models enable predictive maintenance, real-time performance monitoring, and fuel consumption optimization, significantly reducing downtime and environmental impact. Furthermore, the transition to advanced hydraulic systems presents a strategic advantage in achieving environmental compliance by improving energy efficiency and reducing hydraulic fluid waste. By analyzing operational efficiencies, emissions reductions, and overall sustainability, this study identifies the advantages and disadvantages of servo and hydraulic pilot actuated control valves in dozers. The findings suggest that while both systems offer benefits, servo control systems may provide a superior balance of efficiency and environmental performance, particularly in reducing fuel consumption and emissions. The chapter concludes with recommendations for future research and implementation strategies for enhancing sustainability through advanced dozer technologies, with a focus on intelligent automation and next-generation hydraulic solutions.

G. Wrat (✉)
Aalborg University, Aalborg, Denmark
e-mail: gyan007wart@gmail.com

J. T. Jose
Coal India Limited, Kolkatta, India

P. Ranjan
NLC India Ltd., Neyveli, Tamil Nadu, India

R. Kumar
Department of Mechanical Engineering, School of Engineering, P P Savani University, Surat, India

S. K. Tiwari et al. (eds.), *AI-Driven Environmental Pollution Management*, Climate Risks and Solutions, https://doi.org/10.1007/978-3-031-96243-1_16

Keywords Environmental mining · Sustainability · Control valve actuation using servo · Hydraulic pilot actuation of control valve · Bulldozer · AI

16.1 Introduction

The mining industry plays a crucial role in the global economy, supplying essential raw materials for various sectors. However, it is also a major contributor to environmental degradation, including habitat destruction, soil erosion, and greenhouse gas emissions. As sustainability becomes a priority, the industry must integrate advanced technologies to minimize its environmental impact while maintaining operational efficiency (Co 2023; Caruana 2024). Among the key heavy machinery used in mining operations, dozers (bulldozers) are indispensable for tasks such as land clearing, material handling, and site preparation. Their environmental impact is closely tied to the efficiency of their control systems, making the choice between hydraulic pilot and servo actuated controls a critical consideration.

Traditional hydraulic systems, while widely used due to their robustness (Wrat and Das 2023) and reliability, pose significant environmental challenges. They are prone to hydraulic fluid leaks, which contaminate soil and water, and their energy losses result in higher fuel consumption and greenhouse gas emissions. Additionally, the high-pressure operation of hydraulic components generates excessive noise, affecting both operators and surrounding ecosystems. In contrast, servo control systems provide a more sustainable alternative, offering improved efficiency, reduced fuel consumption, and lower emissions. Their precise control mechanisms minimize soil disturbance, which helps preserve local ecosystems and reduces erosion. Furthermore, servo systems operate more quietly, mitigating noise pollution in sensitive environments.

The evolution of dozer technology has also led to advancements in operator control mechanisms, transitioning from traditional mechanical levers to pilot controls. These modern control systems not only enhance efficiency by optimizing power consumption in real-time but also improve operator comfort and reduce fatigue, leading to increased productivity (Carter 2012; Wrat et al. 2024). By integrating servo actuated control systems and real-time data analytics, heavy machinery operations can significantly lower their environmental footprint while improving performance and cost-effectiveness. Dozers play a pivotal role in mining operations, particularly in tasks such as pushing heavy overburden, forming haul roads, and preparing excavation phases for other Heavy Earth Moving Machinery (HEMM). Their crawler-mounted design and high engine power enable them to navigate rough terrain efficiently. After blasting, dozers are among the first machines deployed to prepare mining phases for subsequent operations. Figure 16.1 illustrates the schematic diagram of a dozer, highlighting its key functional components. Figure 16.2 shows the Bulldozer schematic hydraulic system.

By evaluating and adopting sustainable dozer technologies, particularly the shift from hydraulic pilot actuated systems to servo-controlled solutions, the mining

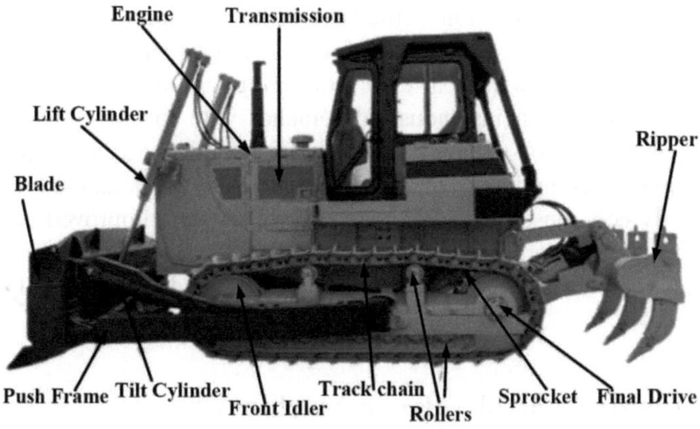

Fig. 16.1 Dozer

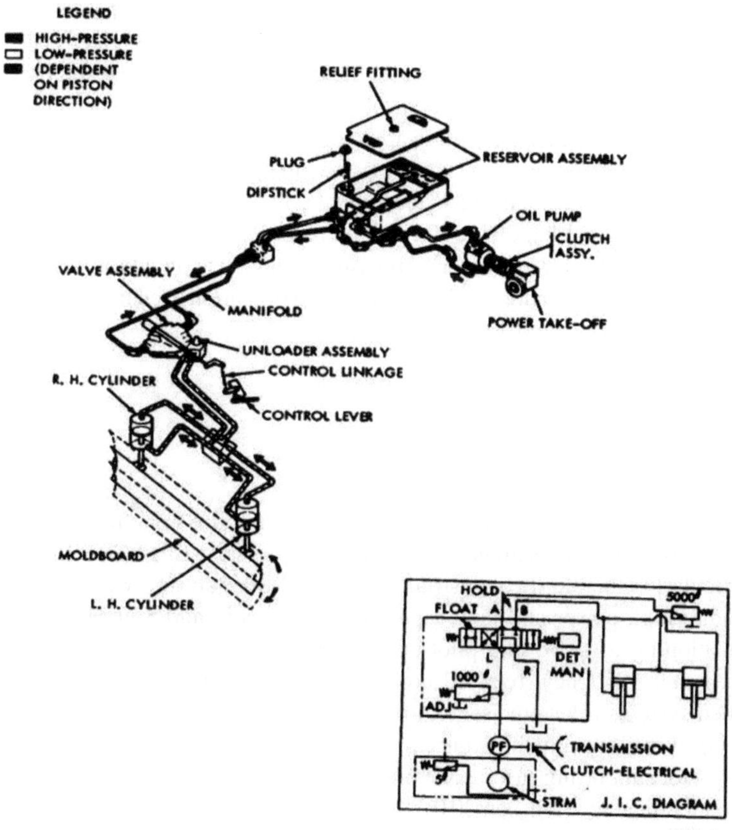

Fig. 16.2 Bulldozer schematic hydraulic system

industry can enhance its operational efficiency while reducing environmental impact (European Commission 2024). As technology continues to advance, integrating AI and ML-driven optimizations, real-time monitoring, and energy-efficient systems will be essential for balancing industrial demands with sustainability goals (Jose et al. 2021). The transition to environmentally friendly dozer technologies not only supports regulatory compliance but also ensures long-term operational resilience in an increasingly eco-conscious world. The ease of use and improved control have made modern dozers more efficient and effective, ultimately benefiting the overall operation and maintenance of heavy machinery (Maas and van Gassel 2003).

16.2 Actuator Control Valve in Machinery

Actuator control valves, such as Servo and pilot-operated control valves are critical components in the hydraulic systems of heavy machinery, such as dozers, where precise movement and control are essential. Servo valves are electrically operated, continuously actuated valves that regulate hydraulic fluid flow and pressure to actuators, offering highly accurate control over actuator position, speed, and force. These valves are particularly useful in applications requiring high precision, such as aerospace, industrial automation, and heavy machinery operations, including dozers. They allow for smooth, controlled movements, making them ideal for tasks like land clearing, material handling, and site preparation in mining. Servo valves are designed to handle high flow rates and pressures, which is crucial for large actuators in heavy machinery, ensuring reliable and efficient operation even in harsh conditions (Anderson et al. 2008; Mouzoure 2023).

16.2.1 Servo Valve Operation

On the other hand, pilot-operated valves use a smaller pilot valve to control a larger main valve, which handles higher flow rates and pressures (Black 2024). The design are particularly suitable for industrial applications like dozers, where large actuators require high power and flow capacity (Karunanidhi and Singaperumal 2010). By transitioning to servo and pilot-operated control systems, dozers can achieve improved fuel efficiency, reduced emissions, and enhanced performance. Servo-controlled systems offer precise, real-time adjustments, reducing energy waste and improving overall sustainability in heavy machinery operations. These advancements are crucial for supporting sustainable practices in the mining industry while ensuring operational efficiency (Nilsson 2016).

From Fig. 16.3a for the servo valve in neutral position; input lever (2) in the "HOLD" position closes the passage between ports A and C by the rotor (3) which is directly connected to the input lever cylinder (5) is then sealed off and the piston

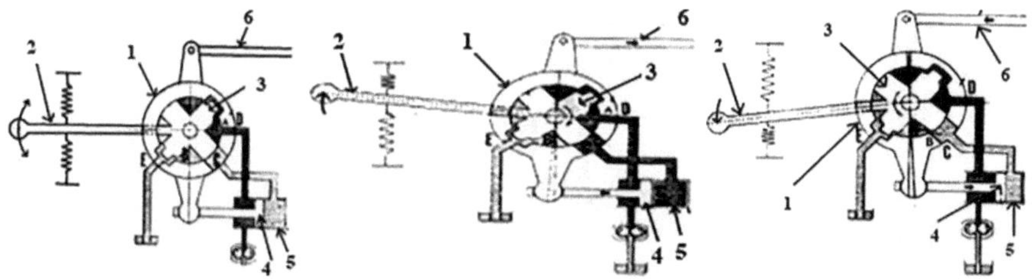

(a)Servo Valve in neutral position *(b)servo valve, Input lever rotated clockwise* *(c)servo valve, input lever rotated counterclockwise*

Fig. 16.3 Servo valve operation

(4) returns to neutral position and stops sleeve (1) and the output shaft are also at neutral position.

In Fig. 16.3b, the servo valve's input lever had been rotated clockwise, shifting the lever (2), and turning the rotor (3) clockwise. This action had opened the passage between ports A and C, allowing pressurized oil to flow into the bottom of the cylinder (5). Due to the difference in area between the bottom and head, the piston (4) had been pushed to the left (←), which in turn had moved the sleeve's lever and output shaft (6) to the right (→). The rotor (5) had continued to rotate clockwise until the sleeve, pushed by the piston, reached a position where the passage between A and C had closed, shutting off the pressurized oil.

In Fig. 16.3c, the input lever had been rotated counterclockwise, turning the rotor counterclockwise and opening the passage between ports B and C. This had connected the drain circuit, allowing oil from the cylinder bottom to flow out via port E, dropping the oil pressure at the bottom. The piston (4) had been pushed to the right (→), moving the sleeve and output shaft (6) to the left (←) until the passage had closed, and the sleeve had stopped. This process had repeated until the control valve spool had reached the designated position. Figure 16.4 had illustrated the hydraulic circuit of the dozer with a servo-actuated control valve.

Figure 16.4 depicted a detailed schematic of the hydraulic circuit used in a dozer, illustrating the flow and control mechanisms for lifting, tilting the blade, and steering. The hydraulic system had started with the hydraulic tank (1), which had stored the hydraulic fluid. The hydraulic pump (PAL 200) (2) had pumped the fluid, creating the necessary pressure for system operation. To maintain safe pressure limits, the main relief valve (3) had regulated the maximum pressure, while the check valve (4) had ensured fluid flow in only one direction, preventing backflow. The lift control valve (5) had managed the flow to the blade lift cylinders (7), responsible for raising and lowering the blade. The quick drop valve (6) had facilitated fast lowering in specific situations. For tilting, the tilt control valve (9) had directed fluid to the tilt cylinders. Check valve (8) had ensured smooth and controlled fluid flow, and the system also included oil filters (11) and a magnetic strainer (15) to keep the hydraulic fluid clean. The steering system had been managed by the steering case (12) and steering pump (FAR 100) (13), while the tilt servo valve (16) had controlled tilt movements (Tables 16.1, 16.2 and 16.3).

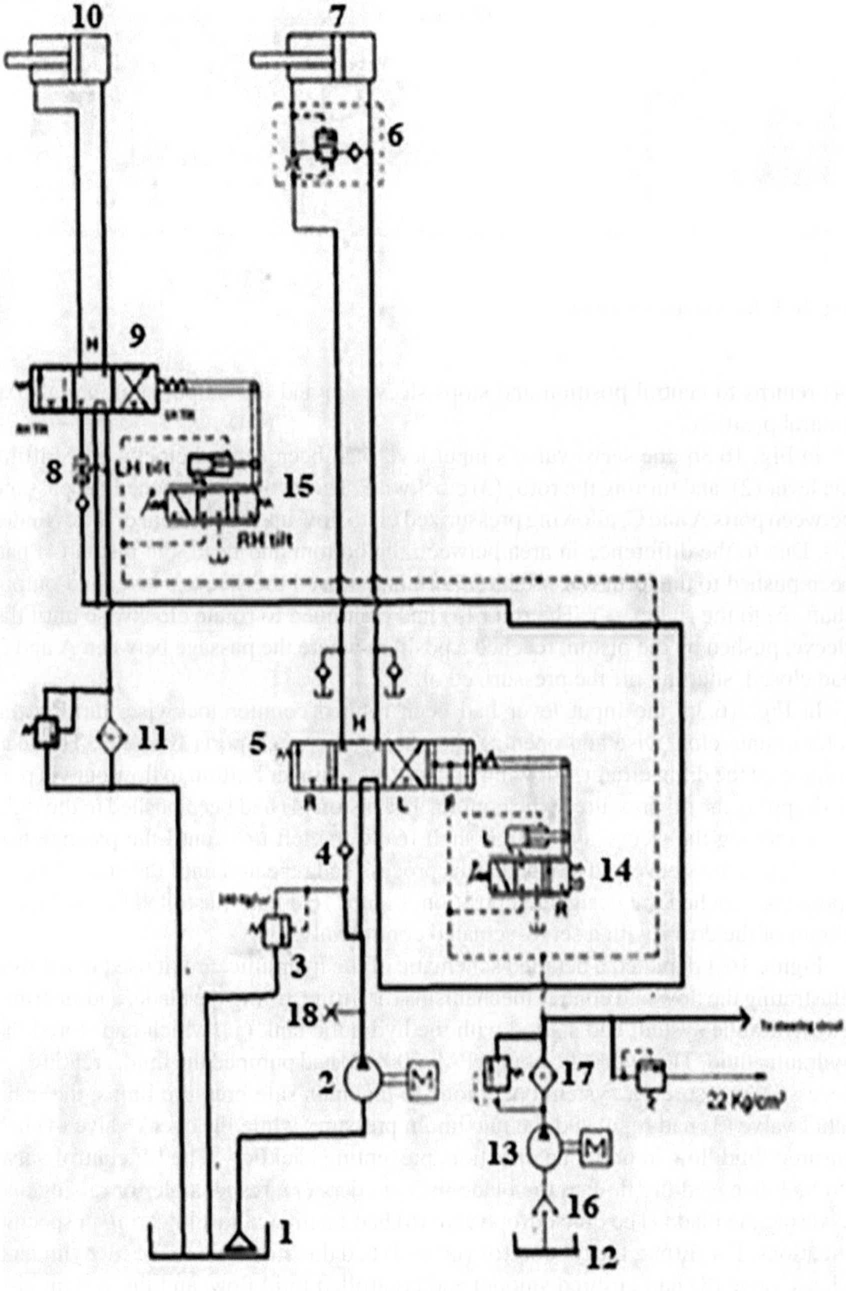

Fig. 16.4 Hydraulic circuit diagram of dozers with servo actuated control valve

Table 16.1 Dozers with servo actuated control valve

Item No.	Component	Item No.	Component
1	Hydraulic tank	10	Blade lift cylinder
2	Hydraulic pump (PAL 200)	11	Oil filter
3	Main relief valve	12	Steering case
4	Check valve	13	Steering pump (FAR 100)
5	Lift control valve	14	Lift servo valve
6	Quick drop valve	15	Tilt servo valve
7	Blade lift cylinder	16	Magnetic stainer
8	flow check valve	17	Oil filter
9	Tilt control valve	18	Pressure pick up plug

Table 16.2 Dozers with servo actuated control valves

Components	Type/specification
Engine	242 KW at 2000 RPM
Hydraulic pump	PAL200
Blade lift control valve (located inside hyd tank)	5l4 control valve, spool actuated by servo link
Blade tilt control valve (located inside hyd tank)	5l3 control valve, spool actuated by servo link
Servo valve	4 position servo valves, operating using transmission fluid

Table 16.3 Lift and tilt cylinder

	Blade lift	Blade tilt
OD of piston rod	75	90
ID of cylinder	140	225
Stroke	1346	160
Max stroke	2380	1468
Min stroke	1030	1308

16.3 Dozers with Hydraulic Pilot Operated Control Valves

Dozers utilize hydraulic pilot-operated control valves to enhance precision and reduce environmental impact. These valves allow operators to control functions like the blade and attachments with minimal physical effort, incorporating proportional control and pressure compensation for efficient performance. Many models feature electro-hydraulic actuation, enabling AI and ML-driven inputs for optimized operation and reduced energy waste. Pilot valves, designed to handle high pressures, ensure reliable function while minimizing hydraulic fluid leaks and noise pollution, contributing to safer and more sustainable dozer operations (Fig. 16.5; Tables 16.4 and 16.5).

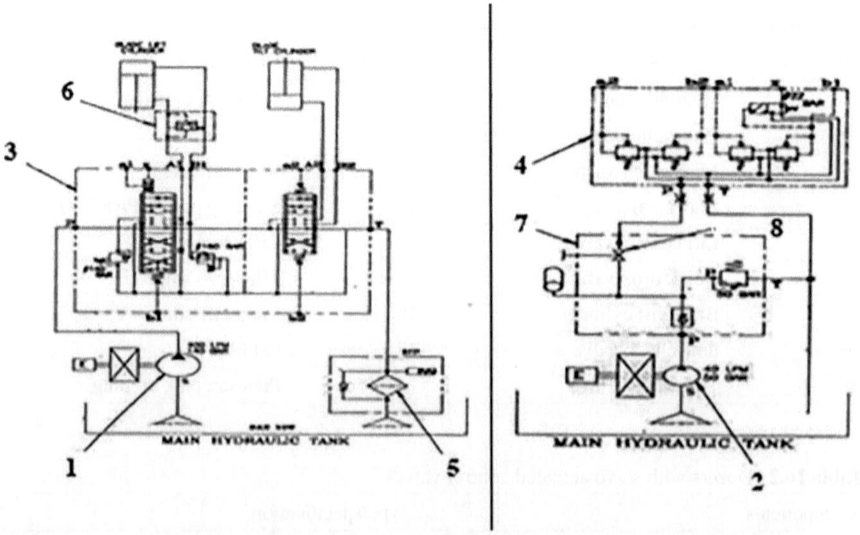

Fig. 16.5 Hydraulic circuit diagram of dozers with hydraulic pilot actuated control valves

Table 16.4 Components of Fig. 16.4

Item No.	Component
1	Main hydraulic pump
2	Pilot pump
3	Main control valve assembly
4	Pilot valve assembly (JOYSTICK)
5	Quick drop valve
6	Return line filter
7	Manifold block assembly
8	Shutoff valve

Table 16.5 Engine and pump specification of dozers with hydraulic pilotactuated control valves

Item No.	Component	Specification
1	Engine	242 KW at 200 RPM
2	Hydraulic pump (Main)	Q = 100 lpm; P_{max} = 140 bar
3	Hydraulic pump (Pilot)	Q = 40 lpm; P_{max} = 50 bar
4	Blade lift control valve	6/4 control spool actuated
5	Blade tilt control valve	6/3 control spool actuated
6	Pilot valve (Joystick)	
7	Accumulator (for blade lifting in emergency)	Volume = 1.1 L; P_{max} = 35 bar; P_{min} = 12.5 bar

The hydraulic system consists of pilot pump (2) which caters to the joystick (4) operations to control the main control valve (3) for desired functions. The flow is directed to a manifold block assembly (7). Which directs the pressurized oil to joystick. It contains an accumulator which is used as storage of pressurized oil and used whenever the engine is OFF to bring down the blade, if in lifted condition. An additional safety shut-off valve is provided to cut off the supply to joystick whenever the operator is away from the seat with engine in ON condition and the blade is in lifted up position there by preventing the danger of unintended fall down of blade.

The key differences between servo actuated control valves and hydraulic pilot actuated control valves for dozers are significant (Adweb-User 2024; Johnson 2024). Servo actuated control valves are electrically actuated using a digital or analog signal, providing highly precise and dynamic flow control of hydraulic fluid to the dozer's actuators. These valves typically have a single stage where the flow is directly controlled by a solenoid or similar device, allowing for incredibly quick adjustments to the fluid flow with response times under 3 ms. The use of brushless DC motors enables smooth and controlled spool adjustments, and magnetic position sensing provides accurate spool position feedback, enabling advanced control, monitoring, and diagnostics capabilities. In contrast, hydraulic pilot actuated control valves consist of a pilot stage that controls the flow to a main stage, which then directs the fluid to the actuators. This two-stage design has many internal seals that are prone to wear and leakage, leading to more maintenance requirements and a shorter lifespan compared to servo valves. Hydraulic pilot valves also suffer from parasitic loss due to constant high internal pressures, resulting in lower efficiency. Additionally, their spool position control is less precise and responsive than servo valves, as they rely on traditional electronics, which limits their performance, size, and weight (Table 16.5).

The superior performance and capabilities of servo actuated control valves make them better suited for demanding applications like dozers, where precise control and reliability are critical. However, the complexity and higher cost of servo valves must be weighed against the benefits they provide in terms of faster response, higher precision, greater efficiency, and advanced diagnostics (Wollin et al. 2020). Table 16.6 shows the comparison of servo actuator control valve and hydraulic pilot actuated control valve.

16.4 Machine Learning and Deep Learning Models in Dozer Optimization

Now a days Machine learning and deep learning models such as Support Vector Machines (SVM), K-Nearest Neighbors (KNN), Artificial Neural Networks (ANN), and Convolutional Neural Networks (CNN) have shown great potential in enhancing the reliability and efficiency of dozers. SVM effective in classifying the health status of dozer components and optimizing operational parameters, as highlighted

Table 16.6 Servo actuated control valves versus hydraulic pilot actuated control valves

	Mechanical linkage/Servo actuated control valve	Joystick/Hydraulic pilot operated control valve
Comfort for operation	Less comfortable, operator has to exert more effort to actuate the mechanical lever	Comfortable operation, joystick can be operated easily even with a finger
Hydraulic circuit	Simple	Complicated
Hydraulic pump	Only single high-pressure pump for entire hydraulic circuit	High pressure hydraulic pump along with low pressure hydraulic pump for pilot circuit
Control valve actuation	Using servo valves	Using hydraulic pilot lines
Maintenance	Easy	Difficult
Reliability of the machine	More reliable	Highly unreliable after few thousands of hours of operation
Accumulator	NA	Provided for lowering the blade in emergency

by Efeoğlu and Tuna (2022). in their study on predictive maintenance using SVM. KNN is useful for predicting component failures and identifying optimal operating conditions, as demonstrated by Halder et al. (2024) in their review of KNN modifications. ANNs can analyze complex sensor data to predict equipment failures and optimize operational parameters, by providing a comprehensive comparison of ANN-based techniques (Moraes et al. 2013). This method improves fault classification efficiency, reducing downtime and maintenance costs. CNNs are particularly effective in analyzing visual data from cameras and sensors to detect early signs of wear and tear and optimize operational efficiency, as discussed by Kim et al. (2019) in their study on dozer productivity correction using CNNs. Implementing these AI techniques can significantly enhance the performance and reliability of dozers by enabling proactive maintenance and continuous optimization of operational parameters.

The use of AI for real-time monitoring of hydraulic systems has shown significant potential in reducing system failures and associated environmental impacts. By leveraging advanced algorithms and machine learning techniques, AI can continuously analyze data from sensors to detect anomalies and predict potential failures before they occur (Kim and Jeong 2020) This proactive approach allows for timely maintenance, minimizing downtime and extending the lifespan of hydraulic systems. Additionally, AI-driven optimization of hydraulic systems can lead to improved energy efficiency and reduced environmental footprint by minimizing waste and optimizing resource usage (Softwareconsultancy.co.th. 2025). For instance, Kim and Jeong (2020) demonstrated the effectiveness of convolutional bidirectional LSTM models for real-time monitoring of hydraulic states, highlighting the benefits of AI in enhancing system reliability and sustainability.

16.5 Advantages of Transitioning to Advanced Hydraulic Systems for Environmental Compliance

The modernization of hydraulic systems in heavy machinery plays a crucial role in achieving environmental compliance by improving energy efficiency, reducing emissions, and minimizing environmental pollution. Traditional hydraulic systems rely on high-pressure fluid circuits that can lead to excessive energy losses, fluid contamination, and increased noise levels. By transitioning to advanced electro-hydraulic systems, industries can enhance operational efficiency while mitigating environmental impacts.

One of the major environmental concerns associated with hydraulic systems is pollution, including hydraulic fluid leakage and noise pollution. Hydraulic leaks contribute to soil and water contamination, posing a significant environmental hazard. Advanced hydraulic systems address this issue by incorporating high efficiency sealing technologies and biodegradable hydraulic fluids that reduce the ecological footprint of heavy machinery (Wollin et al. 2020).

Noise pollution is a critical issue in hydraulic machinery operations, particularly in mining, construction, and industrial applications. The high-pressure operation of hydraulic pumps, actuators, and valves generates substantial noise, often exceeding permissible occupational exposure limits. Excessive noise levels not only affect machine operators but also contribute to broader environmental noise pollution, impacting surrounding communities (Towers 2001). To mitigate hydraulic noise, modern electro-hydraulic systems incorporate noise-dampening technologies such as variable displacement pumps, electronic controls, and optimized actuator designs. These advancements help in reducing the noise generated by hydraulic components, leading to improved workplace safety, enhanced operator comfort, and lower overall noise emissions (Martin 1984). Additionally, electric and hybrid actuators are being integrated into heavy machinery to replace traditional hydraulic systems, further minimizing noise pollution, and enhancing sustainability.

By adopting advanced hydraulic technologies, industries can not only improve efficiency and reliability but also ensure compliance with environmental regulations, promoting sustainable heavy machinery operations.

16.6 Conclusion

The push for sustainability in the mining sector necessitates the adoption of advanced technologies that enhance both environmental performance and operational efficiency. This study compared servo and hydraulic pilot actuated control systems in crawler-mounted dozers, highlighting their impact on emissions, fuel consumption, and sustainability. The findings indicate that while both systems contribute to improved efficiency, servo control systems offer a superior balance by reducing fuel usage and emissions, making them a more sustainable choice. Furthermore, the integration of machine learning (ML) and deep learning (DL) models in dozer optimization presents a significant opportunity for predictive maintenance, real-time performance monitoring, and fuel efficiency improvements, ultimately reducing environmental impact. In parallel, the transition to advanced hydraulic systems further supports environmental compliance by improving energy efficiency and minimizing hydraulic fluid waste. Moving forward, the mining industry should prioritize research and investment in smart automation, AI-driven optimization, and next-generation hydraulic technologies to enhance sustainability. Future advancements in AI-powered control systems and energy-efficient hydraulic solutions will play a crucial role in shaping a more environmentally responsible and operationally efficient mining sector.

References

Adweb-User (2024) Proportional valve vs servo valve: best 5 ways to explain key differences. Rekith Hydraulics, 28 June. Available at: https://www.rekithhydraulics.com/proportional-valve-vs-servo-valve-key-differences

Anderson CN, Bosserman II BE, Morris CD, Cadrecha C, Lescovich JE, Taylor HW, Vuncannon J (2008) Valves. In: Pumping station design. Butterworth-Heinemann, pp. 1–5

Black K (2024) Hydraulic servo valves explained. Domin, 28 May. Available at: https://domin.co/learn/hydraulic-servo-valves-explained

Carter RA (2012) Tracks and treads: dozers and loaders dig into production support roles. Eng Min J 213(2):63

Caruana L (2024) Capcoal's remote-controlled dozer aims to boost safety. Mining Magazine, 19 June. Available at: https://www.miningmagazine.com/surface/news/4324623/capcoals-remote-controlled-dozer-aims-boost-safety

Co W (2023) A look at sustainable mining efforts. Wagner Equipment Co., 15 September. Available at: https://wagnerequipment.com/blog/a-look-at-sustainable-mining-efforts

Efeoğlu E, Tuna G (2022) Machine learning for predictive maintenance: support vector machines and different kernel functions. J Mach Manuf Reliab 51(5):447–456

European Commission (2024) NACE Rev. 2. Available at: https://ec.europa.eu/eurostat/documents/3859598/5902521/KS-RA-07-015-EN.PDF. Accessed 12 Jul 2024

Halder RK, Nasir Uddin M, Ashraf Uddin Md, Aryal S, Khraisat A (2024) Enhancing K-nearest neighbor algorithm: a comprehensive review and performance analysis of modifications. J Big Data 11(1):113

Johnson JL (2024) What's the difference between a servo valve and a proportional valve? Power & Motion, 5 April. Available at: https://www.powermotiontech.com/hydraulics/hydraulic-valves/article/21882788/whats-the-difference-between-a-servo-valve-and-a-proportional-valve

Jose JT et al (2021) Early detection and classification of internal leakage in boom actuator of mobile hydraulic machines using SVM. Eng Appl Artif Intell 106:104492

Karunanidhi S, Singaperumal M (2010) Design, analysis and simulation of magnetostrictive actuator and its application to high dynamic servo valve. Sens Actuators, A 157(2):185–197

Kim K, Jeong J (2020) Real-time monitoring for hydraulic states based on convolutional bidirectional LSTM with attention mechanism. Sensors 20(24):7099

Kim R-H, Park Y-J, Lee D-E (2019) Dozer productivity correction method for eco-dozing assessment. KSCE J Civ Eng 23:2829–2838

Maas GJ, van Gassel FJM (2003) ISARC2003: the future site: proceedings of the 20th international symposium on automation and robotics in construction, 21–24 Sep 2003. Technische Universiteit Eindhoven, Eindhoven

Martin HR (1984) Noise control in hydraulic systems. SAE Trans 825–832

Moraes R, Valiati JF, GaviãO Neto WP (2013) Document-level sentiment classification: an empirical comparison between SVM and ANN. Expert Syst Appl 40(2):621–633

Mouzoure E (2023) Servo valves: precision motion control for a variety of applications. Domin, 24 November. Available at: https://domin.co/learn/precision-in-motion-how-servo-valves-enable-diverse-automation-solutions

Nilsson J (2016) Position control of an electro-hydraulic servo-valve: identification, modelling, control and implementation

Softwareconsultancy.co.th. (2025) AI hydraulics pressure sensor monitoring. AI/ML Development Solutions Thailand. Available at: https://ai.softwareconsultancy.co.th/services/ai-hydraulics-pressure-sensor-monitoring/. Accessed 30 Jan 2025

Towers DA (2001) Mitigation of community noise impacts from nighttime construction. In: Construction and materials issues 2001, pp 106–120

Wollin K-M, Damm G, Foth H, Freyberger A, Gebel T, Mangerich A, Gundert-Remy U et al (2020) Critical evaluation of human health risks due to hydraulic fracturing in natural gas and petroleum production. Arch Toxicol 94:967–1016

Wrat G, Das J (2023) Energy saving in off-road vehicle using leakage compensation technique. ResearchGate. Available at: https://www.researchgate.net/publication/374118311_Energy_Saving_in_Off-road_Vehicle_using_leakage_compensation_technique

Wrat G et al (2024) Modelling and analysis of the hydraulic system parameters under different loading conditions. In: Dynamics of transportation ecosystem, modeling, and control. Springer Nature Singapore, Singapore, pp 9–21

Index

S. K. Tiwari et al. (eds.), *AI-Driven Environmental Pollution Management*, Climate Risks and Solutions, https://doi.org/10.1007/978-3-031-96243-1

MIX
Papier aus verantwortungsvollen Quellen
Paper from responsible sources
FSC® C105338

If you have any concerns about our products,
you can contact us on
ProductSafety@springernature.com

In case Publisher is established outside the EU,
the EU authorized representative is:
Springer Nature Customer Service Center GmbH
Europaplatz 3, 69115 Heidelberg, Germany

Printed by Libri Plureos GmbH
in Hamburg, Germany